HORIZONS OF BIOCHEMICAL ENGINEERING

HORIZONS OF BIOCHEMICAL ENGINEERING

Edited by Shuichi Aiba

Oxford New York Melbourne
Oxford University Press 1988

0 325-5645
CHEMISTRY

Oxford University Press, Walton Street, Oxford OX2 6DP

Oxford   New York   Toronto
Delhi   Bombay   Calcutta   Madras   Karachi
Petaling   Jaya   Singapore   Hong Kong   Tokyo
Nairobi   Dar es Salaam   Cape Town
Melbourne   Auckland

and associated companies in
Berlin   Ibadan

Oxford is a trade mark of Oxford University Press

Distributed outside Japan by Oxford University Press
Distributed in Japan by University of Tokyo Press
under ISBN 4-13-068135-7

ISBN 0-19-856196-2

Printed in Japan

Contents

Preface

The development of recombinant DNA and hybridoma technologies in recent years has had a tremendous impact not only on the basic sciences of molecular biology and immunology but also on applied and environmental microbiology, particularly since the mid-1970s. The wealth of information that has been accumulated on gene cloning, transformation, transduction, and so forth has made it possible to produce in microorganisms various peptide hormones such as insulin and somatostatin, to cite only two examples. In this respect, species barriers between microorganisms, plants, and animals have been, in principle, eliminated. As a result, the microscopic potential of microorganisms can be used advantageously. From the macroscopic viewpoint, there is increasing demand for the development of microorganisms as sources for food protein, energy, and to clean up the natural environment.

Peoples currently working in life sciences and technology are especially interested in the development of biotechnology, extending from medicine, pharmacology, and veterinary medicine to the fields of agriculture, horticulture, fermentation, brewing, and engineering. Biotechnology offers many opportunities for scientific and technical breakthroughs. It has also renewed awareness of the significance of biochemical engineering, because expertise in it is indispensable whenever a new development and/or discovery in biological science is placed on the market as a commodity.

From April to July 1963 Professors A. E. Humphrey (Department of Chemical Engineering, University of Pennsylvania, Philadelphia, Pa., U.S.A.) and N. F. Millis (Department of Microbiology, The University of Melbourne, Parkville, Victoria, Australia) joined me at the Institute of Applied Microbiology, the University of Tokyo, in holding a series of twenty-four seminars on biochemical engineering. About 50 people from academic and industrial circles attended each three-hour seminar. We compiled our experiences during the seminars into a book titled *Biochemical Engineering* initially published in 1965 by the University of Tokyo Press and later distributed by Academic Press, New York. The second edition appeared in 1973. The fact that a number of excellent books have been published during the past two decades or so reflects the keen interest of numerous researchers and workers in the subject of biotechnology.

The present volume is not intended to be an introduction to biochemical engineering, but to serve as a subreference that looks at the brief history of biochemical engineering and also looks forward to future prospects. Publication was motivated by my retirement after working from July 1954 to the end of March 1987 at both the Institute of Applied Microbiology, the University of Tokyo, and the Department of Fermentation Technology, Faculty of Engineering, Osaka University.

This book is composed of 26 papers contributed by my esteemed friends and colleagues from many countries. It is subdivided into two categories: the first deals with individual activities in biotechnology, and the second is mini-reviews of current topics being studied in various institutions. Hopefully, this small compilation of papers will give readers a glimpse of the past and also an image or silhouette of what may happen in the field of biotechnology, particularly biochemical engineering.

I would like to thank my many friends and colleagues for their participation in the project to publish *Horizons of Biochemical Engineering*. I am also grateful to Dr. T. Imanaka and to the University of Tokyo Press for their help in editing and publishing at the right time and the right place.

Shuichi AIBA

March 1987
Osaka

REFLECTIONS

1

What Can the Microbiologist Learn from the Biochemical Engineer?
Personal Reflections on Aiba, Humphrey, and Millis

John D. Bu'Lock

Microbial Chemistry Laboratory, Department of Chemistry, University of Manchester, Manchester M13 9PL, England

Twenty years ago I was fortunate enough to aquire a new book from Academic Press titled *Biochemical Engineering*. The authors were Shuichi Aiba, Arthur E. Humphrey, and Nancy F. Millis. Today that book is one of the more well-worn volumes on my bookshelves. Over the ensuing years it has been joined by other similar titles, some old, some new, and mostly in considerably better condition because they have been used less. Now it is both relevant and happy for me to reflect on what I have learned from *Biochemical Engineering*. In writing this tribute to Shuichi Aiba in particular, and to biochemical engineering in general, I will begin by stating my own credentials. I started scientific life as a classical organic chemist who was put to work on metabolites of higher fungi, and went on by way of natural product structures to biosynthesis. This brought me into the business of growing microorganisms in order to make microbial products. First from necessity, and then from scientific interest, I became involved in "optimizing" this activity, and by 1965 I was already aware of some of its problems and the possibilities. I knew that the optimum conditions for growing an organism were not the same as the optimum conditions for getting my products, and I was interested in developing better ways of getting from one set of conditions to the other.

Since then, my involvement in what nowadays is called "biotechnology" has developed along with the subject. My colleagues and I have labored at biomass production, at ethanol fermentations, at anaerobic digestion, and at water purification as well as in the fields of sophisticated antibiotics and secondary metabolites. We have done our best with techniques that properly belong to other, established, disciplines like chemistry, biochemistry, microbiology, genetics, and chemical engineering. More recently as an editor I have had to exercise critical ability in even wider fields. Having worked in biotechnology from the start, and holding the guiding conviction that an understanding cooperation with biological systems holds the key to improvements in the human condition, I have not succumbed to the fashion that considers biotechnology as a mere synonym for high-tech bioscience, or restricts it to the latest achievements of genetic manipulation. At the same time I recognize the promise of contributions of modern genetics to our process technology, and, although aware of the practical difficulties, look forward to their practical applications.

However, whether we are trying to exploit the genetic potential of some new construct or the even more mysterious (because largely hidden) abilities of some laboriously-discovered "natural" organism, the process by which we do so remains the main challenge. Moreover, implementing a process using microorganisms often brings us closer to the underlying general biology, requiring the kind of understanding of living organisms to which "pure" biologists constantly aspire. In my own research, excursions into process engineering have always seemed to be quite natural extensions of bioscience, and I hope to illustrate here how some requirements of the engineering approach have enriched our understanding of biology.

The contribution of biochemical engineering is not simply a set of useful techniques but, perhaps more importantly, a mode of thought, which remains most important in describing and designing ways through which we can achieve an objective. I take biotechnology to be the deliberate and controlled exploitation, in laboratory or industrial systems, of the two apparent opposites of biological systems: on the one hand, their seemingly uncontrollable momentum towards internally determined ends; on the other, their flexibility of response towards externally determined influences.

I know that biochemical engineering will enable us to calculate the required horsepower of agitator motors and the superficial area required for air filters, but that is what I would have expected it to do; should I want those things, I will still go to a "real" engineer. What came as a surprise, to me, was the way the engineer's mode of description could add to the languages and the logic I had already learned, first as a chemist and then as a microbiologist.

One of the first things I found in Aiba, Humphrey, and Millis was a convincing and generalizable account of metabolic pathways in fermentation processes, written in terms of both reaction routes and regulatory mechanisms. Perhaps it was strange to discover what was essentially an introduction to process biochemistry in a textbook of biochemical engineering, but it was valuable to me for precisely the reasons that led the authors to include it in their account. At that time the only relevant biochemistry I had read was either so generalized that it dealt only with the mythical "standard" organism and left no room for the special metabolic features that were my particular interest, or else so particular that it gave very little direct help with any problem other than the one to which it had been addressed. Here on the other hand was an account that managed to be coherent while dealing with a considerable variety of mechanistically different process systems, done in such a way that readers could go on confidently to deal with further systems, knowing what different features and what common principles were to be involved. Incidentally, *Biochemical Engineering* was also my first introduction to Japanese work on amino acid fermentations, which illustrates so perfectly how successful technology can go hand-in-hand with first-rate fundamental science.

Using such descriptions of systems various work has been done since 1965 in varied and significant systems including sulphur metabolism in β-lactam producers, nitrogen flux in fungi, the overall carbon flow in methanogenesis, and fermentative catabolism in aerobic yeasts. All of these are biochemical problems, but in 1965 very few biochemists wanted to know about them as systems, even when they were working on them as unit problems. Biochemistry was providing the mechanisms, in terms of induction, repression, and allosteric inhibition, but it was the approach of the biochemical engineer that was turning these "unit processes" of biochemical mechanism into descriptions of integrated systems. Although the biochemical engineer must do this in order to be able to describe the system in terms that can be manipulated, I found the approach worthwhile for its own sake.

In due course we were able to use just this kind of thinking for more than one problem in "pure" microbiology. For example, the pro-hormones and hormones that bring about sexual differentiation in the Zygomycetes were explored in terms of their chemistry, and their effects were observed in terms of morphology and behavior, although the mating process as a whole, including its basic genetics, could only be interpreted by a systems approach, which we tried to follow, with modest success, in our work.

The next thing I found in Aiba, Humphrey and Millis was my first introduction to fermentation kinetics. Once again it would be wrong to give these authors all the credit for what I found there; most of it had been published by others, but until then I had missed it. So for the first time I found such qualitative but wholly generalized pictures as Gaden's classification of fermentation kinetics, into which so much of what I had been observing fitted so neatly. It was only later that I began to work with the more quantitative descriptions of kinetics and to try to make use of them, but already I had received the encouragement that comes from the success of others. I had been given confirmation of some cardinal beliefs: that any description of a microbial process needs to be a kinetic one, with lapsed time as the most important variable and with quantitatively-measured phenomena as the basis of observations. The attention to *process* which comes from kinetic approaches was already important to me; Aiba, Humphrey, and Millis made me realize it was also important to other people.

One of the key things you can measure in a microbiological process is the amount of biomass, and yet thinking about this among non-engineers is still deplorably lax. One of the things microbiologists need to learn from their engineering colleagues is how to answer the question "what do you mean by biomass?" I would enumerate some key considerations as follows:

(i) Microbiologists often describe biomass as self-replicating. This is true if and only if we measure only the self-replicating part of the biomass. It is only true for the whole biomass if the proportion of that total which is self-replicating remains in an unchanging proportion. Any changes amount to a form of differentiation.

(ii) Put into kinetic terms, self-replicating biomass grows exponentially, i.e., "logarithmic growth," if and only if it grows without differentiation. Conversely, any growth that is not truly and strictly exponential necessarily involves some kind of differentiation; the self-replicating part of the biomass makes up a changing proportion of the whole, the replicating parts of its mechanisms then operate at changing rates, and the replication intermediates must have changing pool sizes and turnover rates. Classical morphological differentiation may be apparent or not, depending on the acuity of the observer, but this "kinetic differentiation" is ubiquitous even though many microbiologists are quite unaware of it. It follows, more subtly, that there are very few forms of differentiation that can be described in simple "on/off" terms.

(iii) Operationally, what you actually measure in order to measure "biomass" becomes very important, and why you measure it may be equally critical. Except during strictly exponential growth, measures which depend on cell number are not the same as measurements of cell mass; measures of specific cell components may be equivalent to either, or to neither. Measurements of "specific rates" may seem to get around this but they have the same pitfalls in a concealed form; always, an observed rate is divided by some measure of "biomass," and the specific rate in terms of cell numbers or DNA or total protein will follow quite a different course, through a fermentation, to the "same" specific rate in terms of total cell mass.

To a biochemical engineer all these comments will seem obvious and perhaps trivial; however they contain lessons most microbiologists have yet to learn. All too often authors of scientific papers of an otherwise excellent standard do not observe processes, but only end-results, do not know the difference between cell mass and cell number, think all vaguely sigmoid curves are the same, do not understand the significance of changes in specific growth rate, and use expressions like "late log phase" quite shamelessly.

Conversely, however, engineers may need a deeper understanding of why their clear exposition in these areas is important for microbiologists. An outstanding example of failure to understand this, dating from the early days of my scientific education, is the fate of Hinshelwood's pioneering work on bacterial kinetics. As an excursion into unknown territory by a great physical chemist, that work was of real significance, and remains so because most of its conclusions arise from the overall dynamics of the systems treated and not from any particular set of underlying mechanisms. Unfortunately the whole work was described at the time in terms of hypothetical mechanisms which contemporary biology was just at that time proving incorrect. What should have been fruitful advances became lost in sterile controversy. Biologists did not comprehend the mathematics, and the physical chemists scarcely attempted to comprehend the biologists.

The lesson of this episode remains useful today. For instance, until the importance of the distinction is brought home to them, microbiologists do not readily distinguish between the inhibitory effect of ethanol on yeast growth and its inhibitory effect on yeast metabolism; however when properly instructed they will not merely see the point, they will quite happily add to the process description measurements of the effect of ethanol on yeast viability. Even more, they will define viability by one or several distinct criteria, and having done so, will legitimately expect the kineticist to incorporate these into his account. The differential equations may become quite hard to handle as a result.

As a more general example, microbiologists can learn that the only kind of exponential growth they can normally see is what happens in the early (post-log) phase of a typical batch culture, when they will see exponential growth at the maximum specific growth rate for the system. They will further learn that they can observe exponential growth at *other* specific growth rates in a chemostat. The engineer must now learn that this is in fact a worthwhile exercise quite irrespective of the process applications (or otherwise) of continuous culture, and that the microbiologist is not merely being obtuse when he duly reports that the phenomena observed at a given growth rate in a fully equilibrated, steady-state chemostat are not exactly the same as those observed transiently at that same growth rate in a batch culture. If the engineer's kinetic model tells him that they should be the same, then the model requires changing.

In fact both my examples underline a contemporary need in kinetic modelling: the need to develop models which are as convincing, as convenient, and as credible as those now widely available, but which no longer rely on an "instantaneous" description. Both in interpreting general data and through very specific experiments, we are learning that microbiological systems can and frequently do show marked hysteresis in their regulatory loops, but we are often driven to ignore this fact by lack of simply understood, generally applicable kinetic models that will take this into account.

In fermentation research there are a growing number of systems whose behavior seems to include becoming "locked in" to particular regulatory modes, with a greater or a lesser degree of reversibility. In visibly differentiating systems such as sporulating bacteria the phenomena have been interpreted, with difficulties, in terms of some process of "commitment," but similar phenomena in fact occur within chemical differentiation. The effect of toxins like ethanol on yeast functions seems to include some "instantaneous" and reversible effects and some in which duration is as important as intensity. In many commercial fermentations the situation where a batch has "gone wrong" irreversibly at some early stage is recognized as a regulatory effect which it is not easy either to investigate or to describe, at least in part because lacking a model we do not know what to try to observe. Now all such effects may well be quite accessible to certain types of dynamic modelling techniques; the trouble is that no one has yet described those techniques in terms I can understand. So here my lesson from Aiba, Humphrey and Millis is still minimal; I have learned how to recognize what I do not know.

A further lesson from the interaction of microbiological and engineering approaches has had to do with non-homogeneous systems. Here the practical microbiologist's work provides just that touch of reality needed in order to transform theoretical elegance into something much more useful. Most of the better-known kinetic models of fermentation systems are conveniently and simply developed by assuming that the control volume is well mixed. This simplifies the relationship between extensive and intensive variables, allows mass balances to be drawn up in terms of concentrations, and allows continuous systems to be characterized by a single flow rate or residence time.

Most real systems are not ideally well-mixed. Once this is accepted as a feature that many systems actually exploit very advantageously, the engineering models that do not assume a single well-mixed control volume are of course available, and properly handled they are not very difficult to use. Suddenly, heterogeneous systems with variable residence times, different for different components, become susceptible to the same precise and elegant treatment as homogeneous systems. They can now be thought about with a correspondingly greater clarity; real systems which exploit their special features can be designed, installed, and tested. It is common knowledge that such reactor systems applied in a great variety of process contexts constitute a new generation of bioreactors, with the development of both fluidized and fixed-bed reactors, using both free and trapped biocatalysts, in almost every kind of fermentation process.

So here is another area where biochemical engineering has enriched microbiology, and *vice versa*. I am sure that this is the lesson that Shuichi Aiba has been trying to teach us; at least, it is the one I have been trying to learn. But the whole lesson, so far as I have been concerned, can be summarized by the following: try to describe the whole system; always think dynamically; don't be afraid of simple mathematics; talk to the engineers, but don't surrender.

One further lesson for me was entirely personal. After 1965, I had an additional name on the list of people I wanted to meet. When I eventually visited Professor Aiba, I was not disappointed, and he has remained one of my most valued aquaintances, always helpful and conscientious, always a friend. This article may not contain much science, but I hope it will direct at least a few people along pathways of thought I have found enlightening, along which Shuichi Aiba has been my respected guide.

2

Biochemical Engineering—Roots and Blossom

Armin Fiechter

Department of Biotechnology, Swiss Federal Institute of Technology, CH 8093 Zürich, Switzerland

1. Introduction

Aiba began his scientific work when the discipline of biochemical engineering had reached its first peak. The glamor products of microbiology, e.g., antibiotics, were being produced in large-scale batches under sterile conditions; penicillin was produced in considerable amounts. The technology of those days was based on the development of stirred tank reactors with flat blade turbine impellers designed for sterile operations.

Numerous papers have been published on stirred containments displaying good mixing and mass transfer properties. The characteristics of fluid motion and its consequences have been investigated mostly by using aqueous sulfite as a model solution and copper or cobalt as catalysts.[1] The ultimate goal of the efforts was to increase the mass—particularly the oxygen-transfer capacity of the reactor with a simultaneous decrease in specific power input. Agitation and aeration of reaction tanks under sterile conditions were common topics among engineers.

Appropriate rotating shafts, ports, closures, air filters, and sampling devices had to be designed and constructed, although the technology for process control was rather primitive. The development of pH sensors was considered a great step forward, but reliable automatic control of this important growth factor became possible only a short time ago. Temperature and hydrogen ion concentration were for quite a while the only controllable parameters. The detection of dissolved oxygen tension by membrane-covered sensors appeared to be a difficult and expensive task. Nevertheless it was chemical engineers who established gradually a sound basis for engineering in an otherwise empirical field.

The biological and chemical problems of antibiotic production, however, were solved elsewhere. Microbiologists focused their attention on the cell itself and left the rest to the "technicians." In this manner biology and engineering continued to develop separately.

Aiba was a chemical engineer in the early days of biochemical engineering who contributed substantially to the development of the field in the 1960s and 1970s. He devoted great effort to notoriously less investigated subjects crucial to industrial work and rationalization of production. Typical examples of his farsighted research programs are downstream processing and scaling-up criteria. Two review papers[2,3] published in the 1970s show his high scientific quality and capacity. The broad interest of Aiba becomes obvious, when we recognize that as a chemical engineer he did not hesitate to tackle problems of a biological nature. His exemplary work has served as a foundation for both the theoretical and the empirical approach to solve biological problems (see T. Imanaka, *Adv. Biochem. Eng.* **33**, 1 (1986)).

2. Development of Biochemical Engineering

The technical capabilities of the 1960s were inadequate for sophisticated biological experiments. Although the purpose of most bioreactors was continuous operation under sterile conditions, these objectives were not met, and hence systematic studies of the metabolic parameters affecting cell growth and product formation, of the metabolic sequences for the purpose of regulation, and of control of bioprocesses could not be made. Continuous operations were handicapped by poor operational reliability. Due to the lack of extended on-line detection of changes taking place during the course of growth, control of reactions was restricted to control through stirrer speed, temperature, and pH. Consequently, microbiologists gave up the application of this efficient culture method and reconsidered the rather simple batch operations. Only a few groups continued their efforts to develop the equipment necessary for sophisticated, continuous operation devices.

Digital techniques were introduced into this field in the 1960s and agitation and aeration of cultures gradually became more efficient and reliable. On-line detection of cell respiratory activity was achieved by the more advanced groups. Nevertheless, contributions of chemostat methods to the knowledge of biological control mechanisms were only minor because process control accuracy of 10–20% was still inadequate. Computer control of bioprocesses

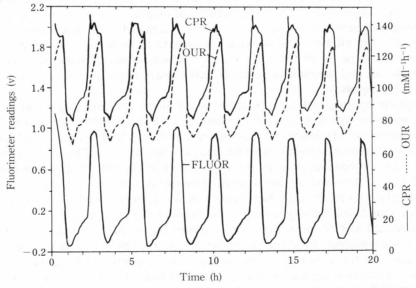

FIG. 1. Oscillation of CO_2-release and NADH in continuous cultivation of *Saccharomyces cerevisiae* during steady state (D = 0.17 h^{-1}), where:
CPR, volumetric carbon dioxide production rate;
OUR, volumetric oxygen uptake rate;
FLUOR, fluorescence signal reflecting intracellular NADH.
Meyer, C., Beyeler, W.: Control strategies for continuous bioprocesses based on biological activities. *Biotechnol. Bioeng.*: **26**, 916 (1984).

turned out to be expensive and had to be based on rather complex concepts. In the mean-time application of recombinant methods for genetic alteration of cells became feasible and attracted many scientists. The importance of development of efficient cultivation methods became even more obvious when industrial-scale production of recombinant strains was contemplated. It was recognized that the physical and chemical parameters of growth and their optimal control are decisive as well. Medium-design for mass production of host strains appeared to be most relevant for the understanding of regulatory factors with a significant bearing upon maximum product synthesis. It became obvious that final product formation was dependent on extracellular as well as on intracellular parameters.[4]

An example of the progress made in biotechnology is a chemostat with direct digital control (DDC) developed in the early 1980s that has made a significant contribution to overall process control of bioreactors. Strict control of feed rate, reaction volume, gas hold-up, and good mixing of reactor contents regardless of viscosity, when combined with the use of media with defined concentrations of certain trace elements, will extend the capacities of the chemostat as a tool for the study of biological reactions.

Figure 1 shows steady oscillation of the intracellular NADH-pool and CO_2-release at a constant dilution rate. The regular peaks can be related to the activity of two subfractions of a growing population which is essentially divided into mother cells with a shorter G_1-phase of the cell cycle and daughter cells bound to a prolonged G_1-phase. This type of oscillation occurs spontaneously without any inducing effector and can be reached only by perfect control of growth parameters. (compare Sonnleitnev et al.: *Proc. 4th Europ. Congr. Biotech.* **3**, 76 (1987))

3. Progress in Biology

The scientific contribution of Aiba during his later career demonstrates his profound insight into biochemical engineering. He was one of the few chemical engineers who focused attention on the cell and included genetic and metabolic methods in his investigations on bioproduct formation. With these tools in hand, he was able to include in his great work investigation of complex biosystems such as photosynthetic microorganisms and protozoa. These investigations are of highest importance in terms of environmental protection. Waste water treatment has traditionally been an area neglected by science and additional effort will be necessary for future improvements.[5,6]

Substantial progress in applied areas like effluent treatment can be made only if basic knowledge of substrate/growth relationship, growth kinetics, control of metabolic path-ways, and product formation can be extended. Investigations on the control mechanisms of substrate uptake and its degradation responsible for the delivery of energy and formation of building blocks for the synthesis of cell constituents are especially important. In the case of glucose, it has been shown recently that the controlling step is substrate flux rather than catabolite repression of glucose in yeasts. An overflow concept explaining the release of ethanol was developed and confirmed by carefully designed chemostat experiments. It was shown that the cell respiratory capacity is the limiting factor for purely oxidative de-gradation of glucose. The uptake of this substrate by the so-called glucose-sensitive yeasts is not subject to any limitation and hence carbon skeletons in excess of the maximum cell respiration potential are excreted on the level of pyruvate.[7,8] Formed metabolites such as ethanol in yeast (and possibly acetic acid in the case of *Escherichia coli*) act as inhibitors on a variety of synthetic pathways limiting the full yield of biomass (Fig. 2).

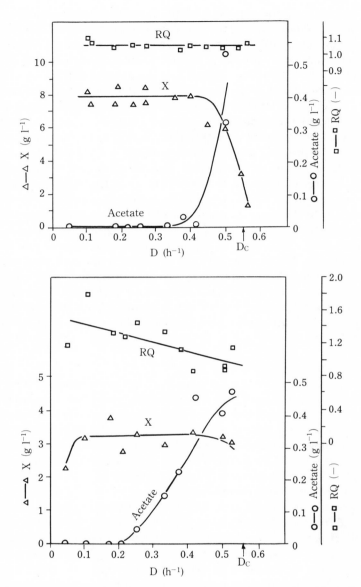

FIG. 2. Metabolic control of central pathways (glycolysis, respiration) is an essential aspect in process development. Chemostat methods are important to evaluate typical strain parameters and appropriate medium design.

The formation of ethanol from glucose is the result of an overflow of carbon skeletons on the C_3-level rather than a primary product of catabolite repression. Ethanol acts as an inhibitor reducing cell yield. Such a control mechanism also exists in some bacteria such as *E. coli*, as shown here, leading to the excretion of acetic acid (top diagram for defined medium, and bottom, complex medium).

Meyer, H.-P., Leist, C., Fiechter, A.: Acetate formation in continuous culture of *Escherichia coli* K12 D1 on defined and complex media. *J. Biotechnol.*: **1**, 355 (1984).

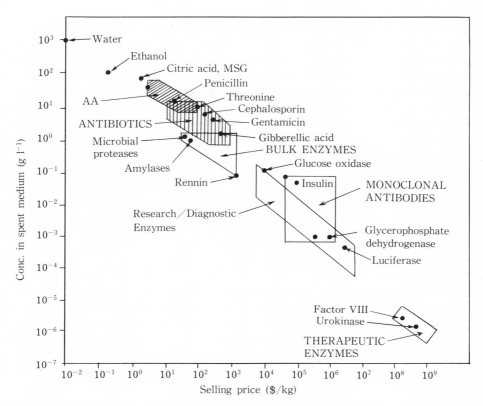

Fig. 3. Product concentration in bioprocesses. Concentrations are generally low especially in processes for high-cost products. In spite of the process made in biochemical engineering, higher product concentrations and better production rates are important goals for the future.

Nystrom, J., Dwyer, J.L.: Int. Congr. of Biotechnol. Munich, 1984.

The consequences of this concept are of great importance for commercial processes as well. Efficient biosynthesis takes place only under conditions where the cells' oxidative potential can be maintained at the maximum by selecting nonlimiting environmental conditions including proper medium design. This prerequisite is not met in many cases when ethanol formation is observed and is ascribed simply to "catabolite repression." It has been shown however, that ethanol formation depends upon the respiratory capacity and the feeding rates of sugar.[7,8] Glucose fed in excess leads therefore to ethanol excretion, irrespective of the oxygen pressure. Consequently, large amounts of ethanol can be obtained aerobically at maximum substrate feed rates, as shown by Lorencez-Gonzalez.[9]

Increasing the productivity of bioprocesses is still an important task in biochemical engineering although much progress has been made in the past (Fig. 3). Further improvements however, are necessary and can be made only by exploiting fully the kinetic potential of the cell and by increasing the actual substrate concentrations applied. Profound knowledge of metabolic potential and advanced equipment with the highest productivity are therefore important prerequisites for future progress.

4. Prospects

Despite the impressive progress made during the last ten years, high expectations have yet to be fulfilled. Technical methods used in research on biochemical engineering should be improved and made suitable for application in complex biological processes. Processing of microbes which grow under extreme and inverse conditions, such as thermophilic, barophilic, acidophilic, or strictly anaerobic cells, as well as fastidious mammalian and recombinant cells, is difficult and demanding. The conventional equipment and methods for growing cells need to be improved and adapted to these special systems.

Mammalian cells have no walls, are very sensitive to mechanical stresses, and their growth and multiplication are relatively slow. These characteristics as well as the typical genetic instability of mammalian cells have hindered investigations of biological regulations of these cells. In other words, for the lack of suitable techniques, the potentials of these cells in the synthesis of useful substances such as hormones and interleukin, has not yet been fully utilized. Cultivation of plant cells has also given rise to similar problems. There is hardly an industrial process in which a known or a new active agent can be produced from plant cells economically. In spite of the progress made in the cultivation of certain plant cells, including cell immobilization, no major industrial breakthrough has been attained.

In addition to the need for improved cultivation techniques, current analytical and measurement possibilities are also limited and need to be extended. Biotechnology requires sensors and detectors to measure factors never before controlled industrially, as well as operation under sterile conditions for which there is little experience in the chemical industry. Expenditures for biological tests and analyses are high. Many bioassays are still carried out manually. Enzyme sensors for routine work are available only in rare cases. On-line mechanization of these assays needs to be carried out, and this is where robotics can be applied.

Separation and purification must make significant advances for biotechnology. The market demands faster speed and greater purity, as well as more versatility. A bacterial cell contains about 5,000 different proteins, all or any one of which may be the desired objective of a biotechnology process. Separation of individual cell fractions requires more subtle technology. At present, costs for the separation and purification of bioproducts can amount to 60% of the production cost. This significant cost factor has been a main obstacle in the commercialization of many bioproducts. The present downstream processes need therefore to be improved to make various biosystems commercially attractive. All these problems present a real challenge for scientists and engineers.

A sound basis for the future has been created by a generation of scientists like Shuichi Aiba, our good colleague and friend who has devoted his efforts and time to educate a future generation of scientists to develop the field of biotechnology in a manner that has opened several new and promising research areas. He has given his best to the Japanese and international scientific community. All of us are thankful for his many contributions and for his friendship, and we are looking forward to his future work.

REFERENCES
1. Cooper, C.M., Fernstrom, G.A., Miller, S.A., "Performance of agitated gas-liquid contactors." *Ind. Eng. Chem.*: **36**, 504–509 (1944).

2. Aiba, S., "Separation of cells from culture media." *Adv. Biochem. Eng.*: **1**, 31–54 (1971).
3. Aiba, S., "A complementary approach to scale-up." *Adv. Biochem. Eng.*: **7**, 111–130 (1977).
4. Fiechter, A., "Physical and chemical parameters of microbial growths." *Adv. Biochem. Eng./Biotechnol.*: **30**, 7–60 (1984).
5. Aiba, S., "Growth kinetics of photosynthetic microorganism." *Adv. Biochem. Eng./Biotechnol.*: **23**, 85–156 (1982).
6. Sudo, R., Aiba, S., "Role and function of protozoa in the biological treatment of polluted water." *Adv. Biochem. Eng./Biotechnol.*: **29**, 117–141 (1984).
7. Käppeli, O., Sonnleitner, B., "Regulation of sugar metabolisms in *Saccharomyces*-type yeasts: experimental and conceptual considerations." CRC (in press) (1986).
8. Sonnleitner, B., Käppeli, O., "Growth of *Saccharomyces cerevisiae* by limited respiratory capacity: Formulation and verification of a hypothesis." *Biotechnol. Bioeng.*: **28**, 927–937 (1986).
9. Lorencez-Gonzalez, I., "Aerobic ethanol production with a flocculent yeast in a biomass recycling system." Thesis ETHZ No.: 7781, Zürich (1985).

3
The Contribution of Microbial Physiology to Biotechnology

Nancy F. Millis

Department of Microbiology, University of Melbourne, Parkville 3052, Australia

In my contribution to this volume honoring Professor Shuichi Aiba, I should like to set out some thoughts on the area of science encompassed by biotechnology and on Professor Aiba's contribution to it. It is interesting to note that when I was first associated with Prof. Aiba in Tokyo in 1962, the word biotechnology was not yet widely used; indeed, it has only recently been coined.

The first course in Japan devoted to this topic was organized by Prof. Aiba in 1963 and Prof. Arthur Humphrey and I took part in it. The importance of this initiative on the part of Prof. Aiba would be hard to overemphasize. It was the beginning of a serious appreciation of the fact that to exploit biological processess, three major disciplines, genetics, physiology, and biochemistry (particularly that of microorganisms), and chemical engineering must be jointly applied to the problem. The various specialists need to understand sufficient of the other's discipline to work as a team to develop a biological process. This is a commonplace today, but it was a very novel concept at that time and it owes much to Prof. Aiba's flexibility of mind that, from his strictly chemical engineering background, he saw this need very clearly and exploited it brilliantly. Undergraduate degrees in biotechnology are advocated by some, but I believe such graduates would be equipped for only the less sophisticated tasks within the industry. The very breadth of the field dictates that undergraduate biotechnology courses cannot have great depth in any one area and its graduates will thus be ill equipped to undertake research into novel problems. For research and development in biotechnology, I believe it is important for practitioners to acquire first a sound basic knowledge of a relevant discipline and later acquire the breadth necessary to enable them to appreciate the knowledge and skills of other experts in the team.

Professor Aiba is an excellent example of a rigorously trained chemical engineer who has applied that rigor to every biological problem he has tackled, and who has had great success in his many collaborations with biological scientists.

Since the 1960s, knowledge of the genetic material of cells and the control of its expression has expanded at a very fast rate. The possibility of introducing and expressing foreign genes in other hosts, particularly in bacterial hosts, put stars in the eyes of venture capitalists and managers of large pharmaceutical houses. This powerful technique has also fired the imagination of research workers and the brightest young graduate students. Consequently molecular biology and genetic engineering have become the focus of many groups in universities, research institutes, and government-sponsored promotion schemes.

The importance of the contribution of molecular biology to expanding the horizons of biotechnology is undeniable. But the breeding of a novel strain is only the first step on the path to a successful industrial process. It is essential to establish the conditions that will enable the cell to express the introduced character in high yield, to grow vigorously in large vessels, and to have high stability. Good graduate students must be trained to address these problems so that geneticists can know which traits should be incorporated into production strains, and chemical engineers can know the limits of the physical environment that can be tolerated and still achieve high yields. To provide these data, microbial physiologists are essential members of the team. I should like to indicate some of the contributions microbial physiologists can make to biotechnology. It will be clear, I believe, that there are challenges here to stretch the ingenuity of the most intelligent scientists.

1. Chemostat Studies

The examination of the behavior of production strains in chemostats is perhaps the physiologist's most powerful tool. It enables single parameters to be examined under otherwise fixed conditions. Batch growth does not permit this. For example, a change in temperature or pH will usually change the growth rate; if the environmental change affects the yield of product, is this due to the test parameter, or to the consequent change in growth rate which cannot be controlled in a batch system? Although the possibilities of the chemostat were amply demonstrated by the workers at Porton in England some thirty years ago, the fermentation industry has not used chemostats as extensively in their development work as might have been expected. The morphology of many organisms producing antibiotics is different during active growth from the form associated with antibiotic production. This phenotypic change is clearly complex but if a chemostat culture with the appropriate morphology can be grown, changes in single parameters can be made to explore their effects on productivity. This cannot be as incisively addressed in batch cultures.

The information obtained in the chemostat allows rational programs to be developed for managing fed-batch fermentations. Although continuous culture methods would clearly be inappropriate for commercial antibiotic production since this method of cultivation selects so strongly for faster-growing variants, the research and development phases can be greatly assisted by such studies. Dr. S.J. Pirt's pioneering work[1] on the effect of growth rate on penicillin production in a glucose-limited chemostat showed quite conclusively that slow growth rates were not a necessary condition for penicillin production, but rather that the organism could grow at fast rates and maintain fast rates of penicillin production, provided glucose limitation prevailed. The optimum rate at which glucose should be fed could thus be established.[2] The production of other secondary metabolities and of specific enzymes is amenable to similar analysis.

The price of fermentation products such as biomass, or the end-products of energy production such as ethanol, lactic acid, or acetone butanol, is extremely sensitive to the cost of the major substrate in the growth medium. Production strains should have high affinity for substrate and low maintenance requirements, and for biomass, a high yield. To select strains with fast growth rates and good yields of biomass, a continuous flow system can be run at low cell density as a turbidostat with a high flow rate. By contrast, a chemostat run with limiting substrate at relatively slow flow rates will tend to favor variants with low K_s

values, and will select for constitutive variants if, in the wild type, the substrate is metabolized by inducible enzymes.

Although the chemostat has not been used as much as it might have been for this purpose, it can also be used to simulate the adverse environmental conditions that inevitably occur in very large vessels where mixing is imperfect. These effects can be potentially serious, for example, if poor oxygen transfer develops in parts of the vessel (this can occur readily with viscous broths) or if a substrate added at a point is imperfectly mixed in fed-batch fermentations. The likely deleterious effects on productivity were illustrated by basic studies of organisms in chemostats where the partial pressure of oxygen was carefully controlled. It was established that the rate of respiration of cultures at very low oxygen tensions was very unstable and that transitions from near anaerobiosis to well-aerated conditions frequently caused respiration to uncouple from ATP production.[3] Other chemostat studies have established that variation in the concentration of the nutrient limiting growth may often cause changes in cellular metabolism to invoke "energy spilling" reactions. Clearly, if mixing times in large vessels are known, the production strain can be grown in a chemostat under optimal conditions and then stressed to simulate the likely events in large vessels and thus determine their importance to the productivity of the process.

The chemostat also provides an excellent tool for determining conditions favoring the retention of wanted genetic traits, particularly for strains with inserted plasmids. Basic studies have shown that different limitations and different growth rates influence the retention of plasmids in cells where there is no specific selection for the plasmid.[4,5]

2. Organisms from Extreme Environments

Interest in enzymes capable of operating at high temperatures and alkaline pH was stimulated by the use of proteases in laundry detergents, but work to explore the further potential of organisms from extreme environments has only recently been seriously supported by the biotechnology industry. Organisms from extreme environments (pH, temperature, salinity, hydrocarbon contaminated environments) clearly have physiological properties of interest industrially, both as a source of genes for insertion into other production strains and as a means of uncovering the mechanism by which enzymes from such organisms withstand extreme conditions. Already certain principles are emerging concerning the balance of charged amino acids which improves the heat resistance of enzymes, and concerning the importance of the position of cystine residues in peptide chains which permit disulphide bridges to form and so improve stability. The exploitation of this knowledge linked with site-directed mutagenesis will enable enzymes in existing production strains to be altered to improve their stability, and hence widen their conditions of use and lengthen their life. This has particular application for reactions with isolated enzymes or for whole cells used as a source of enzymes.

3. Cells Grown on Solid Surfaces

There is a need for further study of the effect of immobilization on cell growth. This technique has such wide applications that it is important to establish the basic kinetics of growth, substrate affinity, etc. which will enable immobilized cells to produce optimally. It is predictable that these conditions will not be identical to those which apply to free cells.

The possibilities of conducting reactions with immobilized enzymes in solvents have changed the prospects for the manipulation of many water-insoluble substrates. There is much basic work to be done here on solvent systems and on multiphase enzyme reactions.

4. Anaerobes

I also feel that the anaerobes in nature are a much neglected group. The very nature of their environment requires them to use electron acceptors other than oxygen which occur in the environment, or to develop such acceptors as part of their metabolism. This offers the opportunity of exploring the ability of anaerobes from natural environments to carry out reductions of interest. Success with elective culturing and enrichment techniques elsewhere in microbiology would seem to offer encouragement to put effort into probing this sector of the microbial world for interesting reactions and metabolites of possible value.

5. Mixed Cultures

With the exception of wastewater treatment, industrial processes predominantly use pure cultures and this will of course remain the case for many processes, especially those using strains which have been extensively manipulated genetically. However, pure cultures may not invariably be the most appropriate catalysts. A known mixture of organisms may provide the ideal way of removing unwanted by-products as was found by the workers at Shell who grew organisms on methane for single-cell protein.[6] Here the minority organisms removed free methanol (which was toxic to the methane oxidizer) and the end-products of the growth of the majority organism. The cell yield of the majority organism from methane was greatly enhanced in mixed culture compared with its productivity as a pure culture, and as a further benefit, foaming problems were eliminated in the mixed culture fermentation. Basic work to develop complementary mixed cultures and to understand their interactions needs to be fostered, particularly for processes where toxic by-products reduce the performance of the producer strain.

In waste treatment little attention has been directed toward developing specific mixtures of organisms with enhanced ability to metabolize recalcitrant substrates. Some effort has been put into incorporating extra catabolic genes into organisms, particularly pseudomonads, to widen their ability to degrade hydrocarbon wastes. However the full potential of elective culturing to establish known mixtures of robust organisms with complementary metabolic ability has hardly been addressed. The use of two-stage continuous flow systems in which the selected mixture is grown in the first stage, as a fully adapted inoculum for the second stage, which receives the main waste stream, has yet to be applied widely in the treatment of difficult wastes. This approach would allow higher concentrations of inhibitory wastes to be treated in the second stage along with other waste streams, without fear of losing the selected organisms.

6. Professor Aiba and Biotechnology

The ideas outlined above by no means exhaust the contributions that a knowledge of cellular physiology can make to biotechnology, and when I considered the publications of Prof. Aiba, I was immediately struck by the fact that his work and that of his students has indeed exploited most of those avenues and many more besides.

Professor Aiba's work has been characterized by the clear aim of providing reliable, kinetic data to describe biological processes. This is true whether he was discussing dissolved oxygen and growth yields, the role of protozoa in waste treatment, the relationship of bacteria to cyanobacteria in algal blooms, the rates of enzyme reactions, or areas in which kinetics are more conventionally applied, such as the sterilization of media, the removal of organisms from air streams, mass transfer, and the scaling up of fermentations. Most recently he has become interested in working with genetically manipulated bacteria for the production of amino acids and the manipulation of the genetic material of thermophilic bacilli. This very incomplete survey of the work of Prof. Aiba illustrates a most important feature of his approach to biotechnology. It underscores the important principle that biological processes have many features in common and can be analyzed and quantified using established methods, although the particular systems involved may at first glance appear very different. Professor Aiba's faith that his primary training in chemical engineering could be profitably applied to biological systems has been amply demonstrated in his own work and this approach will continue to contribute much to biotechnology as a multidisciplinary branch of science.

REFERENCES

1. Pirt, S.J., Righelato, R.C., "Effect of growth rate on the synthesis of penicillin by *Penicillium chrysogenum* in batch and continuous culture." *Appl. Microbiol*: **15**, 1284–1290 (1967).
2. Ryu, D.D.Y., Hospodka, J., "Quantitative physiology of *Penicillium chrysogenum* in penicillin fermentation." *Biotechnol. Bioeng.*: **22**, 289–298 (1980).
3. Harrison, D.E.F., "Growth, oxygen and respiration." *Crit. Rev. Microbiol.*: **2**, 185–228 (1973).
4. Primrose, S.B., Derbyshire, P., Jones, I.M., Robinson, A., Ellwood, D.C., "The application of continuous culture to the study of plasmid stability." In *Continuous Culture* 8, A.C.R. Dean, D.C. Ellwood, C.G.T. Evans (eds.), p. 212–265. Ellis Horwood Ltd., Chichester, 1984.
5. Goodwin, D., Slater, J.H., "The influence of the growth environment on the stability of a drug resistance plasmid in *Escherichia coli* K12." *J. Gen. Microbiol.*: **111**, 201–210 (1979).
6. Wilkinson, T.G., Topiwala, H.H., Hamer, G., "Interactions in a mixed bacterial population growing on methane in continuous culture." *Biotechnol. Bioeng.*: **16**, 41–59 (1974).

TECHNICAL PAPERS

I. Physiology and Kinetics

Reflections on the Dynamics of Growth and Product Formation in Microbial Cultures

S. John Pirt

Department of Microbiology, King's College (London), Kensington Campus, Campden Hill Road, London W8 7AH, England

1. Introduction

Shuichi Aiba and his school have done much not only to advance the subject of biochemical engineering and fermentation science but also to stimulate others to do so. The text[1] that emanated from his laboratory has become a classic in the field. I am honored by the invitation to contribute to this volume of studies collected under the aegis of the distinguished school from which Aiba is about to retire.

Microbial process dynamics is concerned with the control of biomass action by environmental factors such as substrate concentration. This involves manipulation of the genes, the contents, and the activities of enzymes in the biomass, and the structure and environment of the biomass. Of these aspects, the manipulation of the genome has become highly fashionable and the other aspects have been relatively neglected, largely, in my view, because of both theoretical and technical difficulties. This paper elaborates upon some of the problems involved and suggests possible solutions from the microbial physiologist's point of view.

2. Technical Difficulties

The fermenter or bioreactor is the apparatus required for the control of the biomass in its environment. Batch culture, fed-batch culture, and chemostat modes of operation may be used. One difficulty is that the bioreactor and its instrumentation, of necessity, have evolved into a highly complex and expensive form and access to it is much restricted by the present economic recession and consequent retrenchment in science. The big gap between the simple shake-flask and the sophisticated bioreactor needs to be bridged.

The theory of culture dynamics seems to defeat too many workers in the field. This points to a neglect of teaching the subject. With understanding, much more could be done with simple continuous culture apparatus and the choice of conditions to circumvent the limitations of the equipment.

The necessity for careful control of a continuous culture, possibly for weeks or months, is another difficulty because it calls for high reliability of the apparatus and its control, which is tedious as well as exacting on the apparatus. The multiplicity of factors which influence the process points to the need for computer control.

Continuous culture of the chemostat type with constant culture volume is the method *par excellence* to determine the biomass function-environment relations because it permits time-independent steady states to be obtained. Much of the tedium of continuous culture stems from the practice of allowing an arbitrary number of generation times of the organism under a particular set of conditions, in order to arrive at the steady state. The time required may be minimized and more information gained if the experimenter observes the changes in the transient state between two steady states. This rarely used approach is typified by its application to characterize bacterial cell wall changes associated with the phosphate supply[2] and to determine the trace element requirements for bacterial growth.[3]

3. Mathematical Models of Growth

The object of a mathematical model of growth is to express the dynamics of the process as quantitatively and generally as possible. The dynamics are based on material and energy balances and insight into the biological function involved. Insight comes only by much study of microbial physiology. Conversely, the microbial physiologist often lacks the chemical engineer's ability to manipulate mathematically the material and energy balances. The text by Aiba, Humphrey, and Millis[1] provided the first major bridge between the engineering and the physiological approaches.

A successful model is one which usefully describes a significant area of process behavior. It is not a failure if there are exceptions to the model or remote areas of interest in which it is invalid. A classical criterion of the success of a model is its predictive value and the significance of deductions that can be made from it. Also, a simple premise can express a basic principle which may be elaborated upon to extend the utility of the model, a possible example of which is the extension of the maintenance energy model to slow growth rates (see below).

Mathematical models of microbial growth of necessity start with the "black box" or "unstructured" model, which ignores the structure of the organism. In the "structured" model the structure considered can only be partial because of the complexity of the organism and the lack of knowledge about the structure. Alternative terms for the structured and the unstructured model are, respectively, the phenomenological and the "mechanistic," also, in energetics, macro- and microenergetics. Macroenergetics deals with the energetics of the whole unstructured biomass, whereas microenergetics refers to energetics at the molecular level. The field of macro- and microeconomics provides an interesting analogy of both terminology and some concepts, for instance, yield and growth rate. Where the unstructured model will do, it is pointless to adopt a more complex structured one; Occam's razor should be used to limit the number of assumptions. However, the ultimate goal must be to accommodate in the model the fullest possible description of the biomass structure and function. An excellent new synthesis of ideas in this field has been provided by Roels.[4]

The kinetics of a microbial process are expressed in terms of three basic rates: the rate of substrate consumption $(-ds/dt)$, the rate of biomass growth (dx/dt), and the rate of product formation (dp/dt). Growth on a carbon and energy source (C and E) may be expressed as

$$x + s = (1 + Y)x + (1 - Y)p \tag{1}$$

where s, x, and p represent 1 c-mol of substrate, biomass, and product, respectively, and Y is the growth yield (c-mol biomass/c-mol substrate; one c-mol of substance contains one mol c. For example, if glucose $(CH_2O)_g$ is the C and E source and lactate $(CH_2O)_L$ is the product, then we have

$$(CH_2O)_g + x = (1 + Y)x + (1 - Y)(CH_2O)_L.$$

The kinetic expressions for consumption or accumulation of material are: $-ds/dt = q_s x$; $dx/dt = \mu x$; $dp/dt = q_p$ where s, x, and p are the concentrations of substrate, biomass, and product, respectively; μ is the specific growth rate; and q_s and q_p are the specific rates of substrate consumption and product formation, respectively. One can also write $dx/dt = -Y ds/dt$ and $dp/dt = (Y - 1)ds/dt$, and then from this set of equations it follows that

$$-q_s = \mu/Y; \quad q_p = (1 - Y)\mu; \text{ and } q_p = (Y - 1)q_s.$$

During the last four decades or so attention has been focused on the way in which μ depends on s, that is, the function $\mu = f(s)$. In the simplest case it is assumed that μ is dependent on one s value only and is independent of all other nutrients. Monod[5] basically solved the problem of the $\mu - s$ relationship when he noted that his experimental results could be fitted by the equation

$$\mu = \mu_m s/(s + k_s) \tag{2}$$

where k_s is termed the "saturation constant" and μ_m is the maximum specific growth rate obtained when $s \gg k_s$ (see Fig. 1a). Monod[5] specifically disclaimed that his model could have a mechanistic or structural basis. He also omitted any mention of the analogy between his model and that of Michaelis-Menten for enzyme kinetics. Chemists use a similar model based on the Langmuir isotherm for chemically catalyzed reactions.

4. Role of Substrate Uptake Enzyme

One can take a step toward a structured model by considering that the reaction of the growth limiting substrate with the substrate uptake enzyme (E), is the rate limiting step or "bottleneck." Thus, the rate limiting reaction is represented as

$$E + S \underset{k_{-1}}{\overset{k_{+1}}{\rightleftharpoons}} ES \xrightarrow{k_2} E + P \tag{3}$$

where ES is the enzyme-substrate complex, P is an intermediary metabolite used in the synthesis of biomass, and the k values are the velocity constants of the reactions. Suppose that the substrate uptake enzyme is a fraction α of the biomass. Applying Michaelis-Menten kinetics, the velocity of substrate consumption is given by

$$-ds/dt = k_2 c = q_s x \tag{4}$$

where c is the concentration of ES (c-mol l^{-1}). On saturation of the biomass with substrate we assume that

$$c = \alpha x \text{ and } -(ds/dt)_{max} = k_2 \alpha x = q_{sm} x, \tag{5}$$

where $-(ds/dt)_{max}$ is the maximum possible rate of substrate uptake. From Eqs. (4) and (5) we obtain

$$q_s/q_{sm} = c/\alpha x. \tag{6}$$

We put $\alpha x = c + e$ where e is the concentration of E (c-mol l^{-1}) and $k_s = e.s/c =$

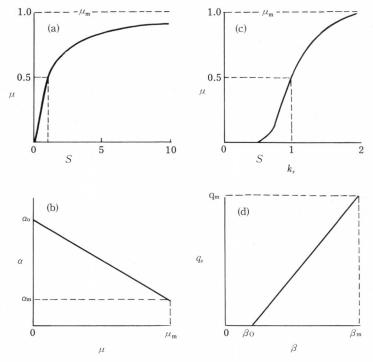

FIG. 1.   Biomass growth kinetics. Specific growth rate, μ and substrate conc. s (arbitrary units).
(a) Monod kinetics. (b) Linear variation of substrate uptake enzyme content (α) with μ, e.g., PTS
enzyme of *Klebsiella aerogenes*.[7] (c) Cooperative effect (Hill number, $n = 5$). (d) Linear variation
of specific rate of conserved element uptake (q_s) with conserved element content of biomass (β),
e.g., phosphate uptake by *Chlorella*.[10]

k_{-1}/k_{+1} on the assumption that the reaction of E with S is practically at equilibrium. By
means of these simultaneous equations, c is eliminated from Eq. (6) and one obtains

$$q_s = q_{sm}s/(s + k_s). \tag{7}$$

From Eq. (7) it follows that

$$\mu = \mu_m s/(s + k_s). \tag{8}$$

It is implicit in Eq. (8) that Y is constant and independent of μ. The limited validity of
that assumption is considered below. Another implicit assumption is that the enzyme content α is constant, but there is mounting evidence that the amount of substrate uptake
enzyme in the biomass varies inversely with the specific growth rate. An example is the
PTS (phosphotransferase system) for glucose uptake enzyme in *Klebsiella aerogenes*, the
amount of which, during glucose limitation, has been found to increase linearly with decrease in μ[7] as shown in Fig. 1b. In that case it may be derived (see Appendix) that

$$\mu = \mu_m s/(s + k_s') \tag{9}$$

where $k_s' = (\alpha_m/\alpha_o)k_s$ and α_m and α_o are the enzyme (E) contents of the biomass when

$\mu = 0$ and $\mu = \mu_m$, respectively. Thus, the Monod equation may still be applied and the apparent saturation constant takes account of the variation in the amount of the substrate uptake enzyme.

The quantitative description of the factors affecting μ requires the action of inhibitors to be taken into account. Inhibitors may be either a product, a substrate, or an inhibitor added with the substrate. The addition of inhibitors to cultures appears to offer almost infinite variation in the growth kinetics through the competitive and non-competitive modes,[8] to which must be added the uncompetitive mode.[9] So far the effects of inhibitors on growth dynamics have been largely neglected and remain a wide open field for future investigation.

5. The Cooperative Effect

If the substrate uptake enzyme is subject to a cooperative effect whereby uptake of substrate increases the affinity of the enzyme for the substrate, then the substrate uptake reaction is considered to be

$$E + nS \rightleftharpoons ES_n \rightarrow E + P. \tag{10}$$

Application of Hill kinetics to this system leads to the following expression[10]

$$\mu = \mu_m s^n / (s^n + k_s) \tag{11}$$

which gives a sigmoid response of μ to s (Fig. 1 c). Panikov and Pirt[10] found that urea-limited growth of *Chlorella* conformed to this model with $n = 5.5$. This model of the cooperative effect may be regarded as a more general one since it contains the Monod model when $n = 1$.

6. Conserved Elements

A conserved element is defined as one which is entirely assimilated into the biomass; usually the elements N, P, S, K, Mg, and trace elements come into this category. In contrast, examples of unconserved elements are C, H, and O. The conserved element is characterized by the fact that the content of the element in the biomass is $\beta = 1/Y$ where Y is the growth yield. Growth limitation by a conserved element seems to conform to the relation illustrated in Fig. 1 d, that is, q_s is a linear function of β. Experimental evidence for this relation is found in the studies on N- and P-limited growth of algae[10,11] and N-, P-, K-, or Mg-limited growth of bacteria.[12,13] Organic growth factors also are conserved and the uptake of vitamin B_{12} by an alga conforms to the relation given in Fig. 1 d.[11]

The observed relation of q_s to β (Fig. 1 d) can be represented as

$$q_s / q_m = (\beta - \beta_o) / (\beta_m - \beta_o) \tag{12}$$

where β_o is the threshold value of β when $q_s = 0$ and β_m is the maximum nutrient content of the biomass occurring when $q_s = q_m$. Putting $q_s = \mu\beta$ and $q_m = \mu_m\beta_m$ we obtain for the μ-s relation,[10]

$$\mu = \mu_m s / \left(s + \frac{\beta_o}{\beta_m} k_s \right). \tag{13}$$

Thus, the effect of the variation in the growth yield may be accounted for by a simple modification of the k_s value. For phosphate limitation of *Chlorella*,[10] $\beta_o/\beta_m = 0.25$, and for

B_{12} limitation of an alga,[11] $\beta_o/\beta_m = 0.33$. Such an effect on the saturation constant, of course, would be crucial in determining the survival of the species in competition with other organisms.

7. Multiple Substrate Limitation

Multiple substrate limitation of growth with different C and E sources may occur.[14] For nutrients with different functions, for instance, N and P, there appears to be a sharp transition from dependence of μ on one nutrient to dependence on the other.[15] Examples are change from urea to phosphate dependence of *Chlorella*[15] and from NH_3 to sugar dependence in *Penicillium chrysogenum*.[16] However, the work of Panikov[15] showed that although a single substrate (urea) was controlling μ, the growth yield from a second substrate (phosphate) varied over a wide range of phosphate concentrations above the limiting level. The production of penicillin is also dependent on a wide range of NH_3 levels above the growth limiting level.[16] The study of Tempest et al.[2] suggested that the composition of the polymers in *Bacillus* cell walls could vary progressively with the increase in the phosphate concentration above the growth limiting level. In general, it appears that both the biomass structure ånd its metabolic activity can be functions of a number of substrate concentrations apart from that of the growth limiting substrate. From this it follows that the practice of adding the substrate in arbitrary amounts, apart from the growth limiting substrate, can be inadequate to achieve full control over the process by environmental factors.

8. Trace Element Requirements

The influences of the trace elements (TE) on microbial growth are among the most obscure aspects of growth dynamics. The salient TE is Fe because of the multiplicity of functions it serves, and also because the demand for it is usually sufficiently large to make it easy to demonstrate an Fe requirement for growth. However, general knowledge of the quantitative requirements for the TE is totally lacking. A TE deficiency may have a surprisingly novel effect, for instance, Mn deficiency in a culture of *Agrobacterium* induces formation of the novel enzyme glucose-3-dehydrogenase.[17]

A limit to the supply of TE may be set by some simple stoichiometry as follows. The enzyme (protein) content of biomass is taken to be 50% of the dry wt. The average mol. wt. of an enzyme is assumed to be 25,000. The TE is assumed to be present to the extent of one mol/mol enzyme. Then the mol of enzyme per g dry wt. will be 2×10^{-5}. If the TE were Fe, then the requirement would be 1.12 mg/g dry wt. This is consistent with the Fe requirement for growth of *Chlorella*.[18] This requirement was dependent on μ like that of the conserved elements. In all cases the quantitative dynamics and control of TE metabolism are among the least understood aspects of fermentation science. Understanding and quantifying the TE requirements could prove to be the most crucial problem in microbial process dynamics.

The technical difficulties in the study of the TE requirements are not trivial. One problem is the insolubility of the trace metals except at low pH values. This may be overcome by a continuous feed of an acid solution of the TE. Alternatively, a chelating agent, usually EDTA, may be introduced to act as a metal ion buffer. However, EDTA is inadequate to prevent precipitation of ferric hydroxide except at acid pH values. This precipitation of iron in the presence of EDTA can be overcome by the use of ferrous ions, although access of O_2

will slowly oxidize the ferrous ions. The other outstanding problem is to eliminate or stand-ardize the amount of TE present in the medium constituents as contaminants. The solution to this problem is *ad hoc*. Use of the highest possible concentration of biomass will mini-mize this difficulty.

9. Macroenergetics

Maintenance energy. According to the concept of maintenance energy[19] the specific rate of uptake of the energy source (q_E) by a growing culture is the sum of two terms

$$q_E = \mu/Y_{GE} + m \tag{14}$$

where Y_{GE} is the "true" or "maximum" growth yield from the energy source and m is the maintenance energy coefficient. Also, since $q_E = \mu/Y_E$ where Y_E is the actual growth yield it follows that

$$1/Y_E = 1/Y_{GE} + m/\mu. \tag{15}$$

If the substrate is the source of both carbon and energy (C and E) and q_s is the specific rate of this consumption then

$$q_s = \frac{\mu}{Y_c} + \frac{\mu}{Y_{GE}} + m \tag{16}$$

where Y_c is the growth yield from the substrate used to provide cell C only. If the substrate used to provide cell C were completely conserved then $Y_c = 1$ (c-mol/c-mol). Actually, in the case of a yeast it has been found that $Y_c \approx 0.9$, according to Professor Kuenen of Delft (personal communication). From Eq. (16) putting $q_s = \mu/Y$, it follows that

$$1/Y = 1/Y_G + m/\mu \tag{17}$$

where $1/Y_G = 1/Y_c + 1/Y_{GE}$. Rearrangements of Eqs. (15) and (17) give

$$Y_E/Y_{GE} = \mu/(\mu + a_E) \tag{18}$$

and

$$Y/Y_G = \mu/(\mu + a) \tag{19}$$

where $a_E = mY_{GE}$ and $a = mY_G$. The a terms are called the "specific maintenance rates." The term $a = mY_G$ corresponds to the "endogenous metabolism" rate of Herbert.[20] Thus, there is a subtle difference between the a and a_E terms. One consequence is that when comparing maintenance energy by means of the a term one must ensure that either the a or a_E term is used consistently. It may be argued that the m term is the more fundamental because it is independent of the definition of the Y_G term.

Meaning of maintenance energy. Our recent study[21] of the growth dynamics of strictly anaerobic digestion of glucose associated with methanogenesis showed that the maintenance energy term is surprisingly small compared with that of aerobic bacteria if m is viewed as the ATP equivalent (m_{ATP}). For instance, the m_{ATP} of the GD organism, a strict anaerobe that ferments glucose, is about one-tenth of that of the aerobic *Aerobacter cloacae*. To account for this difference in m_{ATP} values, it was suggested that the major role of maintenance energy in aerobic cultures is not to supply ATP but rather reducing equivalents to reverse oxidation of reduced molecules in the cell or to "mop up" toxic O_2 radicals such as superoxides. According to this hypothesis the maintenance energy term could increase with the dissolved O_2 tension. This, of course, seems to be the case in the nitrogen fixing bacteria.[22] The results of Maclennan et al.[23] suggest that hyperbaric O_2 increases the maintenance energy. Another cause of the difference in m_{ATP} values could be that the aerobes, unlike the anaerobes, depend on the chemiosmotic mechanism and that leakage of protons across the osmotic barrier dissipates some of the proton motive force with consequent loss of energy. This latter mechanism might explain why the maintenance energy terms for the eukaryotic yeasts and fungi[8] and algae[24] tend to be significantly lower than those of the aerobic prokaryotes, since in the eukaryotes the chemiosmotic membranes are in internal organelles and are thus protected to some extent against the external environment of the cell.

Slow growth rates. Numerous studies over two decades have confirmed that over most of the growth rate range (from μ_m down to about 0.05 μ_m) the maintenance energy concept holds. However, at slow growth rates (less than about 0.05 μ_m) the maintenance energy appears to decrease and the growth yield is greater than that expected.[25] This growth yield deviation at slow growth rates may be accounted for by postulating that a part of the population differentiates into a dormant or non-viable state with zero maintenance energy.[25] According to this model, the uptake of the carbon and energy source is given by

$$q_s = \mu_o/Y_G + \gamma m \qquad (20)$$

where γ is the growing fraction of the culture, the dormant fraction being $(1 - \gamma)$. The observed specific growth rate is $\mu_o = \gamma\mu$ where μ is the specific growth rate of the growing fraction. In a chemostat culture in the steady state $\mu_o = D$. Knowing the value of m, it is possible by means of Eq. (20) to calculate the value of the growing faction γ. Several ways in which this extension of the maintenance energy model to cover slow growth rates may be tested have been suggested.[25]

Secondary metabolism such as antibiotic formation is frequently associated with slow growth rates. The dynamics of secondary metabolism constitutes another little understood problem and the industrial processes involved are consequently highly empirical and largely based on trial and error. Control of the dormant state could be highly relevant to the control of secondary metabolite formation.

10. Conclusion and Summary

The dynamics of growth and product formation in microbe and cell cultures are still in an early state of development that is just leaving behind the "black box" approach. There is unlimited scope for the development of both the technical methods and theoretical approaches used in the field.

The Monod equation for growth kinetics (Eq. 2) is remarkably flexible and able to accommodate the complexities of structured models such as the variations in the growth yield of conserved substrates and in the amount of substrate uptake enzyme in the biomass. Also, by means of the Hill variation on the kinetics and the cooperative effect the sigmoidal response of μ to s can be introduced. It is sometimes suggested that these variations in the growth kinetics are fine distinctions of little significance. That is not true because they are supremely important in determining the ability of the organism to compete in its environment against genotypic or phenotypic variants.

Microbial dynamics present unlimited opportunities for research of both fundamental importance and potential to advance biotechnology. Outstanding problems are: biomass behavior at slow growth rates, the influence of inhibitors, the effects of trace elements, and the dynamics of secondary metabolite formation. An attack on the complexities of the effects of slow growth rates could be facilitated by the concept of dormant cell formation.

Microbial growth dynamics are now on the crest of a wave, the creation of which owes much to the contributions of the school of Shuichi Aiba.

Appendix

The influence of linear variation in the substrate uptake enzyme content of biomass (α) as shown in Fig. 1b is modelled as follows. The specific rate of substrate uptake is given by

$$q_s = q_T s/(s + k_s). \tag{21}$$

The maximum enzyme activity is given by

$$q_T = k\alpha, \; q_o = k\alpha_o, \text{ and } q_m = k\alpha_m \tag{22}$$

where k is a constant. From the linear relation (Fig. 1b) it follows that

$$\alpha = \alpha_o - \frac{(\alpha_o - \alpha_m)}{\mu_m} \mu. \tag{23}$$

From relations (22) and (23) and assuming $q_s = \mu/Y$, $q_m = \mu_m/Y$ where Y is a constant it follows that

$$q_T = q_o - \frac{(q_o - q_m)}{q_m} q_s. \tag{24}$$

Substitution for q_T in Eq. (21) leads to

$$q_s = q_m \frac{s}{s + (q_m/q_o)k_s} = q_m \frac{s}{s + (\alpha_m/\alpha_o)k_s}. \tag{25}$$

Consequently, assuming $q_s = \mu/Y$, it follows that

$$\mu = \mu_m s \Big/ \Big(s + \frac{\alpha_m \, k_s}{\alpha_o}\Big) = \mu_m \, s \Big/ \Big(s + \frac{q_m \, k_s}{q_o}\Big). \tag{26}$$

The values of q_T may be obtained by decryptifying the enzyme.[7]

*Nomenclature with S.I. Units**

$a = mY_G$	specific maintenance rate, s^{-1}
$a_E = mY_{GE}$	specific maintenance energy rate, s^{-1}
c	conc. of enzyme-substrate complex, mol m^{-3}
e	enzyme conc., mol m^{-3}
k	reaction velocity constant (units defined by law of mass action)
k_s	saturation constant, mol m^{-3}
k_s'	modified value of saturation constant, mol m^{-3}
m	maintenance energy coefficient, mol substrate (c-mol biomass)$^{-1}$s^{-1}
n	Hill number

* Departures from this nomenclature are noted in the text.

p	product, mol, or product conc., mol m^{-3}
q	specific metabolic rate, mol (c-mol biomass)$^{-1}$s^{-1} subscripts: m, maximum value; p, product; s, substrate
q_m	maximum specific rate of substrate utilization at a particular value of μ, mol (c-mol biomass)$^{-1}$s^{-1}
s	substrate, mol or substrate conc., mol m^{-3}
t	time, sec
x	biomass conc., c-mol m^{-3}
Y	growth yield, c-mol biomass, mol substrate^{-1}, subscripts: E, from energy source; G, maximum value
Y_c	growth yield from c substrate used to provide cell c only, c-mol, mol substrate^{-1}
α	enzyme content of biomass, c-mol, c-mol biomass^{-1} subscripts: o, value at $\mu = 0$; m, value at $\mu = \mu_m$
β	conserved element content of biomass, mol c-mol biomass^{-1}
γ	growing fraction of biomass
μ	specific growth rate, s^{-1}
μ_m	maximum specific growth rate, s^{-1}
μ_o	observed specific growth rate

REFERENCES

1. Aiba, S., Humphrey, A.E., Millis, N.F., In *Biochemical Engineering* (2nd ed.), Academic Press, New York, 1973.
2. Tempest, D.W., Herbert, D., Phipps, P.J., In *Microbial Physiology and Continuous Culture, Third International Symposium*, Powell, E.O., Strange, R.E., Tempest, D.W. (eds.), p. 240–255. Her Majesty's Stationery Office, London, 1967.
3. Mateles, R.I., Battat, E., "Continuous culture used for media optimization." *Appl. Microbiol.*: **28**, 901–905 (1974).
4. Roels, J.A., In *Energetics and Kinetics in Biotechnology*, Elsevier, Amsterdam, 1983.
5. Mond, J., *Recherches sur la Croissance des Cultures Bactériennes*, Herman, Paris, 1942.
6. Hinshelwood, C.N., *The Chemical Kinetics of the Bacterial Cell*, Clarendon Press, Oxford, 1946.
7. O'Brien, N.W., Neijssel, O.M., Tempest, D.W., "Glucose Phosphoenol-pyruvate Phosphotransferase activity and Glucose uptake rate of *Klebsiella aerogenes* growing in chemostat culture." *J. Gen. Microbiol.*: **116**, 305–314 (1980).
8. Pirt, S.J., *Principles of Microbe and Cell Cultivation*, Blackwell Scientific, Oxford, 1975.
9. Miles, R.J., Pirt, S.J., "Inhibition by 3-deoxy-3-fluoro-D-glucose of the utilization of lactose and other carbon sources by *Escherichia coli*." *J. Gen. Michobiol.*: **76**, 305–318 (1973).
10. Panikov, N., Pirt, S.J., "The effects of cooperativity and growth yield variation on the kinetics of nitrogen or phosphate limited growth of *Chlorella* in a chemostat culture." *J. Gen. Microbiol.*: **108**, 295–303 (1973).
11. Droop, M.R., "The nutrient status algal cells in continuous culture." *J. Mar. Biol. Ass. U.K.*: **54**, 825–855 (1974).
12. Nyholm, N., "A mathematical model for microbial growth under limitation by conservative substrates." *Biotechnol. Bioeng.*: **18**, 1043–1056 (1976).
13. Nyholm, N., "Kinetics of phosphate limited algal growth." *Biotechnol. Bioeng.*: **19**, 467–492 (1977).
14. Harder, W., Dijkhuizen, L., "Physiological responses to nutrient limitation." *Ann. Rev. Microbiol.*: **37**, 1–23 (1983).
15. Panikov, N., "Steady state growth kinetics of *Chlorella vulgaris* under double substrate (urea and phosphate) limitation." *J. Chem. Tech. Biotechnol.*: **29**, 442–450 (1979).
16. Court, J.R., Pirt, S.J., "Carbon- and nitrogen-limited growth of *Penicillium chrysogenum* in fed batch

culture: The optimal ammonium ion concentration for penicillin production." *J. Chem. Tech. Biotechnol.*: **31**, 235–240 (1981).

17. Kurowski, W.M., Fensom, A.H., Pirt, S.J., "Factors influencing the formation and stability of D-glucoside 3-dehydrogenase activity in cultures of *Agrobacterium tumefaciens*." *J. Gen. Microbiol.*: **90**, 191–202 (1975).

18. Pirt, S.J., Walach, M., "Biomass yields of *Chlorella* from Iron ($Y_{x/Fe}$) in iron-limited batch cultures." *Arch. Microbiol.*: **116**, 293–296 (1978).

19. Pirt, S.J., "The maintenance energy of bacteria in growing cultures." *Proc. Roy. Soc.*: **163**, 224–231 (1965).

20. Herbert, D., "Some principles of continuous culture." In *Recent Progress in Microbiology*, p. 381–396, VII Intern. Congr. for Microbiology, 1958.

21. Pirt, S.J., Harty, D.W., Salmon, I., Lee, Y-K., "Methanogenic digestion of glucose plus yeast extract by a defined bacterial consortium: carbon balances and growth yields in chemostat culture." *J. Ferment. Technol.*: **65**, 159–172 (1987).

22. Nagai, S., Aiba, S., "Reassessment of maintenance and energy uncoupling in the growth of *Azotobacter vinelandii*." *J. Gen. Microbiol.*: **73**, 531–538 (1972).

23. Maclennan, D.G., Ousby, J.C., Vasey, R.B., Cotton, N.T., "The influence of dissolved oxygen on *Pseudomonas* AM1 grown on methanol in continuous culture." *J. Gen. Microbiol.*: **69**, 395–404 (1971).

24. Pirt, S.J., Lee, Y-K., Richmond, A., Watts-Pirt, M., "The photo-synthetic efficiency of *Chlorella* biomass growth with reference to solar energy utilisation." *J. Chem. Tech. Biotechnol.*: **30**, 25–34 (1980).

25. Pirt, S.J., "The energetics of microbes at slow growth rates: maintenance energies and dormant organisms." *J. Ferment. Technol.*: **65**, 173–177 (1987).

I-2

Suggestions Concerning Measurement of the Concentration of Actually Effective Cells in Fermenting Media

Walter Borzani

Instituto Mauá de Tecnologia, Centro de Pesquisas, Estrada das Lágrimas 2035, 09580 São Caetano do Sul, SP, Brazil

Present methods for the measurement of cell concentration (Calam, 1969; Kubitschek, 1969; Mallette, 1969; Postgate, 1969;[1-5] Stouthamer, 1969[7]) do not provide information on the fraction of the microbial population effectively responsible for the reactions observed in a fermenting medium. Even the number of viable cells cannot be considered, in this respect, an evaluation of the fraction mentioned above, because in the plate counting method the microbial cells are subjected to conditions very different from those in the fermentation process. Thus, quantifying the cells actually responsible for a given transformation in a fermenting mash is one of the major problems to be solved by those interested in the study of microbial processes.

Active cells are those microbial cells effectively responsible for a given transformation A → B in a fermenting mash. Other cells present in the fermenting medium are referred to as *inactive* cells in the transformation A → B. It must be pointed out that an *inactive cell* is not necessarily a dead one. The transformation could be, for example, cell growth, substrate consumption, or product formation.

We may then write:

$$X = X_a + X_i. \tag{1}$$

The *activity* of the microbial population regarding the transformation A → B will be defined by:

$$a = \frac{X_a}{X}. \tag{2}$$

To show only the influence of X_a and X_i on kinetic parameters, let us consider the very simple case of a continuous culture in a chemostat (Sinclair and Topiwala, 1970).[6] In this case, X_a is the concentration of growing cells, and X_i is the concentration of non-growing cells. We may then write:

$$\frac{dX_a}{dt} = \mu \cdot X_a - D \cdot X_a - k \cdot X_a \tag{3}$$

$$\frac{dX_i}{dt} = k \cdot X_a - D \cdot X_i. \tag{4}$$

39

Equations (1), (3), and (4) lead to:

$$\frac{dX}{dt} = \mu \cdot X_a - D \cdot X. \tag{5}$$

In the steady state $(dX/dt = 0)$ Eq. (5) permits us to write:

$$\mu = D \cdot \frac{X}{X_a} \tag{6}$$

In other words, the specific growth rate will be equal to the dilution rate only when $X_i = 0$. Otherwise, the steady-state defined by $dX/dt = 0$ may represent the following two different situations. First,

$$\frac{dX_a}{dt} = \frac{dX_i}{dt} = 0. \tag{7}$$

Obviously, this situation leads to constant values of X_a, X_i and μ, respectively.

Secondly, we may assume:

$$\frac{dX_a}{dt} = -\frac{dX_i}{dt} \neq 0. \tag{8}$$

Then the respective X_a, X_i, and μ values will not be constant during the steady state.

The above example clearly shows that the measurement of the active cell concentration may substantially affect the interpretation and the consequences of experimental results.

It is possible, however, to demonstrate that certain parameters do not depend on the presence of inactive cells. In fact, the mass balance of the limiting substrate in the continuous culture considered above, assuming that the substrate consumption for the cell maintenance is negligible, leads to:

$$\frac{dS}{dt} = D(S_0 - S) - \frac{1}{Y} \cdot \mu \cdot X_a. \tag{9}$$

Combining Eqs. (6) and (9), we obtain at steady state $(dS/dt = 0)$:

$$X = Y(S_0 - S). \tag{10}$$

Hence, the value of Y obtained from a steady-state chemostat experiment does not depend on the active cell concentration.

Several important results have already been obtained from fermentation experiments in spite of the impossibility of evaluating the actual active cell concentration in fermenting media. We cannot predict what the results would have been if methods of measuring active cell concentration were available. Efforts in this direction will most likely lead to a more correct interpretation of experimental results and also to new theoretical and/or practical conclusions. This paper presents some suggestions based on experimental results published elsewhere.

Let us comment, first of all, on a paper by Sinclair and Topiwala (1970).[6] Defining the *viability* of a microbial population by the ratio of viable to total cells as

$$v = \frac{X_v}{X} \tag{11}$$

those authors observed that the value of v, in a steady-state, continuous culture experiment, depends on the dilution rate according to the following equation:

$$v = \frac{D}{D + \gamma}.$$
(12)

Figure 1 shows typical experimental results.

From a qualitative point of view, the influence of dilution rate on cell viability would be expected, because as D decreases, the culture medium in the chemostat will become more and more depleted and, consequently, less favorable to cell maintenance and growth. At relatively low dilution rates the medium composition in the reactor will negatively affect the viability of the microbial population.

The results presented by Sinclair and Topiwala (1970),[6] despite the known limitations of the method used to measure the total cell concentration and the number of viable cells, clearly indicate that the maximum viability value will be attained in a steady-state, continuous culture test at the maximum dilution rate. In other words, if the dilution rate is sufficiently high (the condition $D < \mu_{max}$ must be fulfilled in order to avoid wash-out) we may assume that 100% of the microbial cells are active.

This leads to the first suggestion regarding measurement of the active cell concentration in microorganism culture (in this case, active cells and growing cells are synonymous).

Assume that microorganism M is batch cultivated on culture medium C in fermentor F; that microorganism M is continuously cultivated in a chemostat, using culture medium C as feeding mash; and that a steady state is established at a very high dilution rate ($a \cong 1$). To measure the activity of the microbial cells in fermentor F at a given time t, we may proceed as follows:

1. At time t, suitable volumes of media are sampled both from fermentor F and from the chemostat, and then centrifuged (or filtered) in order to obtain the respective microbial cells.
2. A flask F_1, containing a volume V of the fresh culture medium C is inoculated with mass m_1 (dry matter) of cells obtained from the batch fermentor F.
3. A flask F_2, containing also the same volume V of fresh culture medium C, is inoculated with mass m_2 (dry matter) of cells obtained from the chemostat.
4. Both flasks, F_1 and F_2 are then maintained under the same conditions for optimal cell growth.
5. The experimental growth data obtained from flasks F_1 and F_2 permit us to calculate the respective initial growth rates, R_1 and R_2.

Because the only differences between flasks F_1 and F_2 are the masses of the inocula and their respective activities, we may write:

$$R_1 = \alpha \cdot m_1 \cdot a_1$$
(13)

$$R_2 = \alpha \cdot m_2$$
(14)

where α is a constant and a_1 is the activity of cells obtained from batch fermentor F.
Equations (13) and (14) yield:

$$a_1 = \frac{R_1}{R_2} \cdot \frac{m_2}{m_1}.$$
(15)

This method will be called Method 1.

The problem will certainly be more complex if we intend to measure microbial cell activity in terms of the production of a given substance during a fermentation test. In this case we may write:

$$\mu_P = \frac{1}{X} \cdot \frac{dP}{dt} \qquad (16)$$

Equations (2) and (16) give rise to:

$$\mu_P = a \cdot \frac{1}{X_a} \cdot \frac{dP}{dt} = \beta \cdot a \qquad (17)$$

where:

$$\beta = \frac{1}{X_a} \cdot \frac{dP}{dt}. \qquad (18)$$

Under a given set of experimental conditions, however, β will be constant, whatever the value of cell activity may be. Equation (17) then shows that the specific production rate is proportional to the activity of the microbial population. In other words, the value of μ_P may be adopted as an indirect evaluation of the cell activity in terms of the production of a particular substance, although in certain continuous fermentations the specific production rate increases when the dilution rate increases (Wang et al., 1979).[8]

Figure 2 shows values calculated from experimental results obtained in steady-state, continuous ethanol fermentation of blackstrap molasses (Perego Jr., et al., 1985)[4] and shows that the correlation between μ_E and D is similar to that proposed by Sinclair and Topiwala (1970)[6] in studying the influence of D on cell viability (Fig. 1).

It therefore seems acceptable to assume that when a Monod-like equation correlates the specific production rate of a given substance to the dilution rate in steady-state, continuous fermentation tests (Fig. 2), the maximum value of cell activity (regarding the production of the substance) will be attained in a steady-state, continuous fermentation experiment carried out at a dilution rate as high as possible.

Based on the above, a method similar to Method I can be applied to the evaluation of cell activity in terms of the production of a given substance in a fermenting medium. This new method will be called Method II.

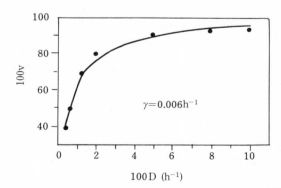

FIG. 1.   Influence of dilution rate (D) on cell viability (v) in steady-state, continuous culture (Sinclair and Topiwala, 1970).

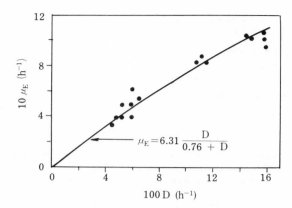

FIG. 2. Influence of dilution rate (D) on specific production rate of ethanol (μ_E) in steady-state, continuous ethanol fermentation. Values are calculated from the results published by Perego Jr., et al. (1985).

In order to avoid unnecessary repetition only the main differences between Methods I and II will be noted.

1. In Method I the experimental conditions favor cell growth, while the conditions adopted in Method II favor the production of a particular substance.
2. R_1 and R_2 in Method II are the initial production rates of the substance in flasks F_1 and F_2, respectively.

Based on the values of m_1, m_2, R_1, and R_2, the cell activity may be calculated from Eq. (15).

A great deal of experimental work must be done to verify the possibility of applying the above suggestions in practice, but I am sure that the problem discussed here deserves the attention of workers interested in fermentation processes.

Nomenclature

a, a_1	activity of microbial cells
D	dilution rate
k	specific rate of cell inactivation
m_1, m_2	mass of cells (dry matter)
P	product concentration
R_1, R_2	initial rates
S	concentration of limiting substrate
S_0	value of S of feeding mash
t	time
v	viability of microbial cells
V	volume of culture medium
X	cell concentration (dry matter)
X_a	active cell concentration (dry matter)
X_i	inactive cell concentration (dry matter)
X_v	viable cell concentration (dry matter)
Y	yield factor
α	empirical constant; see Eqs. (13) and (14)

β	actual specific production rate; see Eq. (18)
γ	specific death rate
μ	specific growth rate
μ_{max}	maximum value of μ
μ_E	specific production rate of ethanol
μ_P	specific production rate

REFERENCES

1. Calam, C.T., "The Evaluation of Mycelial Growth." In *Methods in Microbiology* (J.R. Norris, D.W. Ribbons) (eds.) **1**, p. 567–591. Academic Press, London (1969).
2. Kubitschek, H.E., "Counting and Sizing Micro-Organisms with the Coulter Counter." In *Methods in Microbiology* (J.R. Norris, D.W. Ribbons) (eds.), **1**, p. 593–610. Academic Press, London (1969).
3. Mallette, M.F., "Evaluation of Growth by Physical and Chemical Methods." In *Methods in Microbiology* (J.R. Norris, D.W. Ribbons) (eds.) **1**, p. 521–566. Academic Press, London (1969).
4. Perego Jr., L. Dias, J.M.C. de S., Koshimizu, L.H., Cruz, M.R. de M., Borzani, W., Vairo, M.L.R., "Influence of Temperature, Dilution Rate and Sugar Concentration on the Establishment of Steady-State in Continuous Ethanol Fermentation of Molasses." *Biomass*: **6**, 247–256 (1985).
5. Postgate, J.R., "Viable Counts and Viability." In *Methods in Microbiology* (J.R. Norris, D.W. Ribbons) (eds.), **1**, p. 611–628. Academic Press, London (1969).
6. Sinclair, C.G., Topiwala, H.H., "Model for Continuous Culture which Considers the Viability Concept." *Biotechnol. Bioeng.*: **12**, 1069–1079 (1970).
7. Stouthamer, A.H., "Determination and Significance of Molar Growth Yields." In *Methods in Microbiology* (J.R. Norris, D.W. Ribbons) (eds.), **1**, p. 629. Academic Press, London (1969).
8. Wang, D.I.C., Cooney, C.L., Demain, A.L., Dunnill, P., Humphrey, A.E., Lilly, M.D., In *Fermentation and Enzyme Technology*, p. 131. John Wiley & Sons, New York (1979).

I-3
Studies on Flow Bioreactors for Microbial Reaction

Kiyoshi Toda

Institute of Applied Microbiology, University of Tokyo, Bunkyo-ku, Tokyo 113, Japan

1. Introduction

A new methodology for investigating the biological activity of microorganisms growing in a constant environment was devised in the 1950s and has been used since the classic studies of continuous culture.[1,2] Nowadays, we cannot do anything without the technique of continuous culture in the quantitative assessment of microbial physiology and bioreaction kinetics. Experimental batch culture procedures are easier to perform than continuous culture and have been carried out extensively in scientific and engineering studies. However, the reproducibility of the results is often unsatisfactory for quantitative analysis or model simulation of the microbial reaction. This partly originates from the difficulty in preparing inoculum cultures of the same quality as used in previous experiments. Moreover, chemical as well as physiological conditions of the environment surrounding microorganisms (concentrations of substrate and product, pH, osmotic pressure of nutrient medium, etc.) change in batch culture during the time course of microbial growth. This profile of change is quite hard to reproduce, because the nature of microbial growth is stochastic or probabilistic rather than deterministic.

In the experiments using continuous culture, such problems as the variability of inoculum activity and transient changes in the microbial environment can be minimized. However, a spontaneous mutation in continuous culture may occasionally exert a serious influence on the characteristics of the culture population so far as the phenotypic expression of the mutation is evident. Table 1 summarizes the advantages and disadvantages of continuous culture.

Usually continuous culture cannot be operated at a high rate of incoming flow of fresh medium whenever the growth rate of microorganisms cannot compete with the dilution rate of the inflowing medium. This phenomenon of "wash out" limits the applicability of continuous culture as a bioreactor. Since studies on immobilized microorganisms initiated in the 1960s were extended during the 1970s, a new aspect has appeared in the study of continuous culture of microorganisms. Entrapment of microbes in an appropriate support not only enhances the upper limit of space velocity in the continuous culture but also excludes the serious defect of accidental contamination by foreign microbes in the culture. The latter defect has been the bottleneck to industrial and large-scale applications of continuous culture. Many engineers and technicians are now interested in manufacturing

46   K. Toda

TABLE 1. Advantages and disadvantages of continuous culture.

Advantages
   (1) Cell physiology can be studied under constant environmental conditions.
   (2) Microbial reaction at low substrate concentration (such as decomposition of toxic substances and derepression of biosynthesis) can be investigated.
   (3) Nutrients that limit the microbial growth rate can be selected arbitrarily.
   (4) Data reproducibility is better than in batch experiments.
   (5) Strains can be selected for specific qualities, i.e., resistance to product inhibition, susceptibility to substrate inhibition, or low maximum specific growth rate but high affinity to substrate.

Disadvantages
   (1) The probability of contamination is high over the long term.
   (2) The distribution of cell maturity is broad due to short retention time of liquid in the vessel.
   (3) The influence of wall growth is conspicuous.
   (4) Spontaneous mutants may become predominant.

microbial products using this new type of continuous culture utilizing immobilized microorganisms.

A short review on the application of continuous culture as a tool for investigating microbial physiology and production of microbial metabolites will be attempted and our studies on both chemostat culture and immobilized microorganisms will be introduced.

2. Flow Bioreactors

Figure 1 shows several flow bioreactor types. The single-stage (SS) bioreactor has been widely used for chemostat culture of microbes. In continuous culture using the SS fermentor, enzyme derepression can be realized, because a low concentration of substrate can be maintained at a low dilution rate. This phenomenon of improving enzyme production

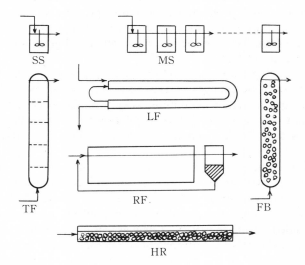

FIG. 1. Flow bioreactors.
   SS: Single-stage continuous fermentor; MS: multistage continuous fermentor; TF: tower fermentor with perforated plates; LF: tubular loop fermentor; RF: fermentor with cell recycling; FB: fluidized bed of immobilized cells; HR: horizontal bioreactor of immobilized cells.

in continuous culture in comparison with batch culture will be discussed later on as pertaining to glucose and phosphate effects on specific invertase activity of yeast.

The basic prototype of a single-stage vessel for continuous culture is assembled horizontally in series (multistage bioreactor, MS in Fig. 1) or vertically (tower fermentor, TF) in order to change the mixing characteristics from the flow of back-mix in the SS bioreactor to that of the non-back-mix. In both the loop fermentor (LF) and the cell-recycle fermentor (RF), a fraction of the exit culture of the bioreactors is returned to the inlet with and/or without concentration of cells. The effect of the culture recycling rate and mixing characteristics on the cell-mass concentration in the exit of the LF reactor will also be discussed later on.

The RF reactor is used in the activated sludge process for domestic and industrial wastewater treatment. Recently, this type of reactor equipped with cell separation by membrane filtration or centrifugation is being used for pure cultures of microorganisms at high cell densities.

All the reactors mentioned above can be used for reaction by immobilized microbial cells. The simplest type is a column reactor, in which gel beads of immobilized microbial cells are fluidized (fluidized bed, FB) by virtue of liquid or gas flow.

Various configurations of reactors have been proposed. One sophisticated bioreactor that will be discussed keeps the cell concentration at the steady-state level even when the dilution rate exceeds the maximum specific growth rate of microorganisms.

3. Enzyme Derepression in Chemostat Culture

The result of a continuous culture of *Saccharomyces carlsbergensis* is shown in Fig. 2[3]. The specific enzyme activity (SEA) of invertase of the yeast (closed circle) and the remaining glucose concentration (open circle) in the culture medium at steady state are plotted against the dilution rate (D) of the continuous culture.

The profile of specific enzyme activity against the dilution rate shows a bell-shaped

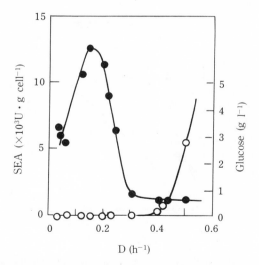

FIG. 2. Specific enzyme activity (SEA) of invertase against dilution rate (D) of continuous culture of *Saccharomyces carlsbergensis* LAM 1068.[3] SEA (closed circle) and glucose concentration (open circle).

curve. When the dilution rate exceeded the critical value of 0.3 h^{-1}, the specific enzyme activity became low and constant (1.2 kU · g cell^{-1}). Thus, the glucose concentration of the culture was more than 100 mg l^{-1}, which was clearly the cause for the repression of enzyme activity in the yeast cells. At the lower dilution rate, meanwhile, the glucose concentration was too low to be detected accurately by the usual method of glucose analysis. In accordance with the disappearance of glucose from the culture medium due to the decreasing dilution rate, the specific enzyme activity of invertase reached a peak value of 12.5 kU · g cell^{-1} at $D = 0.15$ h^{-1}. The peak value of SEA was 8 times as large as that of 1.5 kU · g cell^{-1} for yeast cells grown in batch culture with the same medium. However, the SEA value decreased again with the decrease of the dilution rate below $D = 0.15$ h^{-1}. This peculiar phenomenon cannot be explained by the concept of enzyme repression by glucose. Unknown substances required to induce invertase synthesis might have been supplied insufficiently at this low feed rate of fresh medium.

The reduction of specific enzyme activity (SEA) at dilution rates higher than 0.3 h^{-1} as seen in Fig. 2 was not observed when a mutant strain resistant to glucose repression was used. Figure 3 shows the result of the spontaneous mutant strain A3 that was isolated from batch culture of *S. carlsbergensis* LAM 1068 that continued for two weeks in the presence of 2-deoxy glucose.[3] The maximum enzyme-specific activity of 25 kU · g cell^{-1} was observed at every dilution rate higher than 0.15 h^{-1}. The maximum value was about twice the highest value ever recorded with the wild-type strain LAM 1068.

Enzyme derepression occurred not only in conditions of glucose starvation but also in the depletion of inorganic orthophosphate. Figure 4 indicates the dilution rate effect on the SEA value of invertase of *S. carlsbergensis* LAM 1068 grown in phosphate-limited continuous culture.[4] At the lowest dilution rate tested, the value of SEA was estimated to be 40 kU · g cell^{-1}; the value was 1.6 times the maximum value attained with the glucose

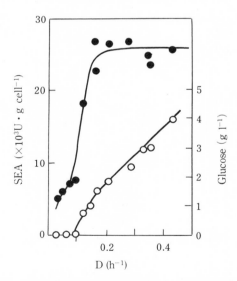

FIG. 3. Specific enzyme activity (SEA) of invertase against dilution rate (D) of continuous culture of glucose repression-resistant mutant strain A3 derived from *Saccharomyces carlsbergensis* LAM 1068.[3] SEA (closed circle) and glucose concentration (open circle).

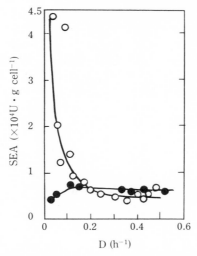

FIG. 4. Specific enzyme activity (SEA) of invertase for *Saccharomyces carlsbergensis* LAM 1068 grown in continuous culture using phosphate-rich and -deficient media.[4] Molecular ratio of potassium phosphate to glucose in feed medium was 0.0029 (open circle) and 0.029 (closed circle).

repression-resistant mutant. The effect of phosphate depletion on the regulatory mechanism of enzyme production is left open for further experimentation and discussion. Specific invertase activity of *S. carlsbergensis* in batch, fed-batch, and continuous cultures with different growth-limiting substrates but with the sole carbon source of either glucose or ethanol is summarized in Table 2.

4. Kinetic Models of Enzyme Regulation in Continuous Culture

Many researchers[5-7] have presented kinetic models of enzyme production by microorganisms in mathematical forms, wherein the intracellular control of protein biosynthesis principally developed by molecular biology has been formulated. Both enzyme induction

TABLE 2. Specific invertase activity of *Saccharomyces carlsbergensis* LAM 1068.

Carbon source	Growth-limiting substrate	Specific invertase activity (kU · g cell⁻¹)	Dilution rate (h⁻¹)
Glucose	Glucose	1.5	[batch]
	(complex	12.5	0.15
	medium)	27.5*	0.15>
	Glucose	4.5 – 8.5	[batch]
	(defined	8.0 – 11.0	[fed-batch]
	medium)	8.5	0.2>
	Inorganic	40.0	0.02
	phosphate		
Ethanol	Ethanol	1.0	0.2<
	Inorganic	7.5	0.02
	phosphate		

* glucose-repression-resistant mutant A3.

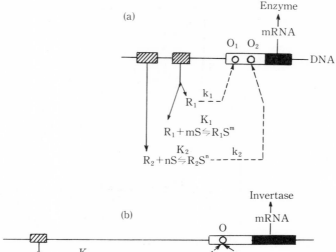

(a)

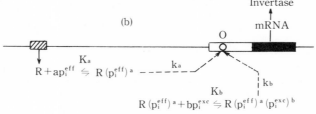

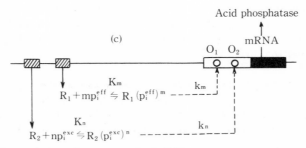

FIG. 5.   Three models for regulation of enzyme synthesis.[10, 11]
    (a) Dual control of induction and repression.
    (b) Competitive repression.
    (c) Non-competitive repression.
    Structural gene (closed rectangle); regulator gene (hatched rectangle); and operator gene (circle).

and repression were first formulated by Yagil and Yagil[8] who were faithful to the operon theory.[9] Figures 5(a), (b), and (c) show schematically three kinetic models that are extensions of Yagil and Yagil's. The models in Fig. 5 (a) to (c) are meant to explain the effect of dilution rate on the enzyme activities of invertase or acid phosphatase in continuous culture of *S. carlsbergensis* (for symbols, see the next paragraph or refer to the list of nomenclature at the end of text). A model of the simultaneous and dual regulation of enzyme synthesis by inductive activation of an operator gene O_1 and repressive inactivation of another operator gene O_2[10] is shown in Fig. 5(a).

What the enzyme produces in either a fully induced, fully repressed, or in between manner with the increase of substrate concentration (S) in the culture medium depends on the relative values of the parametric constants that appear in the model. A high substrate concentration in continuous culture with a large dilution rate would enhance repression of enzyme synthesis. On the other hand, a low substrate concentration at a low dilution rate would reduce the inductive effect of the substrate or its metabolites on enzyme synthesis. The experimental finding that the specific activity of invertase (see Fig. 2) was maximized at an intermediate feed rate to the medium justifies assuming that the enzyme must be subject to the dual controls of induction and repression simultaneously. The final equation derived from the model in Fig. 5(a) is:

$$\mathrm{SEA/SEA_{max}} = \frac{(1 + K_1 S^m)(1 + K_2 S^n)}{(1 + k_1[R_1]_t + K_1 S^m)\{1 + (1 + k_2[R_2]_t)K_2 S^n\}} \cdot {}^{10)} \qquad (1)$$

Inorganic phosphate in the medium is taken up by the cells and is either converted to phosphate compounds intracellularly or remains free in the cell plasma. A fraction of inorganic phosphate transported into the cells is incorporated into structural units of cell constituents such as DNA, RNA, and/or lipids. Another fraction of the phosphate compounds (designated as p^{eff}) determines the cellular growth rate; the third phosphate fraction (p^{exc}) is stored intracellularly where it contributes to miscellaneous activities in cell metabolism.

The effect of phosphate limitation on the invertase production in *S. carlsbergensis* was modeled (Fig. 5(b)) by assuming competitive repression of enzyme synthesis by repressor complexes produced as a result of the binding between aporepressor (R) and phosphate fractions p^{eff} and p^{exc}.[11]

Figure 5(c) illustrates another model of non-competitive repression of enzyme synthesis. This model was to explain the experimental result that acid phosphatase of *S. carlsbergensis* was derepressed notably as the inorganic orthophosphate concentration in the feed medium decreased.[11]

Genes O_1 and O_2 for the enzyme formation are assumed to be associated non-competitively with the repressor complexes. The equations obtained from the models in Figs. 5(b) and (c) on invertase and acid phosphatase biosyntheses as regulated by orthophosphate are[11]

$$\mathrm{SEA/SEA_{max}} = \frac{1 + K_a[p^{eff}]^a + K_a K_b[p^{eff}]^a[p^{exc}]^b}{1 + (1 + k_a[R]_t)K_a[p^{eff}]^a + (1 + k_b[R]_t)K_a K_b[p^{eff}]^a[p^{exc}]^b} \qquad (2)$$

and

$$\mathrm{SEA/SEA_{max}} = \frac{(1 + K_m[p^{eff}]^m)(1 + K_n[p^{exc}]^n)}{\{1 + (1 + k_m[R_1]_t)K_m[p^{eff}]^m\}\{1 + (1 + k_n[R_2]_t)K_n[p^{exc}]^n\}} \cdot \qquad (3)$$

Simulations based on Eq. (1) and on Eqs. (2) and (3) are shown in Figs. 6 and 7(a) and (b), respectively, where appropriate values are taken in parametric constants.[10,11]

Curve 1 in Fig. 6 simulates the experimental data in Fig. 2. The characteristic bell-shaped relationship between the specific invertase activity and dilution rate is well represented by Eq. (1). Curves 2, 3, and 4 are for simple repression, simple induction of enzyme, and constitutive enzyme synthesis, respectively.[10]

Figure 7 (a) and (b) simulates specific activities of invertase and acid phosphatase of *S. carlsbergensis* grown at steady state in continuous culture. The parameter is the molec-

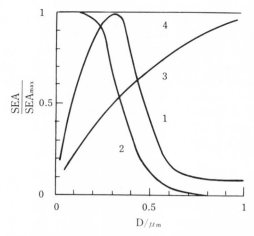

FIG. 6. Specific enzyme activity (SEA) of microbial cells vs. dilution rate in dimensionless correlation.[10] Curve 1 corresponds to enzyme subjected to dual control of induction and repression. Curve 2, repressible enzyme. Curve 3, inducible enzyme. Curve 4, constitutive enzyme.

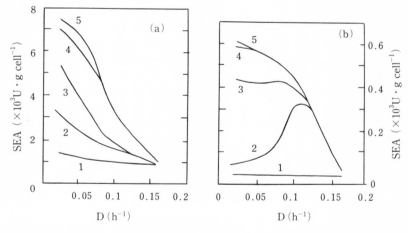

FIG. 7. Simulation of specific enzyme activity (SEA) against dilution rate for invertase (a) and acid phosphatase (b) of *Saccharomyces carlsbergensis* LAM 1068.[11] Molecular ratio of potassium phosphate to ethanol in feed medium for Curves 1 to 5 is 0.015, 0.006, 0.003, 0.0015, and 0.00074, respectively.

ular ratio of inorganic orthophosphate to ethanol in the feed medium.[11] Curves 1 through 5 are for the feed, where concentrations of potassium phosphate decrease consecutively (see caption for Fig. 7).

In contrast to glucose, which is transported into the cells and decomposed irreversibly to intermediate metabolites or carbon dioxide, inorganic phosphate accumulated intracellularly can be released into the medium and then reused by the cells. Accordingly, the dilution-rate effect on specific enzyme activity of cells growing under phosphate limitation cannot be expressed straightforwardly as in glucose-limited growth. Specific activities of both enzymes (invertase and acid phosphatase) increase as the phosphate limitation becomes more stringent and/or the dilution rate decreases, except when the maximum value of the specific activity as in Curve 2 in Fig. 7 (b). These models (Eqs. (2) and (3)) simulate fairly well the experimental results reported elsewhere.[4]

5. Effect of Incomplete Mixing and Culture Recycle Ratio on Cell Productivity in Flow Bioreactors

In most continuous culture studies, complete mixing (back-mix type) reactors have been used. However, the degree of product inhibition is more conspicuous in the continuous culture with back-mix flow than with plug flow, because the concentration of inhibitory product in the former is kept high throughout the reactor whenever one wishes to establish high reaction conversion. On the contrary, a concentration distribution of product occurs along the axis of the flow path if the culture passes through a plug-flow reactor. Microorganisms are subjected to less inhibition than in the back-mix flow reactor, at least in regions where the product concentration is low. However, stable microbial growth in the plug-flow reactor cannot be realized due to the absence of liquid mixing, unless a particular device, for instance, a continuous seed of preculture to the inlet of the reactor, is installed. Recycling of the exit culture serves as a substitute for the continuous seed, thus preventing microorganism wash-out.

A dimensionless correlation between the exit cell concentration (normalized to the maximum cell concentration estimated stoichiometrically) and the dilution rate (normalized to the maximum specific microorganism growth rate), is given in Fig. 8 from a (recycle ratio, $R = 0$) to d ($R = 0.9$). In each diagram, the Péclet Bodenstein number (PeB) is

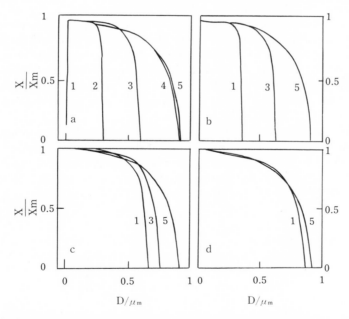

FIG. 8.   Dimensionless correlation between exit cell concentration and dilution rate for various combinations of culture recycle ratio (R) and Péclet Bodenstein number (PeB). Recycle ratio in (a) to (d) is R = 0, 0.1, 0.5, and 0.9, respectively. Curves 1 to 5 in each diagram are for PeB = infinity, 10, 3, 1, and 0, respectively.

taken as a parameter. Curves 1 and 5 in these diagrams are for the two extremes: PeB = infinity (plug flow) and PeB = 0 (complete mixing), respectively, and the rest are in between the two extremes.

Figure 8(a) shows the case of no culture recycling (R = 0); it is obvious that the flow condition as represented by the PeB number considerably affects the correlation between the cell concentration and dilution rate.

It is also evident from these diagrams that the dimensionless dilution rate of cell washout varies depending on whether the flow is completely mixed (Curve 5), a plug flow (Curve 1) or in between the two extremes. Consequently, the productivity of cell mass, that is, the cell concentration multiplied by the ratio of volumetric flow rate of the culture to the reactor's working volume, increases remarkably when the flow condition changes from plug flow to complete mixing, especially in diagrams (a) and (b).

However, if the recycle ratio increases while the PeB number remains unchanged (see Curve 1), the correlation approaches that of the back-mix reactor. Although Curves 5 in Fig. 8 (a) and 1 in (d) are similar, they differ substantially in that cells in the plug-flow reactor with a large recycle ratio experience, while passing through the reactor, a transient environment (in concentrations of substrate, product, and dissolved oxygen as well as pH, etc.). The effect of exposing the cells to such dynamic environmental changes is left open for further investigation, although a favorable effect was reported on the nucleic acid content of *Azotobacter vinelandii* in continuous culture when the cells were exposed to changes in the dilution rate.[12]

6. Dual Fermentors

Non-back-mix flow can be established by using a multistage (MS) fermentor, in which single-stage (SS) reactors are arranged in series. However, a defect of MS fermentors is that microbial cells are washed out at a lower dilution rate than that in SS fermentors.[13] The dilution rate at which the cells are washed out decreases inversely with the increase in the number of SS reactors. We describe here another type of MS fermentor in which microorganisms can apparently be grown at a dilution rate higher than their maximum growth rate.

Figure 9 shows a scheme of a SS fermentor that is coordinated with another equi-volume reactor by the mutual exchange of culture broth. If the medium is fed to Fermentor 1 and the culture is withdrawn from the same fermentor, it is calculated that the microbial cells will not be washed out, even if the dilution rate turns out to be higher than the maximum specific growth rate of the cells. The dilution rate here is defined as the feed rate of fresh

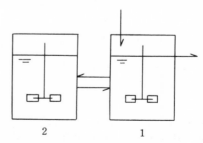

FIG. 9.   Dual fermentors for super high-growth-rate microorganisms.[14]
    1: main fermentor; 2: auxiliary fermentor.

TABLE 3. Cell concentration of *Saccharomyces carlsbergensis* in dual fermentors.[15]

Degree of over-growth rate	Cell concentration (O.D.$_{660}$) in fermentors		Mutual liquid exchange
(dilution rate) (maximum specific growth rate)	Fermentor 1	Fermentor 2	(liquid exchange rate) (feed rate of medium)
1.24	0.152	0.532	0.21
1.58	0.131	0.576	0.18
1.75	0.131	0.578	0.16
2.23	0.114	0.600	0.13
3.15	0.055	0.255	0.20
4.46	0.037	0.142	0.16
5.15	wash out	wash out	0.16

medium divided by the total volume of the fermentors. It was confirmed experimentally (Table 3) that the microbial cells were not washed out from the dual fermentors at a dilution rate that exceeded the maximum value of specific growth rate. The application of this type of reactor has been discussed elsewhere.[14]

7. Horizontal Bioreactors for Alcohol Fermentation

The concentration of microorganisms in a continuous culture at steady state does not exceed the value $X_{max} = Y \times S_i$, where Y is growth yield and S_i is the inlet concentration of substrate. If the cells are immobilized and their flow out of the reactor can be prevented, the cell population density increases and the reaction rate is enhanced. Whatever the ideal

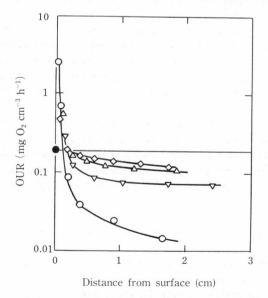

FIG. 10. Axial profile of respiratory activity of *Candida lipolytica* IAM 12188 growing in agar gel support.[15] Incubation time is 0 h (closed circle), 3 h (open diamond), 12 h (open triangle), 24 h (open inverted triangle), and 48 h (open circle).

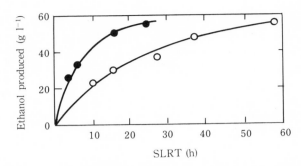

FIG. 11. Exit concentration of ethanol produced by immobilized *Saccharomyces carlsbergensis* LAM 1068 packed in a horizontal reactor (closed circle) and fluidized bed reactor (open circle).[16] Abscissa is superficial liquid residence time (SLRT) defined by the ratio of working volume including volume fraction of beads of immobilized cells to volumetric flow rate of medium.

picture might be, the growth of immobilized cells entrapped in gel support is limited only on and/or near the surface, where sufficient nutrients and/or oxygen are available. Accordingly, the space effective for microbial growth in the reactor is far less than the whole volume of the support. In other words, even if the cell concentration in the gel is supposed to be 10 times that of the surrounding medium, the reaction rate does not necessarily increase 10 times due to diffusion resistance to the transport of nutrients and/or oxygen within the support.

An example of deteriorated activity of yeast cells immobilized in support gel[15] is shown in Fig. 10. Satisfactory cell growth was confined within a narrow space (about 100 μm) from the agar gel surface. The core area far from the gel surface is essentially inert in terms of microbial activity.

The FB reactor in Fig. 1 has been frequently used as a vessel for immobilized microorganisms. When the microbial reaction is accompanied by product inhibition, e.g., alcohol fermentation or vinegar manufacture, the quasi-back-mix characteristics of the FB reactor reduce the fermentation rate significantly for the reason mentioned above. This problem can be solved by using non-back-mix bioreactors, i.e., MS or TF fermentors or the HR reactor (Fig. 1).

The HR reactor can also reduce inhibitory effect of the fermentation product. Figure 11 demonstrates the performance of an HR reactor in the production of ethanol by immobilized yeast cells as compared to FB and/or TF fermentors.[16] When the same superficial liquid residence time (SLRT) was used, the exit ethanol concentration in the three reactors was enhanced in the order of FB, TF, and HR[16] (data for TF not shown in Fig. 11).

8. Conclusion

1. Various flow bioreactors and their performance in terms of enhancing the cell productivity were discussed.

2. Three models of regulation of enzyme biosynthesis were simulated. The experimental data obtained with invertase and acid phosphatase of *Saccharomyces carlsbergensis* in continuous culture were used to demonstrate the validity of the models.

3. A special arrangement of single-stage continuous fermentors that allows a specific growth rate exceeding the maximum value was presented and discussed.

4. The growth rate of gel-immobilized *Candida lipolytica* was illustrated from time-dependent changes of local distribution of cell concentration in the support.

5. The utility of the horizontal bioreactor in ethanol production was demonstrated.

Nomenclature

a, b	stoichiometric constant
D	dilution rate, h^{-1}
FB	fluidized bed
HR	horizontal reactor
K	equilibrium constant for complex formation between substrate and (apo-) repressor
k	equilibrium constant for complex formation between repressor and operator gene
LF	loop fermentor
MS	multistage fermentor
m, n	stoichiometric constant
O	operator gene
OUR	oxygen uptake rate, g O_2 h^{-1} cm^{-3}
p	hypothetical intracellular concentration of phosphate, mol $\cdot$ g $cell^{-1}$
PeB	Péclet-Bodenstein number ($= uL/E$, where u: linear liquid flow rate; L: axial length of reactor; E: dispersion coefficient)
R	intracellular concentration of (apo-) repressor, mol $\cdot$ g $cell^{-1}$ or culture recycle ratio
RF	cell-recycle fermentor
S	substrate concentration in culture medium, g l^{-1}
S_i	inlet concentration of substrate, g l^{-1}
SEA	specific enzyme activity, $U \cdot$ g $cell^{-1}$
SLRT	superficial liquid residence time, h
SS	single-stage continuous fermentor
TF	tower fermentor
X	dry cell concentration, g l^{-1}
Y	growth yield, g dry cell g $substrate^{-1}$
μ	specific growth rate, h^{-1}

Subscripts

a, b	species of complex formation
m, n	species of complex formation
m, max	maximum value
t	total
1, 2	species of repressor or operator gene

Superscripts

eff	effective for growth
exc	excess

58   K. Toda

REFERENCES

1. Monod, J., "La technique de culture continue; theorie et application." *Ann. Inst. Pasteur*: 79, 390–410 (1950).
2. Novick, A., Szilard, L., "Experiments on spontaneous and chemically induced mutations of bacteria growing in the chemostat." *Cold Spring Harbor Symp. Quant. Biol.*: 16, 337–343 (1951).
3. Toda, K., "Invertase biosynthesis by *Saccharomyces carlsbergensis* in batch and continuous cultures." *Biotechnol. Bioeng.*: 18, 1103–1115 (1976).
4. Toda, K., Yabe, I., Yamagata, T., "Invertase and phosphatase of yeast in a phosphate-limited continuous culture." *European J. Appl. Microbiol. Biotechnol.*: 16, 17–22 (1982).
5. Terui, G., Okazaki, M., Kinoshita, S., "Kinetic studies on enzyme production by microbes (1). On kinetic models." *J. Ferment. Technol.*: 45, 497–503 (1967).
6. van Dedem, G., Moo-Young, M., "Cell growth and extracellular enzyme synthesis in fermentations." *Biotechnol. Bioeng.*: 15, 419–439 (1973).
7. Imanaka, T., Aiba, S., "A kinetic model of catabolite repression in the dual control mechanism in microorganisms." *Biotechnol. Bioeng.*: 19, 757–764 (1977).
8. Yagil, G., Yagil, E., "On the relation between effector concentration and the rate of induced enzyme synthesis." *Biophys. J.*: 11, 11–27 (1971).
9. Jacob, F., Monod, J., "Genetic regulatory mechanisms in the synthesis of proteins." *J. Mol. Biol.*: 3, 318–356 (1961).
10. Toda, K., "Dual control of invertase biosynthesis in chemostat culture." *Biotechnol. Bioeng.*: 18, 1117–1124 (1976).
11. Toda, K., Yabe, I., Yamagata, T., "Kinetics of yeast growth and enzyme syntheses in a phosphate-limited continuous culture." *European J. Appl. Microbiol. Biotechnol.*: 16, 10–16 (1982).
12. Nagai, S., Nishizawa, Y., Endo, I., Aiba, S., "Response of a chemostat culture of *Azotobacter vinelandii* to a delta type of pulse in glucose." *J. Gen. Appl. Microbiol.*: 14, 121–134 (1968).
13. Aiba, S., Humphrey, A.E., Millis, N.F., "chapter 5, continuous cultivation," p. 128–162. In *Biochemical Engineering, 2nd ed.*, University of Tokyo Press, Tokyo, 1973.
14. Toda, K., Dunn, I.J., "Continuous culture in combined backmix-plug flow-tubular loop fermentor configurations." *Biotechnol. Bioeng.*: 24, 651–668 (1982).
15. Sato, K., Toda, K., "Oxygen uptake rate of immobilized growing *Candida lipolytica*." *J. Ferment. Technol.*: 61, 239–245 (1983).
16. Toda, K., Ohtake, H., Asakura, T., "Ethanol production in horizontal bioreactor." *Appl. Microbiol. Biotechnol.*: 24, 97–101 (1986).

I-4

Some Comments on the Physiology of Immobilized Cells: A Proposal for Kinetic Modeling

Tarun K. Ghose*

Department of Chemical Engineering, University of Delaware, Newark, DE 19716, U.S.A.

1. Introduction

This review does not intend to discover the wheel. It proposes to offer a few comments on some current literature on the characteristic metabolic shifts and physiological behavior reportedly displayed by some viable immobilized cells (VIC). Based on the observations on the physiological state of *Saccharomyces cerevisiae* cells a proposal of kinetic modeling has been made. Although this communication should not be considered a complete documentation on the subject it may be useful for young biochemical engineers who intend to contribute to this fascinating field. Treatment of immobilized enzyme is not included in this review. At this time very few biological and engineering studies on immobilized recombinants have been reported and therefore they are not dealt with.

The cell has been the prime focus of Professor Shuichi Aiba throughout his life. He began by correlating data on air sterilization in packed fiber filters. His first three papers on the subject appeared between 1949 and 1950, and two subsequent communications on agitation and aeration in chemical engineering appeared in 1951. During the decade of the 1950s, Aiba gradually extended his research interest to subjects like mixing and flow patterns, gamma ray counting, entrainment, cell distribution analysis in fiber filters, oxygen absorption in bubble aeration, etc., all of which attracted the interest of chemical engineers. Until the early 1960s, he was laying the foundations of biochemical engineering sciences in Japan along with a few of his most eminent colleagues, notably at the universities of Tokyo, Osaka, Kyoto, Tsukuba, and Hiroshima. After that time all his research was directly concerned with subjects like microbial kinetics, and simulation, optimization, and mixed culture dynamics in water pollution control systems. During the first five years of the current decade Aiba published a record number of thirty-three original papers on the engineering analysis of recombinant DNA.

Shuichi Aiba never stopped, and his contributions will continue to guide the minds of young biochemical engineers. I have chosen to dedicate this review on immobilized cells to commemorate the retirement of this original man.

---

* *Current address*: Department of Chemical Engineering, National University of Singapore, Singapore 0511.

2. *The State of Immobilized Cells*

It appears that some reported observations lack adequate analysis of the many changes that viable immobilized cells (VIC) undergo in the attached (adsorbed) stage. Vijaya-lakshmi et al. (1979) and Fletcher and Marshall (1982) reported that attachment of microbial cells onto a carrier somehow alters the permeability of the cell membrane. Fletcher and Marshall (1982) also observed permeability changes in some cell membranes due to adsorption on a surface. It has been suggested that adsorption leads to a short-term increase in the respiration rate of the cells at the surface (Novarro and Durand, 1977), while Atkinson and Fowler (1974) pointed out that the adsorption process may initiate a selection among a cell population in which case the behavior of cells near the surface may differ from that of cells further from the surface.

In a recent study, Robinson et al. (1985) reported reduced respiration rate, reduced growth rate, and increased chlorophyll synthesis among immobilized *Chlorella* cells compared to those in the free state. Yeast cells adsorbed on a carrier surface are reported to exhibit a marked change in generation time (Ghose and Bandyopadhyay, 1980). The growth of cells in different generations is reflected by different peaks of O.D. plotted against time (Fig. 1).

It is argued that if the appearance of the cells in the effluent were only due to desorption, the reactor would be washed free of practically all cells and further growth would become impossible. Daughter cells budded out of the parents in the VIC reactor are thrown into the fluid and rapidly carried out of the reactor. The formation of a monolayer of yeast cells on the carrier surface has been demonstrated (Bandyopadhyay and Ghose, 1982) on scanning electron micrographs at × 2,000 and × 1,000 magnifications. The cycle process gives sequential peaks either because no additional sites are left for attachment as monolayer sites become completely covered by adsorbed cells or because interfacial instability occurs (Davies and Ridal, 1961).

This periodic release of cells in the medium followed by their washout with the effluent is attributed to synchronous growth with the release of one offspring following cell division (Hattori et al., 1972a, 1972b; Navarro and Durand, 1980). The results of several independent studies have demonstrated the following significant and quantitative changes in immobilized cells:

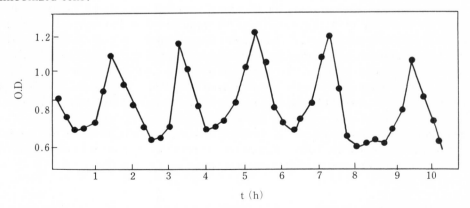

Fig. 1. Frequency of appearance of cells in the reactor effluent (30°C, *S. cerevisiae* cells adsorbed on ligno-cellulose fibers and placed in the reactor).

a) Conditions of optimal growth are often different from those of free cells (Shimizu et al., 1979).

b) The rate of growth may be considerably different from that of cells in the free state (Ghose and Bandyophadyay, 1980; Navarro and Durand, 1980).

c) Significantly different morphological forms may appear resulting from the attachment of cells onto a carrier (Jirkú et al., 1980).

d) The formation of a monolayer of single cells on the matrix surface is seen in scanning electron micrographs (Bandyopadhyay and Ghose, 1982).

e) Product yields and productivities are substantially higher than those of free cells (Tyagi and Ghose, 1982; Hattori and Furusaka, 1959; Ghose and Bandyopadhyay, 1980; Navarro and Durand, 1980; Larsson and Mosbach, 1979).

f) Uncoupling of product formation from cell growth occurs to enable the VIC to use most of the available energy for ion transport (Doran and Bailey, 1984).

g) The chemical make-up, particularly of polysaccharides and polynucleotides, is substantially different in VIC and their offspring from those in the free state (Doran and Bailey, 1986).

Despite many efforts our understanding of the relationship between the immobilization method and the catalytic behavior of fixed cells is not yet adequate. More importantly, the degree to which VIC behavior has been observed to deviate from that of free cells appears to lie far outside the range within which the variations can be attributed primarily to the availability of nutrients as in the case of yeasts (Brodelius et al., 1979; Hattori, 1972). In this respect the capabilities of VIC offspring may exceed those of the mother cells (Larsson and Mosbach, 1979). In the case of immobilized yeast cells, recently published data (Doran and Bailey, 1986) call for further analysis of several significant biochemical events.

Our current knowledge base of glucose transport in VIC is dubious. van Stevenick and Rothstein (1965) and van Uden (1967) showed that uptake of glucose limits the fermentation rate of VIC of *Saccharomyces cerevisiae*. An increase in the rate of ethanol formation, caused by glucose uptake, the rate limiting step for the operation of the Embden-Meyerhof-Parnas (EMP) pathway, as seen in all VIC systems so far, without any improvement in yield, would imply that glucose consumption is much faster than expected. This has been observed despite the fact that substrate transport is intercepted by support surface due to anchorage of a part of the cell surface.

In addition to the report on the change of cell permeability mentioned earlier, some have suggested (Suzuki and Karube, 1979) that damage caused to the VIC cell wall might induce faster entry of glucose into the cell. An increased metabolic rate of ethanol formation might be another reason for the increased glucose demand.

Ghose and Bandyopadhyay (1980) have also mentioned that the increased activities of yeasts are indeed due to improved availability of nutrients on the solid-liquid interface. Hahn-Hägerdal et al. (1982) hypothesized that the presence of polyethylene glycol (PEG) and dextran in the fermentation medium caused a decrease in the water activity and this might have directed metabolism towards ethanol formation. It has been suggested that when suppressed yeast cells are not able to divert glucose to the EMP pathway they accumulate energy in the form of glycogen and trehalose (Küenzi and Fiechter, 1969). In a way, immobilized cells may be considered to be in a state of suppression. Doran and Bailey (1986) demonstrated that there is a substantial increase in the polysaccharide content of VIC which is even higher in their progeny. Neither the reason for the preference for stored glycogen nor for the increased glucose contents in immobilized cells and their offspring is yet clear.

Panek (1963) provided experimental evidence that during the cell cycle of *S. cerevisiae* extrusion of the bud follows the degradation of cell storage reserves and a period of polysaccharide synthesis occurs. In the cell cycle of budded yeasts (Hartwell, 1971) DNA replication is preceded by bud emergence and followed by the separation of the bud from the mother. Results obtained with synchronous cultures of *S. cerevisiae* (Hartwell, 1971) suggest that the CDC-4 gene functions in the initiation and the CDC-8 gene functions in the continuation of DNA replication. Hartwell concluded that (a) nuclear and cell separation in yeasts is dependent upon prior DNA replication; (b) a cellular master timer controls bud initiation and the operation of this timer is independent of other events in the cycle, such as DNA replication, nuclear division, and cell separation; and (c) premature bud initiation is normally prevented as a consequence of the successful initiation of DNA replication.

A year later Mitchison (1972) pointed out that two possible mechanisms exist for directing a fixed sequence of cell cycle events connected to each other. In one, called the "dependent pathway" mechanism, it is necessary for the previous event in the cycle to be completed before the next event can occur. The second possibility calls for a direct casual connection between the two events but these are directed by signals from a master timer. It is therefore implied that there is no need for the earlier event to be completed before the next event occurs and yet the activity of the timer controls the orders of the connected events. The events are (a) initiation of DNA synthesis, (b) bud emergence, (c) continuity of DNA synthesis, (d) nuclear migration, (e) nuclear division, (f) late nuclear division cytokinesis, and (g) cell separation.

According to Hartwell et al. (1974) there is evidence for the existence of a timer that controls bud emergence but there is no evidence that the timer plays any role in the coordination of various events of the cell cycle. They felt that the timer serves to phase bud emergence in sequence with the events of DNA synthesis and nuclear division and argued that the joint dependence of bud emergence and initiation of DNA synthesis is sufficient to explain the coordination between the two possibilities. Their conclusion about the role of the timer is that it either phases successive starts of events (monitoring cell growth) or it phases successive bud emergence events in order to limit the cell to one such event per cycle.

Sloat et al. (1981) experimentally established that temperature-sensitive yeast mutants in gene CDC-24 maintained a continued increase in volume and cell mass at the restrictive temperature of 36°C but failed to form buds. Stained with the fluorescent dye calcofluor, it was seen that mutants were also unable to form discrete chitin rings on the cell wall at budding sites (normally called bud scars) at 36°C. However, substantial amounts of chitin were found deposited randomly over the surfaces of growing unbudded cells. Cell wall mannan of mutants labeled with fluorescein isothiocyanate-conjugated concanavalin A suggested that incorporation of this polysaccharide was also decolonized in cells grown at 36°C. Despite the availability of well-defined execution points normally visible prior to bud emergence, inactivation of CDC-24 gene product in budding cells resulted in selective growth of mother cells (rather than buds) and in shifts in the location of chitin deposition.

It is therefore possible that in immobilized yeast cells budding may be delayed while decoupled DNA replication and polysaccharide synthesis continue. The question therefore is what is the event taking place at the gene level that connects the interrupted cell cycle system of yeasts permitting DNA synthesis and delayed bud emergence with an increased rate of glucose consumption and enhanced ethanol production? The conjecture of Doran and Bailey (1986) and of Takagi et al. (1983) appears to link the biochemical behavior of immobilized yeasts with that of polyploid strains. The former group is reported to have identified polyploid yeast cells being generated due to immobilization. Whether the generation of these cells, which otherwise can reproduce by mating, is an exclusive event of immobilization is not clear.

Combining the well-documented studies of Johnson and Gilson (1966) and Hartwell et al. (1974) it appears quite likely that while additional DNA synthesis and nuclear division cycles have been noticed in certain yeast mutants with damaged budding systems, newly synthesized mannan and glucan are pushed up to the bud tips. Therefore, excess production of these polymers, as observed (Doran and Bailey, 1986) in *S. cerevisiae*, might be due to changes at the gene level of VIC offspring. There is also a substantial loss in double-stranded RNA (the more stable form) in immobilized cells and their offspring. If the synthesis of RNA is decoupled like that of DNA from the process of budding and separation of cells and is thereby able to accumulate, as Hartwell (1971) is reported to have observed in several mutants, it may be argued that a very high rate of DNA replication may follow because the same bank of nucleotide precursors is available to both these polyneucleotides. DNA replication continues to proceed at a faster rate and after a few cell cycles (yet to be conclusively established) the ratio of DNA to RNA may become very high.

Other data (Ghose, unpublished) pertain to the biosynthesis and release of fungal cellulases by a mutant of *Trichoderma reesei* D1–6 in the immobilized state. Preliminary saccharification studies were conducted by using cellulase enzyme obtained from immobilized pellets of the fungus on delignified rice straw cellulose. A control experiment was run with free cells. The results (Fig. 2) indicate a significant difference in the rates be-

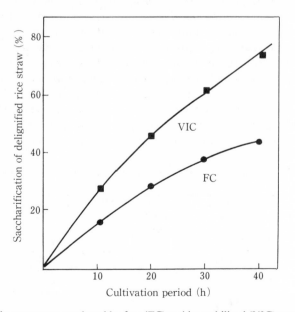

FIG. 2. Effects of cellulase enzymes produced by free (FC) and immobilized (VIC) mycelium of *Trichoderma reesei* D1–6 on the saccharification of rice straw.

tween the free and the immobilized cells. Since the reducing sugar estimation was based on glucose, it is presumed that β-glucosidase is synthesized and released at a higher rate by immobilized pellets than by free cells.

In another study (Ghose, unpublished) pellets of the same fungus were confined inside polymeric sacs similar to those described by Ghose and Subhas Chand (1978) in their experiments with immobilized *Streptomyces* cells. The encapsulated cells, each containing between 6–20 mg of cell mass (conidia, dry basis) were used in a continuous spouted-bed bioreactor using mineral medium (Ghose and Sahai, 1979) with 2% (w/v) microcrystalline cellulose particles to produce the cellulase enzyme. The number of VIC pellets per unit volume of the capsule (VIC-C) and the average cell mass inside the matrix (VIC-X) were taken as parameters. VIC-C effectively represented the growth parameter of the immobilized cells. Control experiments were conducted using free cells of the dispersed mycelium without encapsulation. The results (Table 1) indicate the dramatic effects of fungal cell immobilization. With a nearly 2.4 times increase in the cell mass inside the matrix, there is a three-fold increase in the productivity of the enzyme with an almost 3.3 times increase in the cellulase activity by VIC over free cells (control).

Similarly, with the same cell mass content inside the matrix (VIC-X) but with a 3.75 times increase in the number of VIC in the capsule (VIC-C), the activity increase is almost linear but productivity (expressed per unit weight of VIC) goes up by over four fold. The most dramatic effect is seen when both the parameters VIC-X and VIC-C are increased by 3.0 and 3.7 times over the free cells: The productivity increase jumps by more than 13 times compared to the control values. These results cannot merely be explained either by the increased availability of nutrients and/or by the decrease in the diffusional resistance imposed. There is as yet no valid physiological explanation of these dramatic effects.

A shift in the metabolic pathway in the production of L-malic acid from fumaric acid by immobilized cells of *Brevibacterium flavum* has been reported by Chibata et al. (1983). Such cells have been observed to induce succinate dehydrogenase which diverts part of fumaric acid to succinic acid. The coenzyme of this reducing reaction, $FADH_2$, accumulates inside the cells and as a result the malic acid yield decreases. The success of this important process employing immobilized *Brevibacterium flavum* has been made possible by treating the VIC with bile acid and/or deoxycholic acid, enabling the VIC to denature the intracellular succinic dehydrogenase or to leak out the coenzyme FAD. The optimum temperature of the VIC was observed to be 10°C higher than that of free (intact) cells.

Based on some of the observations mentioned here and some of the results reported

TABLE 1. Performance of *Trichoderma reesei* D1–6 mutant encapsulated pellets in spouted-bed continuous bioreactor.

Experimental conditions	Av. cell mass inside the matrix, mg (VIC-X)	Number of IMC particles per unit volume (VIC-C)	Cellulase activity Filter Paper Activity $l^{-1} \times 10$	Overall productivity FPA $g^{-1}h^{-1}$
Control (FC)	6.02	1200	0.18	30.7
Changed (VIC-X)	14.5	1200	0.59	90.6
Changed (VIC-C)	6.02	4550	0.75	125.5
Changed (VIC-X) + (VIC-C)	18.5	4550	3.0	395.8

Mean sp. productivity within the cell = 5.05 FPA g cell^{-1} h^{-1}; $D = 0.15$ h^{-1}.

earlier, a simple model for adsorbed growing yeast cells is proposed. A substantial part of this analysis has appeared elsewhere (Ghose and Tyagi, 1984).

3. Kinetic Models: Free Cells

Rates of formation of free cells and products (growth associated) and the concurrent disappearence of limiting substrate in a batch process derived from the Monod equation are:

$$\frac{dX}{dt} = \mu X \tag{1}$$

$$\frac{dP}{dt} = \mu X Y_{p/x} \tag{2}$$

$$-\frac{dS}{dt} = \frac{\mu X}{Y_{x/s}} \tag{3}$$

where μ is assumed constant in high substrate concentration ($S \gg K_s$) at the initial phase of growth. Integration of Eqs. (1) to (3) yields:

$$X = X_0 e^{\mu t} \tag{4}$$

$$P = [P_0 - X_0 Y_{p/x}] + [X_0 Y_{p/x} e^{\mu t}] \tag{5}$$

$$S = \left[S_0 + \frac{X_0}{Y_{x/s}} \right] - \left[\frac{X_0}{Y_{x/s}} e^{\mu t} \right] \tag{6}$$

where initial values are designated by the subscript zero.

Equation (1) predicts growth with the assumption that it is neither limited by substrate concentration nor inhibited by products. In most cases of fermentations dealing with high substrate concentrations like ethanol and SCP production, degradation of starch, cellulose, phenol, urea, etc., specific growth rates of the cells are inhibited by the substrate concentration, and the extent of such inhibition is described by a modified Monod equation:

$$\mu = \frac{\mu_m S}{K_{s(g)} + S + \dfrac{S^2}{K_{s(g)} \cdot \omega_{(g)}}} \tag{7}$$

where $\omega_{(g)}$ is the degree of substrate inhibition on growth—a characteristic constant dependent on the chemical mechanism of the inhibition exercised.

When growth is inhibited by the product, the modified specific growth rate for free cells (Aiba et al., 1973) is given by

$$\mu = \left(\frac{\mu_m S}{K_s + S} \right) \left(\frac{K_p}{K_p + P} \right) \tag{8}$$

where K_p is the inhibition constant due to product concentration at which inhibition begins and P is product concentration.

In the case of yeast-ethanol fermentation, when both substrate and product exercise inhibition simultaneously, substrate concentration is considered critical. Ghose and Tyagi (1979) have demonstrated that the inhibitions are multiplicative and proposed the following two equations:

$$r_x = \frac{dX}{dt} = \mu_m \, X \left(1 - \frac{P}{P_{m(g)}}\right) \left(\frac{S}{S + K_{s(g)} + \dfrac{S^2}{K_s \cdot \omega_{(g)}}}\right) \tag{9}$$

and

$$r_p = \frac{dP}{dt} = \nu_m \, X \left(1 - \frac{P}{P_{m(p)}}\right) \left(\frac{S}{S + K_{s(p)} + \dfrac{S^2}{K_{s(p)} \cdot \omega_{(p)}}}\right). \tag{10}$$

Similarly, rate limiting nutrient consumption can be written as

$$- r_s = - \frac{dS}{dt} = \frac{1}{Y_{x/s}} \cdot \frac{dX}{dt} = \frac{1}{Y_{p/s}} \cdot \frac{dP}{dt} \tag{11}$$

provided:

μ_m: maximum specific growth rate of yeast cells, h^{-1};
ν_m: maximum specific rate of ethanol formation, h^{-1};
$P_{m(g)}$: maximum ethanol concentration, above which cells do not grow, g l^{-1};
$P_{m(p)}$: maximum ethanol concentration, above which cells do not produce ethanol, g l^{-1};
$K_{s(g)}$: saturation constant for growth, g l^{-1};
$K_{s(p)}$: saturation constant for product formation, g l^{-1};
$Y_{x/s}$:   yield of cells based on substrate, g g^{-1};
$Y_{p/x}$:   yield of product based on cells, g g^{-1};
$Y_{p/s}$:   yield of product based on substrate, g g^{-1};
$\omega_{(g)}$: degree of substrate inhibition on growth;
$\omega_{(p)}$: degree of substrate inhibition on product formation.

4. Kinetic Models: Immobilized Cells

A steady-state reactor mass-balance of a monolayer of viable immobilized cells in respect of all the events taking place can be written as:

$$\frac{d(X^i V)}{dt} - \frac{dX^i_{(+)}}{dt} + \frac{dX^i_{(-)}}{dt} + \frac{dX^i_{(e)}}{dt} = 0 \tag{12}$$

or

$$\frac{d(X^i V)}{dt} = \frac{dX^i_{(+)}}{dt} - \left(\frac{dX^i_{(-)}}{dt} + \frac{dX^i_{(e)}}{dt}\right) \tag{13}$$

where

$X^i V$:   net viable immobilized cell concentration on the monolayer of the matrix, g l^{-1};
$X^i_{(+)}$: concentration of new immobilized cells produced on the monolayer, g l^{-1};
$X^i_{(-)}$: concentration of old immobilized cells decaying on the monolayer and leaving the surface g l^{-1};
$X^i_{(e)}$: concentration of offspring of immobilized cells leaving the reactor, g l^{-1};
V: volume of cell layer;
and the superscript 'i' denotes viable immobilized cells.

Since in a given volume cells cannot grow unabated, a steady-state condition will determine the signs of the terms on the right-hand side of Eq. (13) with the increase in the value of X^i. In other words, at steady-state conditions the three terms on the right-hand side of Eq. (13) must remain coordinated relative to their point values. The conditions that control the situation are:

a) increased cell concentration varies inversely with increased value of $X^i_{(+)}$,
b) increased cell concentration varies directly with both $X^i_{(-)}$ and $X^i_{(e)}$,
c) uncoupling of product formation from cell growth takes place (Doran and Bailey, 1984); thereby, cells use most of the available energy for ion transport, and
d) two or more cells may cluster together in the bulk fluid due to surface-active agents released by the cells; this situation has not been considered in this analysis because of its negligible effect on the overall system.

The ideal immobilization technique should not permit any significant release of cells from the adsorbed phase. However, the rate of leakage has been observed to increase at very high rates in the majority of immobilization techniques when cells grow while gel entrapped or adsorbed (Förberg and Häggström, 1985). In such cases, a steady state is achieved by the loss of cells with the reactor effluent. It can thus be assumed that the rate at which the offsprings leave the reactor can be set to zero in Eq. (12) and a steady state must therefore arise from a decrease in the rate of adsorption of cells or an increase in the rate of decay of the cells.

It has been discussed earlier that immobilized yeast cells are maintained totally viable, while a part of the offspring budding out of the parents is released into the medium. It is unlikely that additional free space will be available for re-entry of the released cells onto the carrier surface again. According to Eq. (12) spontaneous release of cells from the immobilized phase is possible and therefore the assumed monolayer concept is considered a reversible process relative to all the events described by this equation. It also supports the position that only under the condition of a space being vacated, due to detachment of one or more cells that make their way out of the reactor, can a corresponding number of cells from the fluid phase occupy the vacancy. Scanning electron microscopic studies (Bandyopadhyay and Ghose, 1982) revealed that there is hardly any build-up of cells over the monolayer of VIC. This is considered a state of full-loading and a reference state for the dynamic equilibrium described by Eq. (12).

The corresponding rate expressions for monolayers of viable immobilized cells (Ghose and Tyagi, 1984) are:

$$r^i_x = \frac{dX^i}{dt} = \mu^i_m(X + \beta X^i)\left[1 - \frac{P}{P_{m(p)}}\right]\left[\frac{S}{S + K_{s(g)} + \dfrac{S^2}{K_{s(g)} \cdot \omega_{(g)}}}\right] \quad (14)$$

$$r^i_p = \frac{dP^i}{dt} = \nu^i_m(X + \alpha X^i)\left[1 - \frac{P}{P_{m(p)}}\right]\left[\frac{S}{S + K_{s(p)} + \dfrac{S^2}{K_{s(p)} \cdot \omega_{(p)}}}\right] \quad (15)$$

μ^i_m = maximum specific growth rate of VIC, h^{-1};
ν^i_m = maximum specific productivity of ethanol by VIC h^{-1};
x^i = concentration of VIC on mono-layer volume g l^{-1};

$$\alpha = \frac{\nu^i_m}{\nu_m}, \qquad \beta = \frac{\mu^i_m}{\mu_m}.$$

Thus, the overall rate expressions for growth and product formation for VIC system will emerge as:

$$\mu^i_{(O)} = \frac{r^i_x}{(X + X^i)} = \left[\frac{X\mu_m + X^i \mu^i_m}{X + X^i}\right]\left[1 - \frac{P}{P^i_{m(g)}}\right]\left[\frac{S}{S + K_{s(g)} + \dfrac{S^2}{K_{s(g)} \cdot \omega_{(g)}}}\right] \quad (16)$$

and

$$\nu^i_{(O)} = \frac{r^i_b}{(X + X^i)} = \left[\frac{X\nu_m + X^i \nu^i_m}{X + X^i}\right]\left[1 - \frac{P}{P^i_{m(p)}}\right]\left[\frac{S}{S + K_{s(p)} + \dfrac{S^2}{K_{s(p)} \cdot \omega_{(p)}}}\right] \quad (17)$$

where (o) denotes the overall values for either growth rate or product formation rate. Since the degree of inhibition exercised by either substrate or product on the VIC system is assumed to be less than in the corresponding free cell system based on the monolayer concept, $\omega_{(g)}$ or $\omega_{(p)}$ has not been substituted for by the corresponding expression for VIC Eqs. (16) and (17).

In the development of Eqs. (14) to (17) for the VIC system, the following assumptions were made:

a) the system becomes steady when the first and the second terms on the right-hand side of Eq. (13) become identical and the third term (rate at which the offspring leave the system) is negligibly small and therefore considered zero;

and

b) the yield of cells in the immobilized state is considerably smaller than that of free cells. The mass-balance equation for the energy substrate is written as

$$-\frac{dS}{dt} = -\left(\frac{dS}{dt}\right)_m - \left(\frac{dS}{dt}\right)_g \quad (18)$$

or

$$-\left(\frac{dS}{dt}\right)_m = mX \quad (19)$$

and

$$-\left(\frac{dS}{dt}\right)_g = \frac{1}{Y_G} \cdot \left(\frac{dX}{dt}\right) \quad (20)$$

where

dS_m = energy source consumed in the maintenance work;
dS_g = energy source consumed for growth;
m = maintenance coefficient; and
Y_G = true growth yield.

Combining Eqs. (18)–(20) for a total balance we get:

$$-\frac{dS}{dt} = \frac{1}{Y_G} \cdot \frac{dX}{dt} + mX. \quad (21)$$

If the energy for growth and carbon for cell substance is derived from the same chemical source, then the substrate utilization and product formation rates can be written

$$-\frac{dS}{dt} = \frac{1}{b} \cdot \frac{dP}{dt} \tag{22}$$

where b is the stoichiometric coefficient. Its value is 0.51 for glucose conversion to ethanol by *S. cerevisiae* yielding two mols of ATP per mol of glucose.

$$C_6H_{12}O_6 \longrightarrow 2C_2H_5OH + 2CO_2 + 2ATP \tag{23}$$

Equations (21) and (22) can now be rearranged and combined to provide a definition of the ethanol formation rate, ν, as

$$\nu = \theta\mu + \eta; \tag{24}$$

where

$$\nu = \frac{1}{X} \cdot \frac{dP}{dt};$$

$$\theta = \frac{b}{Y_G};$$

$$\eta = mb; \text{ and}$$

μ = specific growth rate for *S. cerevisiae*.

Consequently, a plot of ν vs μ gives a straight line with a slope equal to the amount of substrate consumed per unit weight of cell produced times the stoichiometric coefficient of the conversion reaction.

The values of θ for immobilized cells in the proposed kinetic model and those observed from the experiments are considerably higher. This is because of the fact that the yield of yeast cells in the immobilized state is very low (0.021 g g^{-1}) compared to 0.117 g g^{-1} for the free cells. As a result, the values of product yield for the immobilized system are higher than for the free cell system. Necessarily, the maintenance coefficient for the VIC system, according to Eq. (21) is nearly 50 % higher than the free cells and thus, maintenance energy needs must be higher for the VIC compared to the free cell system. These results are shown in Table 2. Data points of ν and μ for free cell and VIC systems computed from the kinetic models (Eqs. (11), (12), (14), and (15)) and calculated from experimental data are shown in Fig. 3. Various values of constants used in the computations are given in Table 3 (Ghose and Bandyopadhyay, 1981).

TABLE 2. Maintenance requirements for free and immobilized *S. cerevisiae* cells.

Basis	θ	η	m	Remark
Calculated from ATP production for free cells	4.36	–	–	Eq. 23
Computed from kinetic model for free cells	4.54	0.04	0.085	Eqs. 9, 10, 11
Calculated from experimental data for free cells	4.40	0.025	0.053	Fig. 3
Computed from kinetic model for VIC	17.18	0.065	0.132	Eqs. 16, 17
Calculated from experimental data for VIC	18.38	0.05	0.102	Fig. 3

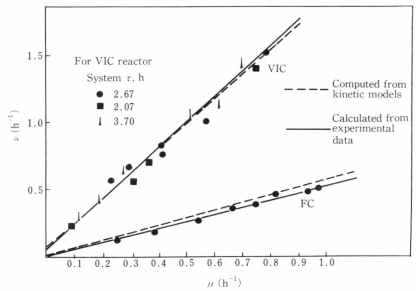

FIG. 3. ν vs μ plots for VIC and FC (Tyagi and Ghose (1982), Ghose and Tyagi (1982), Ghose and Bandyo-
padhaya (1982)).

TABLE 3. Numerical values of constants used in the computations of rates of immobilized cell growth and
product formation.

Constants	Values	Units
P_m	87.0	g l^{-1}
P_m^i	114.0	g l^{-1}
$K_{s(g)}$	0.476	g l^{-1}
$K_{s(p)}$	0.666	g l^{-1}
$Y_{x/s}$	0.09	g g^{-1}
$Y_{p/s}$	0.47	g g^{-1}
$Y_{x/s}^i$	0.02	g g^{-1}
$Y_{p/s}^i$	0.49	g g^{-1}
S_o	150.0	g l^{-1}
μ_m	0.25	h^{-1}
μ_m^i	0.08	h^{-1}
ν_m	1.0	h^{-1}
ν_m^i	1.4	h^{-1}
$\omega_{(g)}$	250	–
$\omega_{(p)}$	30	–

5. Concluding Remarks

From what has been briefly reviewed, it is clear that VIC and their offspring display
considerably different metabolic changes compared with cells in the free state. In summary,
immobilized cells: (a) display a decrease in the generation time of offspring; (b) form monolayers and undergo physical stresses; (c) differ significantly from the chemical make-up of
offspring cells; (d) give very high yields of some polysaccharides as well as ratios of DNA/
RNA turnover rates; (e) shift from fumarase to succinate dehydogenase activities in the
biosynthesis of L-aspartic acid, and immobilized *Trichoderma reesei* cells; (f) undergo
changes in the biosynthesis of β-glucosidase in the cellulase enzymes; and (g) demonstrate

increased productivity of cellulase enzyme from encapsulated fungal conidia. All these observations are considered to have a connection with metabolic shifts or alterations occurring at the gene level in the immobilized state of cells.

Needless to say, more detailed studies with VIC are called for. The kinetic models proposed for the immobilized yeast cells may not be fully applicable to all other kinds of cells but can serve as the basis for a rational approach to the modeling of similar rate processes carried out by VIC.

The basic cause of inhibition of growth of animal cells by what is called "contact inhibition" is not clear, but from the observation of reduced doubling time in VIC of *S. cerevisiae* yeasts and reduced growth rate in contact inhibition in animal cells, it might be implied that the sequence of events occurring in the cell division cycles in yeasts may be affected by these cells coming into contact with each other on the carrier. One may assume, although no experimental evidence is yet available, that at some point in the chain of events connecting the monogenome, DNA replication, and two-genome (nuclear division phase), restrictions are imposed at the genetic level so that delay occurs in the emergence of buds (in yeasts) or daughter cells in bacterial cells. Much more severe inhibitions are exercised on animal cells. The excellent reports on these aspects by Hartwell (1971), Mitchison et al. (1972), and Hartwell et al. (1974) provide substantial motivation for making more detailed analytical enquiries on the subject of viable immobilized cells.

The major questions to which answers must be sought are: Do the immobilized or closely attached cells undergo transformation at the genetic level? Can this process be recognized as random mutation? Is attachment a frequency function? Can these cells be strictly called synchronous if their offspring are washed off the monolayer at regular time intervals? Do two daughter VIC cells divide at different times? What determines the maximum value of X^i? Is it dependent on the cell's physiological state or on the environmental parameters, e.g., the surface morphology of the carrier matrix? Biochemical engineers need to know much more about VIC to be able to make better use of them in large bioconversion systems. The totally virgin area of immobilized recombinants, which may soon have potential applications in many new bioconversion processes, requires special attention.

Acknowledgments

The author expresses sincere thanks to Dr. Gopal Chotani of Eastman Kodak, Rochester, NY, for reading the manuscript and making useful suggestions. Miss Sutini Suratman of the Department of Chemical Engineering, NUS, is also thanked for typing the manuscript.

REFERENCES

1. Aiba, S., Humphrey, A.E., Millis, N.F., In *Biochemical Engineering 2nd ed.*, Univ. of Tokyo Press, Tokyo, 1973.
2. Atkinson, B., Fowler, H.W., "The significance of microbial film in fermenters." *Adv. Biochem. Eng.*: 3, 221–277 (1974).
3. Bandyopadhyay, K.K., Ghose, T.K., "Studies on immobilized *Saccharomyces cerevisiae*. III. Physiology of growth and metabolism on various supports." *Biotechnol. Bioeng.*: 24, 805–815 (1982).
4. Brodelius, P., Deus, B., Mosbach, K., Zenk, M.H., "Immobilized plant cells for the production and transformation of natural products." *FEBS Lett.*: 103, 93–97 (1979).
5. Chibata, I., Tosa, T., Takata, I., "Continuous production of L-malic acid by immobilized cells." *Trends. Biotechnol.*: 1, 9–11 (1983).

6. Davies, J.T., Ridal, E.K., In *Interfacial Phenomena*, Academic Press, New York, 1961.

7. Doran, P.H., Bailey, J.E., *Proc. VII Australian Biotechnol. Conf.*, Brisbane, 1984.

8. Doran, P.M., Bailey, J.E., "Effects of immobilization on growth, fermentation properties, and macro-molecular composition of *Saccharomyces cerevisiae* attached to gelatin." *Biotechnol. Bioeng.*: **28**, 73–87 (1986).

9. Fletcher, M., Marshall, K.C., "Are solid surfaces of ecological significance to aquatic bacteria." *Adv. Microb. Ecol.*: **6**, 199–236 (1982).

10. Förberg, C., Häggström, L., "Control of cell adhesion and activity during continuous production of acetone and butanol with adsorbed cell." *Enz. Microb. Technol.*: **7**, 230–234 (1985).

11. Ghose, T.K., Chand, S., "Kinetic and mass transfer studies of cellulose hydrolyzate using immobilized *Streptomyces* cells." *J. Ferment. Technol.*: **56**, 315–322 (1978).

12. Ghose, T.K., Sahai, V., "Production of cellulases by *Trichoderma reesei* QM9414 in fed-batch and continuous-flow culture with cell recycle." *Biotechnol. Bioeng.*: **21**, 283–296 (1979).

13. Ghose, T.K., Tyagi, R.D., "Rapid ethanol fermentation of cellulose hydrolysate. I. Batch versus continuous systems." *Biotechnol. Bioeng.*: **21**, 1387–1400 (1979).

14. Ghose, T.K., Tyagi, R.D., "Production of ethyl alcohol from cellulose hydrolysate by whole cell immobilization." *J. Mol. Cat.*: **16**, 11–18 (1982).

15. Ghose, T.K., Bandyopadhyay, K.K., "Rapid ethanol fermentation in immobilized yeast cell reactor." *Biotechnol. Bioeng.*: **22**, 1489–1496 (1980).

16. Ghose, T.K., Bandyopadhyay, K.K., "Studies on immobilized *Saccharomyces cerevisiae*. II. Effect of temperature distribution on continuous rapid ethanol formation in molasses fermentation." *Biotechnol. Bioeng.*: **24**, 797–804 (1982).

17. Ghose, T.K., Bandyopadhyay, K.K., "Rapid ethanol fermentation by immobilized whole cells." *Proc. II World Congress of Chem. Eng.*, Canada, p. 293–297 (1981).

18. Ghose, T.K., Tyagi, R.D., "Immobilized whole microbial cells: maintenance requirements, reactor analysis and performance of ethanol production system." In *Recent advances in the engineering analysis of chemically reacting systems*, p. 551–577. Doraiswamy, L.K. (ed.), Wiley Eastern Ltd., New Delhi, 1984.

19. Hahn-Hägerdal, B., Larsson, M., Mattiasson, B., "Shift in metabolism towards ethanol production in *Saccharomyces cerevisiae* using alterations of the physical-chemical microenvironment." *Biotechnol. Bioeng. Symp. Series* No. 12, 199–202 (1982).

20. Hattori, T., Furusaka, C., "Chemical activities of *Escherichia coli* adsorbed on a resin." *Biochem. Biophys. Acta*: **31**, 581–582 (1959).

21. Hartwell, L.H., "Genetic control of the cell division cycle in yeast II. Genes controlling DNA replication and its initiation." *J. Mol. Biol.*: **59**, 183–194 (1971).

22. Hartwell, L.H., Culotti, J., Pringle, J.R., Reid, B.J., "Genetic control of the cell division cycle in yeast. A model to account for the order of cell cycle events is deduced from the phenotypes of yeast mutants." *Science*: **183**, 46–51 (1974).

23. Hattori, R., "Growth of *Escherichia coli* on the surface of an anion-exchange resin in continuous flow system." *J. Gen. Appl. Microbiol.*: **18**, 319–330 (1972a).

24. Hattori, R., Hattori, T., Furusaka, C., "Growth of bacteria on the surface of anion-exchange resin I. Experiment with batch culture." *J. Gen. Appl. Microbiol.*: **18**, 271–283 (1972b).

25. Jirkú, V., Turková, J., Krumphanzl, V., "Immobilization of yeast cells with retention of cell division and extracellular production of macromolecules." *Biotechnol. Lett.*: **2**, 509–513 (1980).

26. Johnson, B.F., Gilson, E.J. "Autoradiographic analysis of regional cell wall growth of yeasts." *Exp. Cell Res.*: **41**, 580–591 (1966).

27. Küenzi, M.T., Fiechter, A., "Changes in carbohydrate composition and trehalase-activity during the budding cycle of *Saccharomyces cerevisiae*." *Arch. Microbiol.*: **64**, 396–407 (1969).

28. Larsson, P.O., Mosbach, K., "Alcohol production by magnetic immobilized yeast." *Biotechnol. Lett.*: **1**, 501–506 (1979).

29. Mitchison, J.H., "The 'DNA-division cycle' and the 'growth cycle'." In *The Biology of The Cell Cycle*, Cambridge University Press, p. 244–249, New York, 1972.

30. Navarro, J.M., Durand, G., "Modification of yeast metabolism by immobilization onto porous glass." *Eur. J. Appl. Microbiol.*: **4**, 243–254 (1977).

31. Navarro, J.M., Durand. G., "Modifications de la croissance de *Saccharomyces uvarum* par immobilisa-

tion sur support solide." Comptes Rendus Des Seances, De L'Acad. Des Sciences, Paris Ser. D. **290**, (6), 453–456 (1980).

32. Panek, A., "Function of trehalose in Baker's yeast (*Saccharomyces cerevisiae*)." *Arch. Biochem. Biophys.*: **100**, 422–425 (1963).

33. Robinson, P.K., Dainty, A.L., Goulding, K.H. Simpkins, I., Trevan, M.D., "Physiology of alginate-immobilized *Chlorella*." *Enz. Microbiol. Tech.*: **7**, 212–216 (1985).

34. Shimizu, S., Tani, Y., Yamada, Y., "Synthesis of coenzyme A by immobilized bacterial cells." *ACS Symp. Ser.*: **106**, 87–100 (1979).

35. Sloat, B.F., Adams, A., Pringle, J.R., "Roles of the *CDC24* gene product in cellular morphogenesis during the *Saccharomyces cerevisiae* cell cycle." *J. Cell. Biol.*: **89**, 395–405 (1981).

36. Suzuki, S., Karube, I., "Production of antibiotics and enzymes by immobilized whole cells." *ACS Symp. Ser.*: **106**, 59–72 (1979).

37. Takagi, A., Harashima, S., Oshima, Y., "Construction and characterization of isogenic series of *Saccharomyces cerevisiae* polyploid strains." *Appl. Environ. Microbiol.*: **45**, 1034–1040 (1983).

38. Tyagi, R.D., Ghose, T.K., "Studies on immobilized *Saccharomyces cerevisiae*. I: Analysis of continuous rapid ethanol fermentation in immobilized cell reactor." *Biotechnol. Bioeng.*: **24**, 781–795 (1982).

39. van Steveninck, J., Rothstein, A., "Sugar transport and metal binding in yeast." *J. Gen. Physiol.*: **49**, 235–246 (1965).

40. van Uden, N., "Transport limited fermentation and growth of *Saccharomyces cerevisiae* and its competitive inhibition." *Arch. Microbiol.*: **58**, 155–168 (1967).

41. Vijayalakshmi, M., Marcipar, A., Segard, E., Broun, G.B., "Matrix-bound transition metal for continuous fermentation tower packing." *Ann. N.Y. Acad. Sci.*: **326**, 249–254 (1979).

II. DNA Technology

II-1
Genetically Structured Models for Growth and Product Formation in Recombinant Microbes

James E. Bailey

Department of Chemical Engineering, California Institute of Technology, Pasadena, California 91125, U.S.A.

The performance of genetically engineered microorganisms in biochemical processes depends upon interactions of key enzymes and regulatory molecules with corresponding binding regions on DNA within the cell. Such interactions are explicitly represented in genetically structured models for cell growth and product formation. Such models, which provide a direct mapping from certain genetic alterations into corresponding host-vector kinetic properties, have been successfully applied to describe λdv plasmid replication, regulation of the cloned *lac* promoter-operator, host-plasmid interactions in recombinant *Escherichia coli*, and product formation in bioreactors with genetically unstable populations. Simulation results obtained from these genetically structured models agree well with available experimental data.

1. Introduction

The first textbooks on biochemical engineering offered little systematic, quantitative treatment of fundamental principles underlying bioprocessing. In a breakthrough from then-traditional approaches based upon empiricism and industrial practice, the book "Biochemical Engineering" by Aiba, Humphrey, and Millis[1] initiated the modern era of biochemical engineering science. Professor Aiba and his coauthors clearly showed in numerous contexts how quantitative description of fundamental transport and reaction processes could be interfaced with known properties of biological molecules and cells to arrive at a rational, systematic description of the interrelationships among different process variables and properties of the biological agents involved.

Extension of these principles to the field of growth and product formation kinetics in cells suggests that kinetic models should represent the central mechanistic processes involved. It has been clearly established in theoretical studies of kinetic model lumping or aggregation that combining or aggregating multiple components into a smaller number of pseudocomponents generally results in modeling errors.[2-4] However, living cells are such complex objects as chemical reactors that careful judgment is required in deciding which components and which fundamental features of the cell should be retained in the kinetic model and which should be treated in more lumped or aggregated form. The nature and depth of model detail retained should be selected based upon the intended application of the model.

Contemporary methods for genetic manipulation of cells provide the opportunity for precise change in nucleotide sequence in cloned DNA sequences. Based upon the mechanisms of plasmid replication, gene expression, and cellular growth and metabolic regulation, it is expected that any change in nucleotide sequence which influences the binding affinity of a catalytic or regulatory species for the DNA is potentially very important. Therefore, kinetic models that are to be used to analyze and eventually to predict the effects of nucleotide sequence changes in regulatory and protein binding sites on the DNA should focus on the interactions between these DNA domains and those components in the cell which bind to them. Kinetic models formulated at this level and with this purpose have been called *genetically structured models*.[5] This review emphasizes the development of genetically structured models in the author's laboratory for description of recombinant cell growth and cloned-gene product synthesis. It is important to emphasize that this research builds upon earlier work in which models with such genetic structure were used to describe enzyme synthesis in wild-type and mutant bacteria and molds. Especially noteworthy in the present context are the detailed models for α-galactosidase production by a *Monascus* sp. mold by T. Imanaka and coworkers[6] and the kinetic model for catabolite repression presented by Imanaka and Aiba.[7]

The examples summarized next are organized in an approximate hierarchical order of complexity. The discussion begins with a model for regulation of a cloned gene using a cloned *lac* promoter-operator, progresses to interaction between gene expression and plasmid replication of the plasmid λdv in *E. coli*, continues to the cellular level to consider partitioning of RNA polymerase and ribosomes between plasmid and chromosomal DNA in a single-cell model for *E. coli* kinetics, and concludes with bioreactor models for recombinant populations which are based upon molecular mechanisms of cloned gene expression. A common thread in all of these examples is description of transcription and translation at a fundamental level involving material balances on messenger RNA and protein derived from a single gene. These central elements of genetically structured modeling are given next.

2. Mathematical Description of Transcription and Translation

Transcription is the production of a single strand of ribonucleic acid (RNA) complementary to a specific deoxyribonucleotide sequence in DNA. The rate of production of a specific messenger RNA (mRNA) by transcription is determined by the number of DNA templates for that mRNA and by the efficiency of transcription, η. The transcription efficiency is regulated by a number of molecules in the cell including the enzyme RNA polymerase. A model for *lac* promoter-operator transcription efficiency will be described below. Messenger RNA is also degraded by certain enzymes, and it is diluted as the cell volume increases.[8] Consequently, a mass balance on mRNA in an individual *E. coli* cell may be written

$$\frac{d[\text{mRNA}]_i}{dt} = k_{p,i}^0 \eta_i [G]_i - k_{d,i}[\text{mRNA}]_i - \mu[\text{mRNA}]_i \tag{1}$$

where η_i is the transcription efficiency at the promoter, and $k_{p,i}^0$ and $k_{d,i}$ represent the overall transcription rate constant and the decay rate constant of mRNA, respectively. The subscript i indicates correspondence with a particular gene.

In a similar fashion, a material balance can be written for the intracellular concentration of the product of translation, namely the protein of interest. In this case, the synthesis term contains the concentration of the specific message coding for the protein i, and the efficiency of utilization of $[\text{mRNA}]_i$ at the ribosomes. The required mass balance is given by

$$\frac{d[P]_i}{dt} = k^0_{q,i}\xi_i[\text{mRNA}]_i - k_{e,i}[P]_i - \mu[P]_i \qquad (2)$$

where ξ_i denotes the translation efficiency and $k^0_{q,i}$ and $k_{e,i}$ denote the overall translation rate constant and the decay rate constant of protein i, respectively. The translation efficiency ξ_i will be determined by the intracellular concentration of active ribosomes, by the level of regulatory elements which control the rate of translation, and by the affinity of the ribosome binding site on $[\text{mRNA}]_i$ with the 30S subunit of the ribosome.[9]

3. Regulation of Transcription of the lac Operon

Expression of the *lac* operon is controlled primarily at the transcription step. Consequently, the molecular mechanism of *lac* operon transcription will be considered in some detail. A schematic diagram of the important DNA sequences and regulatory molecules is given in Fig. 1. The transcription efficiency of the *lac* promoter-operator, which determines the rate of synthesis of *lac* operon gene products such as the enzyme β-galactosidase, is determined by the interaction of the *lac* promoter and operator regions with three regulatory molecules:[10,11] *lac* repressor protein (R), cyclic-AMP receptor protein (CRP) complexed with cyclic AMP (cAMP), and RNA polymerase (RNP) activated by σ factor (σ).

A general expression for transcription efficiency has been derived based upon the assumption that binding of regulatory proteins to binding sites on DNA can be described by mass-action equilibrium equations.[12,13] Using these results and information from the experimental literature on binding affinities and intracellular concentrations of regulatory proteins and regulatory sites, and assuming that the β-galactosidase activity in the cell is proportional to the gene expression efficiency, the values in Table 1 were calculated.[12] Also listed there are corresponding experimental values for a number of different *E. coli* genotypes in the presence and absence of inducer (IPTG). Clearly, the model achieves an

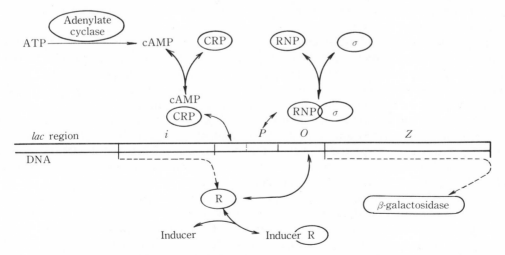

FIG. 1. Schematic diagram of the *lac* operon transcription control system.[12]

TABLE 1. Gene expression of *lac* repressor mutants.[12]

Case	Genotype	Inducer	Enzyme activity	
			Experimental[a]	Calculated[b]
1	I^+	−	<0.1	0.1
2	I^+	+	100	100.0
3	I^-	−	100–140	113
4	I^-	+	100–130	113
5	I^s	−	<0.1	0.1
6	I^s	+	<1	1.8
7	I^q	−	0.01–0.014	0.011
8	I^q	+	65	62
9	I^{sq}	−	0.003–0.004	0.002
10	I^{sq}	+	25	23

[a] Experimental data taken from Jacob and Monod[14] for Cases 1 through 6. Inducer (isopropyl-β-D-thiogalactopyranoside; IPTG) concentrations employed in the experiments were 10^{-4} M for Cases 1 and 2 and 10^{-3} M for other cases. The enzyme activity (β-galactosidase) was normalized to the value of the induced wild-type state. Sources of other experimental data: Cases 7 and 8: Jobe et al.[15] Cases 9 and 10: Miller.[16]

[b] Kinetic parameters used in calculations: Cases 3 and 4: $[R]_0 = 0$; Cases 5 and 6: $K_{C_1}^{mut} = 10^{-3}K_{C_1}^{wt}$; Cases 7 and 8: $[R]_0^{mut} = 10[R]_0^{wt}$; Cases 9 and 10: $[R]_0^{mut} = 50[R]_0^{wt}$ Superscripts *wt* and *mut* represent the wild-type and mutant, respectively.

excellent degree of correspondence with experiment for a large degree of genetic variation and also for environments which differ in IPTG concentration.

3.1   Modeling of lac Promoter-Operator Function in Plasmid-Containing Cells

The model just described has been applied to calculate the synthesis of β-galactosidase encoded in both chromosomal DNA and in F' *lac* episomes. These F' *lac* episomes carry *lac* operon genes and are obtained by sexduction. The cases considered involve F plasmids containing different *lac* region sequences that have been studied experimentally in detail. The major question at this juncture of the modeling effort was the validity of extending the model from description of transcription of the gene in the chromosomal DNA to transcription of gene copies in plasmids.

In calculating gene expression efficiency in plasmid containing cells, the total number of both the regulatory proteins and the regulatory sites must be changed, taking into account the number of plasmids carried by the cell and the makeup of that plasmid. Details of the modeling equations required are presented in Refs. 12 and 13. Table 2 shows calculated β-galactosidase activity for cells containing F plasmids (assumed to be carried at a copy number of 1.7) and experimental data for the corresponding strains. As in the case of chromosomal mutation, it is assumed that I^- and Z^- mutants do not produce active *lac* repressor and β-galactosidase, respectively. Again, the close agreement between model results and experimental observations is striking, spanning several orders of magnitude and accurately describing gene expression behavior in the presence and in the absence of inducer.

The model for plasmid containing cells has also been employed to evaluate the overall transcription rate initiated at the cloned *lac* promoter-operator in the fully induced state as a function of plasmid copy number (defined as the number of plasmids divided by the number of chromosomes).[13] As indicated in Fig. 2, the calculated transcription rate initially increases proportionally to the plasmid copy number, but at high copy numbers, deviates

TABLE 2. Gene expression in diploid systems containing *lac* DNA in F plasmids.[12]

Genotype	Inducer	Enzyme activity	
		Experimental[a]	Calculated[b]
$I^+Z^-Y^+/F'I^+Z^+Y^+$	−	0.10	0.09
$I^+Z^-Y^+/F'I^-Z^+Y^+$	+	156	152
$I^+Z^-Y^-/F'I^-Z^+Y^+$	−	0.41	0.41
$I^+Z^-Y^-/F'I^-Z^+Y^+$	+	150	170
$I^+Z^+Y^-/F'I^+Z^+Y^+$	−	0.14	0.14
$I^+Z^+Y^-/F'I^+Z^+Y^+$	+	239	241
$I^+Z^+Y^-/F'I^-Z^+Y^+$	−	0.61	0.65
$I^+Z^+Y^-/F'I^-Z^+Y^+$	+	233	270

[a] Experimental data from Overath.[17] Activities normalized by induced activity wild-type (i.e., $I^+Z^+Y^+$ = 100). The strain and genetic background of host cell: *E. coli* K12 (rec^- crp^+ cya^+). IPTG concentration (1) = 5×10^{-4} M.

[b] Copy number of F plasmid was assumed to be 1.7.

from linearity, eventually passes through a maximum, and subsequently decreases as the plasmid copy number increases. The two curves shown in Fig. 2 correspond to relatively small (A) and relatively large (B) plasmids, respectively. The transcription rate is not proportional to the number of gene copies at large plasmid content because RNA polymerase is titrated away from *lac* promoters by nonspecific binding on plasmid DNA and other plasmid promoters. The effect of plasmid content on partitioning of RNA polymerase and other intracellular components is a crucial element in the detailed single-cell model for *E. coli* growth and product synthesis to be presented shortly. Based upon the results in Fig. 2,

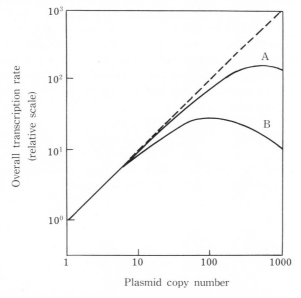

FIG. 2. Overall transcription rate initiated at the cloned *lac* promoter-operator in the fully induced state as a function of plasmid copy number. Ordinate is normalized by the transcritpion rate for a single plasmid. Plasmid genotype is P^+O^+: (A) 5 promoters, 0.1% genome size; and (B) 20 promoters, 1% genome size.[13]

it is clear, qualitatively, that high plasmid copy number will be detrimental to host-cell function also by reducing transcription from chromosomal promoters. However, before considering this application of genetically structured models, the mathematical basis for intracellular equilibrium calculations will be examined, and, subsequently, a genetically structured model for plasmid replication will be examined.

4. Statistical Thermodynamic Analysis of Intracellular Equilibria

In the model just described, interactions between DNA binding proteins and their binding sites on the DNA have been assumed to be in equilibrium, and equilibrium concentrations have been calculated by assuming mass-action kinetics for formation and association of a bound complex. This is equivalent to writing chemical potentials for the components involved in terms of their intracellular concentrations. As noted previously by Harder and Röels[18] and others,[19,20] the validity of this approach is questionable since the interactions considered actually occur in a collection of small closed systems, namely individual cells. Each cell may contain a very small number of the molecules or binding sites involved in the biological mechanism. Therefore, on a single-cell basis, mass-action concepts may not be applicable.

On an individual cell basis, the binding states of a component present in small numbers exist only in discrete values. For example, in a cell containing five molecules of a species S_1 which combines with some other component, the number of bound S_1 molecules is either 0, 1, 2, 3, 4, or 5, and any of these binding states, and only one of them, would be observed if a single cell were examined at a particular point in time. Intracellular concentration of unbound species S_1 in this case is a discrete, not a continuous, variable. The "equilibrium state" of a population of such cells refers to the average of the possible single-cell states, weighted by the probabilities that the molecules are found in these states.

These probabilities are properly calculated applying the method of statistical thermodynamics. The general formalism for accomplishing these calculations has been presented[21] and will be outlined here. Let p_i denote the probability of state i within a cell, where each state corresponds to some particular number of bound and unbound species in a general system of interacting molecules. At equilibrium, the values of p_i should minimize the overall Gibbs' free energy given by

$$\hat{G} = \sum_{i=1}^{N} p_i \hat{G}_i + \sum_{i=1}^{N} p_i \ln p_i. \tag{3}$$

The evaluation of the free energy $\hat{G}_i$ is calculated according to equations given in Ref. 21. The probabilities that minimize $\hat{G}$ must satisfy the constraint

$$p_i + p_2 + \ldots + p_N = 1. \tag{4}$$

The solution to this minimization problem is given by

$$P_i = \frac{e^{\Delta \hat{G}_i}}{\sum\limits_{j=1}^{N} e^{-\Delta \hat{G}_j}} \tag{5}$$

where

$$\Delta \hat{G}_j = \hat{G}_j - \hat{G}_1. \tag{6}$$

It should be noted that p_i has the alternative physical interpretation as the expected fraction of a large population of cells which exist in state i. If S_k^i denotes the number of molecules of species k in state i, then the average number of molecules i in the population is given by

$$\bar{S}_k = \sum_{j=1}^{N} p_j S_k^j. \tag{7}$$

In order to evaluate the quantitative difference between rigorous calculation of intracellular concentrations based upon the statistical thermodynamic approach, which requires significantly more computational effort, relative to the results based on the assumption of classic equilibrium thermodynamics, the *lac* repressor-operator system was analyzed considering the set of reactions given in Eq. (8):

$$R + O \rightleftharpoons RO; \qquad K_1 = 2 \times 10^{12} M^{-1}$$
$$R + I \rightleftharpoons RI; \qquad K_2 = 1 \times 10^{7} M^{-1}$$
$$RI + O \rightleftharpoons RIO; \qquad K_3 = 2 \times 10^{9} M^{-1}$$
$$R + D \rightleftharpoons RD; \qquad K_4 = 1 \times 10^{3} M^{-1}$$
$$RI + D \rightleftharpoons RID; \qquad K_5 = 1.5 \times 10^{4} M^{-1} \tag{8}$$

where R denotes the *lac* repressor protein, O the *lac* operator; I the *lac* inducer (isopropyl-β-D-thiogalactopyranoside; IPTG); D the nonspecific binding sites (the entire genome);

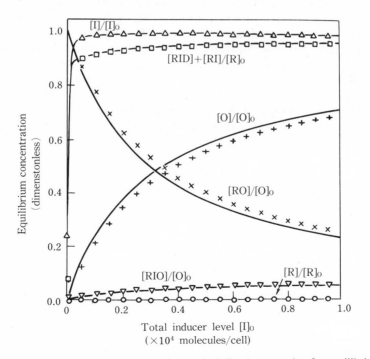

FIG. 3. The results of the statistical (points) and the classical (lines) approaches for equilibrium calculations for the *lac* repressor-operator system.[21]

RO, RI, RIO, RD, and *RID* are the corresponding complexes; and K_m are the corresponding association constants.

Dimensionless intracellular equilibrium concentrations calculated by both methods for this model, after the elimination of reactions involving nonspecific binding sites, are shown in Fig. 3 for a strain that contains a single *lac* operator per cell and ten molecules of *lac* repressor protein.[21] As indicated there, the difference between the classical and the statistical results are relatively minor. Consequently, for this system, assumption of classical equilibrium relationships is a reasonable approximation. Although this comparison for this particular example does not prove in any sense that assumption of classical equilibrium relationships is always a good approximation, this work does provide some indications that the simpler, equilibrium approach may be appropriate, especially in very complex models in which model calculations would be greatly complicated by applying the statistical thermodynamics approach to all the equilibria involved. Accordingly, the classical equilibrium method is adopted in all of the genetically structured models discussed below. However, it should be noted that the statistical thermodynamics framework which has been presented[21] is completely general and can be applied to effect rigorous equilibrium calculations for any set of interacting species of any level of complexity. If questions arise concerning the sensitivity of any model calculations to the application of classical equilibrium methods, the statistical thermodynamic approach can be employed.

5. Replication of Plasmid λdv

Modeling replication of the plasmid λdv introduces additional levels of complexity in biological regulation and interaction and in the corresponding model. As in the synthesis of many biological macromolecules, it is frequency of initiation of synthesis that dictates the plasmid replication rate. Consequently, the mathematical model for plasmid replication must focus on correct calculation of initiation rates in order to calculate properly plasmid synthesis rates. This calculation depends upon consideration of the interactions of a number of proteins with different domains of the λdv plasmid. A detailed mechanistic picture of the replication mechanism, outlined in Fig. 4, can be formulated based upon a wealth of information from the molecular biology literature.[5,22] A critical regulating factor in

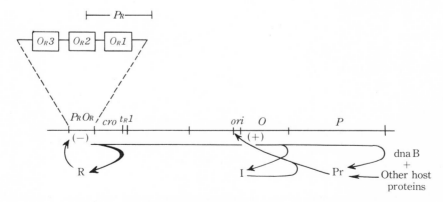

FIG. 4. Schematic diagram of the genetic structure of the λdv replicon.[22]

replication initiation for this plasmid is the frequency of transcription which begins at the promoter P_R upstream of the binding site for initiator proteins at the *ori* site in the origin of replication. Transcription activity at P_R is regulated by binding of the *cro* gene product, here designated R, at three different binding sites, two of which overlap with the promoter P_R and hence, when bound with R, inactivate this promoter.

Calculation of the transcription efficiency at the promoter P_R has been accomplished by assuming equilibrium binding of the repressor R at the three binding sites designated O_R1, O_R2, and O_R3. It is important to note that the equilibrium constants for *cro* repressor binding to each of these operator sites has been experimentally determined, at least for the wild-type plasmid. Transcription efficiency so evaluated can then be used with *cro* protein material balances written in the form of Eqs. (1) and (2) to calculate the *cro* repressor level in the cell and to evaluate the transcription activity from P_R which subsequently passes through the *ori* region, activating it for binding of an initiator complex. The model is too complex to summarize in any detail here; the complete development of model equations, model parameters, and supporting documentation from the molecular biology literature are provided in Refs. 5 and 22. The model satisfactorily simulates the average plasmid content and repressor content for the wild-type plasmid compared to experimental data.

The genetic structure included in this model was used in order to simulate the replication behavior of mutant plasmids in a systematic fashion. The plasmids designated λdvBC2, λdvl, and λdvAJ5 have a single mutation in the O_R2 region, one point mutation in O_R2 and one point mutation in O_R1, and one point mutation in O_R1 and the same mutation in O_R2 as λdvBC2, respectively. Thus, for these mutants, the only model parameters that can and should be altered in accounting for these mutations are K_2, K_1, and K_1, K_2, respectively. Furthermore, the K_2 value used for λdvBC2 and for λdvAJ5 must be identical, although different from the wild-type value. As the binding parameters for the mutant plasmids were not experimentally available, they were estimated approximately by trial and error. The model, with specific parameters so changed, has successfully simulated the trends in repressor level and λdv plasmid copy number for all of the mutant plasmids (see Table 3).[5]

TABLE 3. Effect of mutations of the promoter-operator $P_R O_R$ on the *cro* repressor levels and plasmid copy number.[5]

Case	Plasmid	$P_R O_R$	Relative affinity for repressor binding[a]	Repressor level Units (mg protein)$^{-1}$		Copy number	
				Exp.[a]	Cal.[b]	Exp.[a]	Cal.[b]
1	λdvBB1	wild-type	1	0.14	0.14	25	25
2	λdvBC2	*vir*C34	0.78	0.28	0.27	27	38
3	λdvl	*v1v3*	0.06	1.60	1.23	85	87
4	λdvAJ5	*vir*C34 *vir*R18	0.02	2.54	2.33	125	122

[a] Experimental data were taken from Murotsu and Matsubara.[23]

[b] Except for the following parameters, the same values as listed in Table 2 of Lee and Bailey[22] were used for calculation:

Case 2, $K_2^{mut} = \frac{1}{5} K_2^{wt}$; Case 3, $K_1^{mut} = \frac{1}{10} K_1^{wt}$, $K_2^{mut} = \frac{1}{20} K_2^{wt}$; Case 4, $K_1^{mut} = \frac{1}{250} K_1^{wt}$, $K_2^{mut} = \frac{1}{5} K_2^{wt}$. Superscripts *mut* and *wt* denote the mutant and wild-type, respectively. Calculated repressor contents were converted to activity units using the factor 0.14 units (mg protein)$^{-1}$ obtained from the measured wild-type repressor activity[23] and the calculated wild-type repressor content.[22]

6. A Detailed Kinetic Model for a Single Cell of Escherichia coli with Genetic Structure for Protein Synthesis

The pioneering research by Schuler and colleagues established a base model for growth of an individual cell of *E. coli* using lumped descriptions of cellular metabolites and macro-molecules.[24-26] The model was structured to describe carbon and nitrogen flow approximately as in the real cell, and mechanistic detail was included for chromosomal DNA replication initiation. This model has proven quite versatile and successful in simulating the relationships between different limiting nutrient levels and the corresponding growth rates, population characteristics, and size of the bacterial cell.

In order to apply this type of model to evaluate the intracellular competition between the host chromosome and recombinant plasmids in directing cellular metabolism, substantial extensions have been made in the Cornell model to include genetic structure for protein synthesis.[27] The complete exposition of the model and its extension to plasmid-containing cells is provided elsewhere. However, in the context of the present review, it is interesting to examine the treatment in this model of initiation of transcription. This is a crucial component of the model in order to describe properly allocation of cellular RNA polymerase activity between messenger and stable RNA promoters on the chromosome and promoters on the plasmid. Also, nonspecific binding of core polymerase to the DNA must also be considered, as this is a potentially important interaction which indirectly affects available transcriptional capacity of the cell. (Recall Fig. 2.) In the extended model, three general types of polymerase-DNA binding are considered: nonspecific core enzyme binding and holoenzyme binding to promoter and nonpromoter sites (see Fig. 5). σ subunit attachment to core polymerase causes structural alterations resulting in increased selectivity for transcriptional promoter sites. In addition, nonspecific holoenzyme binding decreases compared with core polymerase binding.

Additionally, two distinct promoter types on the chromosome are used in the calculations of equilibrium distribution of different forms of RNA polymerase. These promoter types are those for stable RNA operons and for an "average" messenger RNA operon. As has been noted experimentally,[28, 29] the sRNA operons have the stronger promoters. Messenger RNA-producing operons on the chromosome are assumed to be evenly distributed whereas the sRNA operons are individually accounted for, and their number is in-

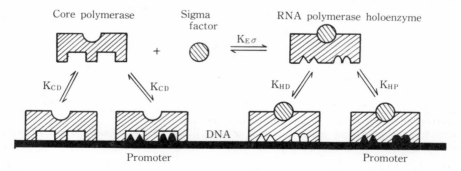

FIG. 5. Schematic diagram of RNA polymerase configurations considered in the single-cell *E. coli* model. Binding of the σ subunit to core polymerase alters binding selectivity of the enzyme. The resultant holoenzyme has much greater affinity for promoter sites and decreased affinity for nonpromoter regions relative to core polymerase.[27]

cremented at the appropriate point as the replication fork passes their position on the chromosome. Furthermore, holoenzyme affinities for m-operon and s-operon promoters are affected by the intracellular concentration of the nucleotide 5', 3'-guanosine tetraphosphate (PG) and by the rate of protein synthesis. PG is produced in a reaction involving idling ribosomes and decreases the affinity of RNA polymerase for s-operon promoters,[30-32] reducing ribosome production. The effect of protein synthesis rate is indirect. Polymerase specificity is affected by translation initiation factor[33] and f-met tRNA,[34,35] the transfer RNA involved in the initiation of protein synthesis. Free initiation factor (IF 2) stimulates s-operon transcription, leading to an increase in cellular protein synthetic capacity. Conversely, excess initiator tRNA signals underutilization of the ribosome and stimulates m-operon transcription. These feedback mechanisms allow the cell continually to monitor the capacity and efficiency of its protein synthesis apparatus and to adjust the levels of message and ribosomes.

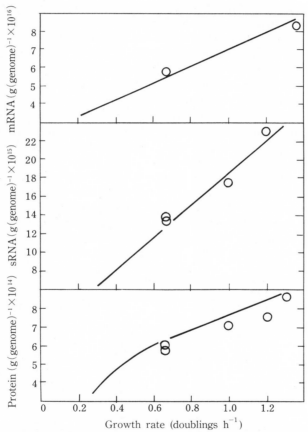

FIG. 6. The data represented (circles) are for *E. coli* B/r grown in minimal medium with different carbon sources utilized to achieve different growth rates.[36] All results are expressed as mass of macromolecule per genome. Shown are the growth rate effect on relative protein content (bottom), stable RNA content, proportional to ribosome content (middle), and mRNA content (top). The lines indicate simulation results, no curves fitted to the data. It is important to note that the model formulation was developed using data for a doubling time of one hour.[27]

A high level of genetic structure is evident in this description. Any genetic change that alters the amount or the biological activity of any of these key compounds in regulating transcription will influence the particular point in the cell model at which this biological activity is physically important, and the overall single-cell model will then evaluate the influence of this change on rates of growth, synthesis of all cellular species included in the model, and cellular morphology. Since little detailed data is available for mutants, the model has been extensively evaluated, for plasmid-free cells, by calculating many different aspects of cellular response and comparing with available experimental data.[27,36] The full evaluation cannot be repeated here, but relative to the previous discussion of modeling of transcription activity and its allocation between synthesis of stable and messenger RNA, it is interesting to examine the model results shown in Fig. 6 and the corresponding experimental data. In these calculations, different growth rates were obtained by changing the external glucose concentration assumed in the simulations—that is, the growth rate is a model output and not something that is fixed or input in the model calculations. Corresponding to these different growth conditions giving different growth rates, the specific cellular content of protein, stable RNA, and messenger RNA have been evaluated and agree very well with the experimental data as shown by the circles in Fig. 6.

When the model is extended to consider plasmids, the binding affinities of promoters on the plasmid relative to the chromosomal promoter are necessary and natural model parameters, as are the affinities of ribosome binding sites encoded on the plasmid relative to average chromosomal ribosome binding sites.[37] Thus, the model is structured in a way to enable simulation of the types of fine scale options that the currently available in constructing expression vectors and to predict the qualitative trends expected from different genetic strategies for increasing cloned-gene expression.

In order to evaluate the quality of the single-cell model including plasmids, the model has been used to evaluate the influence of plasmid number carried by the cell on specific growth rate and cloned-gene product activity in the cell. These quantities have been determined experimentally using a family of closely related plasmids which differ only in their genetic controls for plasmid replication.[38] Model results for recombinant cells compared with the

TABLE 4. Comparison of experimental and simulation results for the relative growth rate as a function of the relative number of plasmids per cell.[37]

Relative number of plasmids per cell	Relative growth rates			
	Experimental			Calculated
0	1			1
1	0.93[a]	0.92[b]	0.97[c]	0.94
2	0.88	0.91	0.94	0.90
5	0.88	0.87	0.88	0.89
10	0.78	0.82	0.84	0.87
20	—	—	—	0.66
34	0.68	0.77	0.82	—

Experimental values are taken from Ref. 38 for *E. coli* HB101 with pMB1 derived plasmids, grown at 37°C in:
  a) M9 minimal medium, enriched with leucine, proline, and thiamine;
  b) Luria broth enriched with leucine, proline, and thiamine; and
  c) M9 minimal medium enriched with thiamine and 0.4% casamino acids.
All media contained 0.2% glucose.

TABLE 5. Comparison of experimental and simulation results for the relative cloned-gene product activity as a function of the relative number of plasmids per cell.[37]

Relative number of plasmids per cell	Relative cloned gene product activity	
	Experimental[a]	Calculated
1	1	1
2	2.1	1.9
5	4.2	3.6
10	5.4	4.1
20	—	5.2
34	7.0	—

experimental data are listed in Tables 4 and 5, indicating excellent qualitative agreement between the single-cell model and the measurements.

Although these are promising results, they should not be taken as conclusive proof that the single-cell model for recombinant cells is in a satisfactory final form. As data accumulate on finer details of recombinant-cell physiology, it is increasingly apparent that these cells can be greatly perturbed relative to wild-type cells in their functions and in their phenotypic responses to different environments. It should be recognized that the presence of a large number of plasmids or a highly active cloned-gene expression system represents a large perturbation relative to the situations for which the cell controls were originally adapted. Extensive experimental research is needed in order to better understand the physiology and metabolic properties of cells that have been subjected to gross genetic perturbations. Then, genetically structured models of the kinds summarized above (and future refinements of these models) can be suitably evaluated, adjusted, and eventually applied to optimize organism design and also to guide bioreactor operating strategies.

7. Application of Genetically Structured Models to Describe Product Formation in Unstable Recombinant Populations

A model that retains certain key elements of genetic structure but forsakes detailed metabolic modeling in order to be able to describe a mixed population situation has been developed for modeling of cloned-gene product formation in unstable recombinant populations.[39] In this model, the growth rate of each representative population is used as a single parameter to indicate the physiological state and macromolecular synthesis activity of the cell.

The basis for formulating a model involving three different populations designated X_1, X_2, and X_3 is the possibility that two different types of genetic instability can arise as a typical expression plasmid (see Fig. 7A) is propagated in a growing cell culture. Such a plasmid must contain a replicon to provide replication functions, carries the product gene and an appropriate promoter, often regulated, and, characteristically, includes a gene which enables the cell containing plasmids to grow in selective medium. Structural rearrangements in the product gene (structural instability) lead to cells which can grow in selective medium but which lose the ability to make product. These are called X_2. Plasmid-containing cells with both product formation and replication functions are grouped together as X_1. Cells that have lost the plasmid due to irregular plasmid partitioning at cell division (segregational instability) are called X_3 (see Fig. 7B).

In the models considered here, the specific growth rate is presumed to depend upon

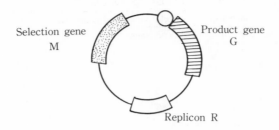

(A) Typical expression plasmid

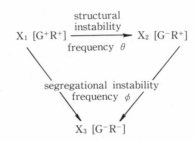

(B) Genotypes and their transitions

FIG. 7. Schematic diagram of basic functional elements in an (A) expression plasmid and diagram of genetic transitions in a (B) recombinant population with structural and/or segregational instability.[39]

limiting substrate concentration s, cloning vector concentration $\hat{G}$, and cloned-gene product concentration $\hat{p}$ as indicated in Eq. (9)

$$\mu = \mu^0 f(s)g(\hat{G})h(\hat{p}). \tag{9}$$

For plasmid-free cells, both g and h are set equal to unity. For cells containing plasmid but lacking the ability to synthesize product (type X_2), the function h is equal to unity. Plasmid-containing cells expressing product are subject to inhibition in growth due both to propagation of the plasmid and to product accumulation. In this work, based upon earlier simulations and available experimental data, the empirical relationship

$$g(\hat{G}) = \left(1 - \frac{\hat{G}}{\hat{G}_{\max}}\right)^m \tag{10}$$

has been adopted for the inhibition function g.

Two different models for product inhibition have been considered. In the first, the inhibitory effect is presumed to depend on the *intracellular* product concentration, $\hat{p}$, according to

$$h(\hat{p}) = \left(1 - \frac{\hat{p}}{\hat{p}_{\max}}\right)^m \text{(Model I).} \tag{11}$$

The second model is based on the total broth concentration, p

$$h(p) = \left(1 - \frac{p}{p_{\max}}\right)^{n'} \text{(Model II)}. \tag{12}$$

As formulated, the model applies to growth in nonselective medium. Extension of this overall growth model to selective medium requires careful consideration of growth rates at small plasmid content and growth of plasmid-free cells. Experiments and model analyses of these questions have only recently begun.[40]

Cloned-gene product formation is described in this model using the material balances for cloned-gene message and for cloned-gene product concentration given earlier by Eqs. (1) and (2). Now, for simplicity and because the response time of messenger RNA concentration is less than that of protein because of more rapid message turnover, the quasi-steady state approximation is applied to Eq. (1) with the result that the material balance for cloned-gene product in the cell may be written as

$$\frac{d\hat{p}}{dt} = f(\mu)\eta\xi\hat{G} - k_e\hat{p} - \mu\hat{p} \tag{13}$$

[The subscript i in Eqs. (1) and (2) has been dropped since the present model treats only one message and protein in this detailed fashion; also, the intracellular cloned-gene product concentration is now denoted by $\hat{p}$.]
where

$$f(\mu) = \frac{k_p^0 k_q^0}{k_d + \mu}. \tag{14}$$

Using the empirical relationships between the specific growth rate and the kinetic parameters in Eq. (14) that were summarized in another paper,[41] Eq. (14) may be written in terms of specific growth rate μ (in units of h^{-1})

$$f(\mu) = \frac{4.5 \times 10^{10} \mu^4}{(78\mu^2 + 233)(145\mu + 82.5)(27.6 + \mu)} \text{ (h}^{-1}\text{)}. \tag{15}$$

Alternatively, it is interesting to consider the implications of a linear approximation to this result in which the growth-rate dependence of the protein synthesis rate is approximated by the linear function given in Eq. (16)

$$f(\mu) = k(\mu + b). \tag{16}$$

It is interesting to note that, with $f(\mu)$ given by Eq. (16), it can be shown that the rate of synthesis of cloned-gene product, expressed in terms of the amount of cloned-gene product per unit volume of medium, can be written as

$$\left.\frac{dp}{dt}\right|_{\text{synthesis}} = \alpha\frac{dx_1}{dt} + \beta x_1 \tag{17}$$

where

$$\alpha = k\eta\xi\hat{G}/\rho_b, \qquad \beta = \alpha b. \tag{18}$$

Here ρ_b denotes the density of the cells. Thus, the genetically structured model parameters can be directly related to the parameters α and β in classical Leudeking-Piret kinetics for product formation.[39]

Using this model, the effect of the number of plasmids carried by the cells can be directly simulated. Figure 8 shows model simulation results based upon the present model for the case of pure segrational instability ($\theta = 0$; $\phi = 0.01$) and without considering product inhibition ($n = 0$). It is interesting to note that the maximum concentration of product obtained, according to these simulations, exhibits a maximum with respect to the number of plasmids per cell. More product is obtained with sixty plasmids per cell than with either ten or 100.

Recently Aiba and co-workers[42] examined experimentally the relationship between the production of tryptophan and the copy number of plasmids using recombinant DNAs with the *trp* operon cloned into four different plasmid vectors RP4, pSC101, RSF1010, and pBR322. They found that accumulation of tryptophan in the culture medium decreased when the plasmid copy number exceeded a certain level. The time trajectories for cell growth, substrate consumption, and product formation shown in Fig. 8 resemble qualitatively the experimental data of Aiba and co-workers for shake flask cultures of the recombinant *trp* system at different copy numbers.

Using this genetically structured model, it has been possible to predict the existence of a maximum intensity of gene expression and also to simulate reasonable operating strategies for batch and continuous bioreactors with regulated promoters and/or replicons.[43] These examples show that judicious application of the principles of genetically structured models contributes to more detailed and more fundamental understanding of cell kinetics and

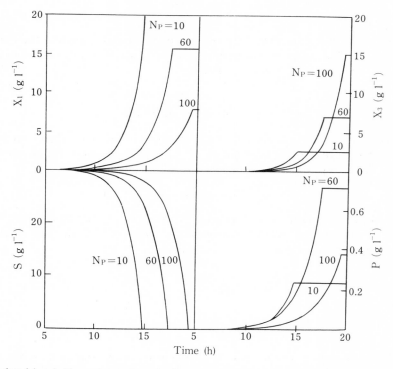

FIG. 8. Simulated batch bioreactor trajectories for pure segregational instability ($\theta = 0$; $\phi = 0.01$) without product inhibition ($n = 0$).[39]

offers the opportunity to formulate kinetic models which directly represent the parametric effects of genetic changes, whether at the level of regulation of a single promoter, description of plasmid replication, modeling the kinetics of a single cell, or describing mixed populations in bioreactors.

8. Concluding Remarks

The overall paradigm involved in genetically structured modeling is summarized in Fig. 9. In the actual physical system, a change in nucleotide sequence at a regulatory locus changes plasmid function and alters overall host-vector properties with respect to growth, product formation, or some other functional attribute of the population of host-vector cells. In a genetically structured model, there is a direct correspondence between specific regulatory genetic loci of interest and corresponding parameters and equations or particular terms in various model equations. Thus, to each genetic modification corresponds an altered model parameter. The model, if it reasonably well approximates other processes in the cell, may then be applied to calculate the altered functional characteristics of the host-vector system containing that specific genetic alteration.

The availability of such models poses fascinating possibilities for host-vector system optimization by collaboration between molecular biologists and biochemical engineers. At some point in the future, it should be possible to predict theoretically or to estimate, based upon accumulated experimental experience, how a particular nucleotide sequence change will alter the binding constant for a binding protein or RNA which interacts with a regulatory domain of the DNA. Then, genetically structured models will enable engineers to establish a mathematical mapping between nucleotide sequence of cloned regulatory regions and overall kinetic properties of the host-vector population.

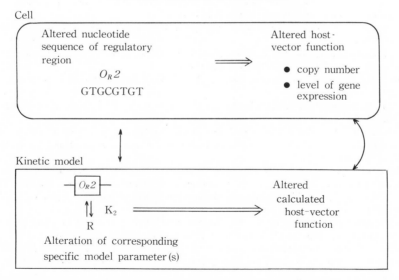

FIG. 9. Schematic diagram summarizing the mapping from nucleotide sequence to predicted host-vector function which is provided by genetically structured models.

Acknowledgments

The author wishes to express his admiration and appreciation to Professor Shuichi Aiba for his leadership in synthesizing engineering and biological science in biochemical engineering research. Professor Aiba has been a pioneer in careful and systematic application of engineering principles and mathematics for the description, analysis, and design of biochemical systems. In addition, his work has inspired many of the current research activities on biochemical engineering aspects of genetically engineered cells, including several research programs in the author's laboratory. The author would also like to thank Dr. Sun Bok Lee for the central role he played in initiating research in the author's laboratory on genetically structured models of recombinant systems. Research on recombinant cell kinetics and genetically structured models in the author's laboratory has been generously supported by the Energy Conversion and Utilization Technology (ECUT) Program of the U.S. Department of Energy and the National Science Foundation.

REFERENCES

1. Aiba, S., Humphrey, A.E., Millis, N.F., *Biochemical Engineering*, University of Tokyo Press, 1965 (2nd edition, Academic Press, 1973).
2. Wei, J., Kuo, J.C.W., "A lumping analysis in monomolecular reaction systems." *Ind. Eng. Chem. Fundl.*: **8**, 114–123, 124–133 (1969).
3. Hutchinson, P., Luss, D., "Lumping of mixtures with many parallel first order reactions." *Chem. Eng. J.*: **1**, 129–136 (1970).
4. Bailey, J.E., "Lumping analysis of reactions in continuous mixtures." *Chem. Eng. J.*: **3**, 52–61 (1972).
5. Lee, S.B., Bailey, J.E., "A mathematical model for λdv plasmid replication: Analysis of copy number mutants." *Plasmid*: **11**, 166–177 (1984).
6. Imanaka, T., Kaieda, T., Sato, K., Taguchi, H., "Optimization of α-galactosidase production by mold. (I). α-galactosidase production in batch and continuous culture and a kinetic model for enzyme production." *J. Ferment. Technol.*: **50**, 633–646 (1972).
7. Imanaka, T., Aiba, S., "A kinetic model of catabolite repression in the dual control mechanism in microorganisms." *Biotechnol. Bioeng.*: **19**, 757–764 (1977).
8. Fredrickson, A.G., "Formulation of structured growth models." *Biotechnol. Bioeng.*: **18**, 1481–1486 (1976).
9. Gold, L., Pribnow, D., Schneider, T., Shinedling, S., Singer, B.S., Stormo, G., "Translational initiation in prokaryotes." *Ann. Rev. Microb.*: **35**, 365–403 (1981).
10. Beckwith, J.R., Zipser, D. (eds.), *The Lactose Operon*, Cold Spring Harbor Laboratory, Cold Spring Harbor, New York, 1978.
11. Miller, J.H., Reznikoff, W.S. (eds.), *The Operon*, Cold Spring Harbor Laboratory, Cold Spring Harbor, New York, 1978.
12. Lee, S.B., Bailey, J.E., "Genetically structured models for *lac* promoter-operator function in the *Escherichia coli* chromosome and in multicopy plasmids: *lac* operator function." *Biotechnol. Bioeng.*: **26**, 1372–1382 (1984).
13. Lee, S.B., Bailey, J.E., "Genetically Structured Models for *lac* promoter-operator function in the chromosome and in multicopy plasmids: *lac* promoter function." *Biotechnol. Bioeng.*: **26**, 1383–1389 (1984).
14. Jacob, F., Monod, J., "Genetic Regulatory Mechanisms in the synthesis of proteins." *J. Mol. Biol.*: **3**, 318–356 (1961).
15. Jobe, A., Sadler, J.R., Bourgeois, S., "*lac* Repressor-operator interaction IX. The binding of *lac* repressor to operators containing O^C Mutations." *J. Mol. Biol.*: **85**, 231–248 (1974).
16. Miller, J.H., "Transcription starts and stops in the *lac* operon." In *The Lactose Operon*, p. 173–188. Beckwith, J.R., Zipser, D. (eds.), Cold Spring Harbor Laboratory, Cold Spring Harbor, New York, 1978.
17. Overath, P., "Control of Basal Level Activity of β-galactosidase in *Escherichia coli*." *Mol. Gen. Genet.*: **101**, 155–165 (1968).

18. Harder. A., Röels, J.A., "Application of simple structured models in bioengineering." *Adv. Biochem. Eng.*: **21**, 55–107 (1982).

19. Berg, O.G., Blomberg, C., "Mass action relations *in vivo* with application to the *lac* operon." *J. theor. Biol.*: **67**, 523–533 (1977).

20. Yagil, G., "Quantitative aspects of protein induction." *Current Top. Cell. Regul.*: **9**, 183–236 (1975).

21. Seressiotis, A., Bailey, J.E., "Intracellular equilibrium calculations based on small systems thermodynamics." *Biotechnol. Bioeng.*: **27**, 1520–1523 (1985).

22. Lee, S.B., Bailey, J.E., "A mathematical model for λdv plasmid replication: Analysis of wild-type plasmid." *Plasmid*: **11**, 151–165 (1984).

23. Murotsu, T., Matsubara, K., "Role of an autorepression system in the control of λdv plasmid copy number and incompatibility." *Mol. Gen. Genet.*: **179**, 509–519 (1980).

24. Domack, M.M., Leung, S., Cahn, R.E., Cocks, G.G., Shuler, M.L., "Computer model for glucose-limited growth of a single cell of *Escherichia coli* B/r-A." *Biotechnol. Bioeng.*: **26**, 203–216 (1984).

25. Shuler, M.L., Leung, S., Dick, C.C., "A mathematical model for the growth of a single bacterial cell." *Ann. N.Y. Acad. Sci.*: **326**, 35–55 (1979).

26. Ho, S.V., Shuler, M.L., "Predictions of cellular growth patterns by a feedback model." *J. theor. Biol.*: **68**, 415–435 (1977).

27. Peretti, S.W., Bailey, J.E., "Mechanistically detailed model of cellular metabolism for glucose-limited growth of *Escherichia coli* B/r-A." *Biotechnol. Bioeng.*: **28**, 1672–1689 (1986).

28. Hamming, J., Gruber, M., Ab, G., "Interaction between RNA polymerase and a ribosomal RNA promoter of *E. coli*." *Nuc. Acid Res.*: **7**, 1019–1033 (1979).

29. Kajitani, M., Ishihama, A., "Determination of the promoter strength in the mixed transcription system, II. Promoters of ribosomal RNA, ribosomal protein S1 and *recA* protein operons from *Escherichia coli*." *Nuc. Acid Res.*: **11**, 3873–3888 (1983).

30. Sokawa, Y., Sokawa, J., Kaziro, Y., "Regulation of stable RNA synthesis and ppGpp levels in growing cells of *Escherichia coli*." *Cell*: **5**, 69–74 (1975).

31. Travers, A., "Modulation of RNA polymerase specificity by ppGpp." *Molec. Gen. Genet.*: **147**, 225–232 (1976).

32. Ryals, J., Little, R., Bremer, H., "Control of rRNA and tRNA syntheses in *Escherichia coli* by guanosine tetraphosphate." *J. Bacteriol.*: **151**, 1261–1268 (1982).

33. Travers, A.A., Debenham, P.G., Pongs, O., "Translation Initiation Factor 2 alters transcriptional selectivity of *Escherichia coli* ribonucleic acid polymerase." *Biochemistry*: **19**, 1651–1656 (1980).

34. Pongs, O., Ulbrich, N., "Specific binding of formylated initiator-tRNA to *Escherichia coli* RNA polymerase." *Proc. Nat. Acad. Sci. U.S.A.*: **73**, 3064–3067 (1976).

35. Travers, A.A., Buckland, R., Debenham, P.G., "Functional heterogeneity of *Escherichia coli* ribonucleic acid polymerase holoenzyme." *Biochemistry*: **19**, 1656–1662 (1980).

36. Dennis, P.P., Bremer, H., "Macromolecular composition during steady-state growth of *Escherichia coli* B/r." *J. Bacteriol.*: **119**, 270–281 (1974).

37. Peretti, S.W., Bailey, J.E., "Simulations of host-plasmid interactions in *Escherichia coli*: copy number, promoter strength and ribosome binding site strength effects on metabolic activity and plasmid gene expression." *Biotechnol. Bioeng.*: **29**, 316–328 (1987).

38. Seo, J.-H., Bailey, J.E., "Effects of recombinant plasmid content on growth properties and cloned gene product formation in *Escherichia coli*." *Biotechnol. Bioeng.*: **27**, 1668–1674 (1985).

39. Lee, S.B., Seressiotis, A., Bailey, J.E., "A kinetic model for product formation in unstable recombinant populations." *Biotechnol. Bioeng.*: **27**, 1699–1709 (1985).

40. Srienc, F., Campbell, J.L., Bailey, J.E., "Analysis of unstable recombinant *Saccharomyces cerevisiae* population growth in selective medium." *Biotechnol. Bioeng.*: **28**, 996–1006 (1986).

41. Lee, S.B., Bailey, J.E., "Analysis of growth rate effects on productivity of recombinant *Escherichia coli* populations using molecular mechanism models." *Biotechnol. Bioeng.*: **26**, 66–73 (1984).

42. Aiba, S., Tsunekawa, H., Imanaka, T., "New approach to tryptophan production by *Escherichia coli*: Genetic manipulation of composite plasmids *in vitro*." *Appl. Environ. Microbiol.*: **43**, 289–297 (1982).

43. Seressiotis, A., Bailey, J.E., "Optimal Gene Expression and amplification strategies for batch and continuous recombinant cultures." *Biotechnol. Bioeng.*: **29**, 392–398 (1987).

II-2
Development and Optimization of Recombinant Fermentation Processes

Dewey D. Y. Ryu* and Sun Bok Lee[2]*

*Department of Viticulture and Enology; Department of Chemical Engineering, University of California, Davis, CA 95616, U.S.A.
[2]*The Korea Advanced Institute of Science and Technology, Seoul, Korea

1. Introduction

Recent developments and advances in the area of genetic engineering and biotechnology have shown that there are unlimited possibilities for their potential applications to various gene products by means of: (1) transferring genes to bacterial host organisms from other gene sources; (2) improving gene expression efficiencies; (3) amplifying gene copy numbers; (4) altering metabolic pathways; (5) improving fermentation yield; and (6) improving purification efficiency.

Within less than a 10-year period immediately following the first demonstration of the recombinant DNA (rDNA) technique by Cohen, Chang, and Boyer,[1] there have been some exciting developments in the area of pharmaceutical and medical applications. Some of these gene products are already available on the market and others are well on the way. The important front runners among these new gene products include: insulin, growth hormones, interferons, interleukin-2, tissue plasminogen activator, vaccines, diagnostics, amino acids, and enzymes.[2-4] Significant progress has been made also in the areas of more traditional "engineering biotechnology" or bioprocess engineering which include fermentation technology, enzyme technology, and bioseparation technology.

Successful commercial development of gene products requires not only recombinant DNA technology and bioprocess engineering but also many other strategic planning tasks related to clinical tests, handling of regulatory affairs, marketing, and legal and business aspects involved in the commercial development of new products and processes. In recent years we have gained considerable experience in these areas and they should not be underestimated, although this paper deals with only the scientific and technological aspects of biotechnology process development.

Recombinant DNA technology enables us to identify, clone, and transform the desired genes, and to have the host cell express the genes very efficiently. On the other hand, bioprocess technology, if and when properly developed and optimized, enables us to produce gene products economically by making good use of recombinant fermentation, purification, and bioseparation technologies. Although many had underestimated the importance of bioprocess technology, during the early period of biotechnological development, it is now widely recognized and many of us, especially in the biochemical engineering community, are now focusing on these problems—namely, scale-up problems related to re-

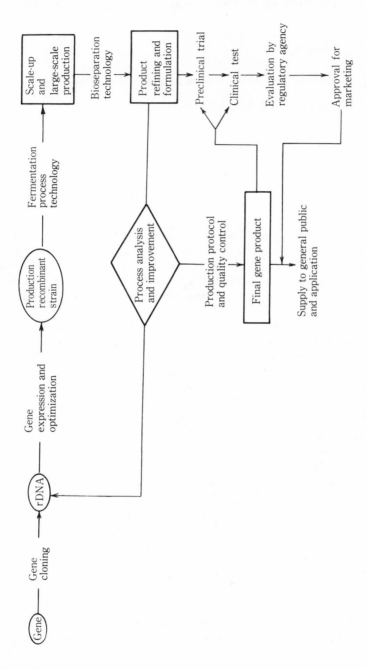

FIG. 1. Commercialization of gene products: development strategy and pathway.

combinant fermentation and bioseparation of gene products. Figure 1 highlights the important strategic planning task components required for the commercial development of gene products.

It is also appropriate to consider the "systems engineering" approach to the biotechnology process optimization task, since the rDNA technique used in the "front end" of the research and the bioprocess technology used in the "rear end" of the process development are highly interactive, and a well-coordinated and concerted research and development endeavor could bring about very efficient commercial development of the gene products. The selection of host organism, gene expression system, and excretion system, and their product separation processes, are good examples.

Recently, as more and more of the second- and third-generation gene products are well on their way to large-scale production and commercialization, the importance of biotechnology process engineering and the scarcity of related knowledge base have been recognized.

In view of this background, Professor S. Aiba and his colleagues at Osaka University have been conducting some pioneering research in this important area. In recognition of the very important contributions that they have made in this field, we felt that it would be a fitting occasion for us to consider the topic of our choice and dedicate our contribution to Professor Shuichi Aiba on the occasion of his retirement from Osaka University. This paper will focus on the effects of important bioprocess variables on the performance of recombinant fermentation and biotechnology process engineering.

2. Bioprocess Design Parameters for Genetically Engineered Recombinant Cells

2.1   Microbial Process Variables

Both genetic and environmental factors (or parameters) affect cell growth and product formation.[5] The genetic factors that render unique characteristics to a given recombinant strain include DNA replication, transcription, and translation. The environmental factors include temperature, aeration, agitation, pressure, apparent viscosity, osmotic pressure, pH, carbon and nitrogen sources, medium composition, growth factors, precursors, inducers, metabolites with toxic and/or inhibitory effects, etc.[5,6] These parameters are also highly interactive and their relationships to overall productivity are illustrated in Fig. 2.

For the development of the recombinant fermentation process, the effects of gene amplification and gene expression on productivity are especially important, in addition to the process design, scale-up, and optimization.[7]

In order to maximize gene product productivity, the strategy for gene cloning, construction of rDNA, and the gene expression system must be carefully planned. The selection list of important microbiological process design parameters that deserve consideration in optimizing recombinant fermentation processes are presented in Table 1. The importance of these parameters is discussed below.

2.1.1   Host cell selection

The selection of the host cell is one of the most important tasks in biotechnology and the production of gene products. The host cells provide all the facilities and precursors required for the biosynthesis of the product, including DNA replication, transcription, and translation. It is also important to understand the mechanism and regulation of gene expression in the host cell.

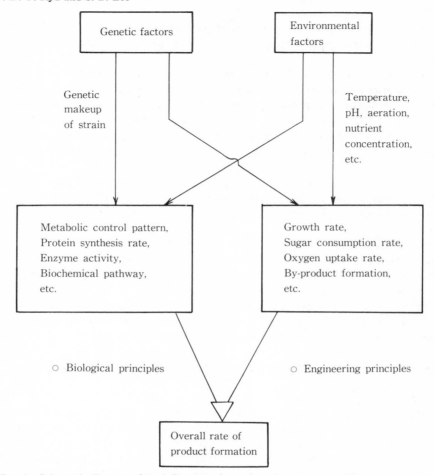

Fig. 2. Schematic diagram of the influences of genetic and environmental factors on overall
product synthesis.

So far, *E. coli* has been the most widely used host organism in recombinant DNA technology. Since its genetic characteristics and gene expression mechanism are relatively well
known, gene manipulation, construction of rDNA, and transformation are relatively easy
with *E. coli* as compared with other organisms. However, *E. coli* as a host organism has
the following disadvantages: (1) the bacterial endotoxin has to be carefully eliminated by
purification; (2) *E. coli* does not carry out post-translational glycosylation of proteins or
gene products; (3) it does not eliminate methionine from the product as in the case with
the human growth hormone (met-hGH); and (4) it does not normally excrete gene products
and they accumulate intracellularly.[8] The selection of the host cell significantly influences
the fermentation conditions, medium composition, cell growth rate, production rate,
stability of mRNA, stability of gene product, and the product recovery process.

Although there are some advantages in employing yeast and bacilli as host organisms,
so far we know relatively little about their genetic characteristics and the gene expression
mechanisms and plasmid stability in these host organisms.

TABLE 1. Microbiological process design parameters.

1. Gene dosage
   ○ plasmid copy number
   ○ regulation of replication
2. Transcription efficiency
   ○ promoter strength
   ○ regulation of transcription
3. Translation efficiency
   ○ nucleotide sequence of ribosome binding site (Shine-Dalgarno sequence)
   ○ distance between Shine-Dalgarno sequence and ATG or GTG initiation codon
   ○ secondary structure of mRNA
   ○ codon usage

4. Stability of recombinant DNA
   ○ stability of cloned gene
   ○ stability of plasmid
5. Stability of mRNA
   ○ structure of mRNA
   ○ nuclease
6. Stability of protein
   ○ structure of protein or gene product
   ○ protease
7. Host cell
   ○ host/vector interaction
   ○ metabolic activity of host cell
8. Others
   ○ protein secretion efficiency
   ○ proper termination between genes

2.1.2   Gene dosage

The gene dosage depends on the plasmid copy number since the desired gene is very often incorporated into the plasmid vector system when rDNA is to be constructed. The plasmid copy number ranges from a low copy number (1 to a few) to a high copy number (on the order of 100). Multicopy number plasmids like pBR322 and ColEl are very widely used in practice.[9]

In some cases, "runaway" plasmids are used.[10] More recently, a method by which the copy number can be controlled has been developed by linking the RNAII gene, which regulates the copy of the ColEl plasmid, to λP_L, *trp*, or to the *lac* promoter.[11]

In principle, the productivity of a gene product is expected to be proportional to the gene dose or plasmid copy number. However, productivity does not increase beyond a certain level when the gene dose is too high, since the primary metabolism of the host organism is severely affected.[12,13] Thus, a means of controlling plasmid copy number to an optimum is very important in certain cases.

2.1.3   Transcription efficiency

Transcription efficiency depends, in large part, on the promoter strength and regulatory mechanism in a given host organism. The ideal promoter system is a very strong one; it should be easy to control or its transcription should be easily turned on or off, and it should be inducible under specific induction conditions. For maximum productivity, a very strong promoter that can easily be controlled is highly desirable.

The rationale for having an easily controllable promoter system is that: (1) usually plasmids containing uncontrollable strong promoters are highly unstable; (2) some foreign proteins produced in large quantities are often very toxic to the host organism; and (3) the degradation or instability of gene products can be minimized when the products are made at a higher rate with controllable promoters as compared to a situation where the protein products are made at a lower rate. Examples of good promoters that have been widely used for *E. coli* are: *lac*, *trp*, λP_L, *tac*, and *recA*.[9,14,15]

2.1.4   Translation efficiency

Translation efficiency depends on the ribosome binding site nucleotide sequence (or Shine-Dalgarno sequence), the distance between the initiation codon (ATG or GTG) for protein biosynthesis and S-D sequence, the secondary structure of mRNA, and the codon usage of the cloned gene.[16-19] Recently, a ribosome binding site nucleotide sequence (TTAAAAT̲T̲A̲A̲G̲G̲A̲G̲G̲, the barred portion of nucleotides represents the SD sequence) with very high translation efficiency was synthesized and applied successfully to the production of interferon, interleukin-2, and other gene products. A product yield of up to 10–20% of intracellular protein was achieved using this approach.[20]

The fact that the distance between the SD sequence and the ATG initiation codon influences the gene expression efficiency has been well recognized. When the SD sequence is not in a single-strand form, the mRNA forms a stem-loop and the translation efficiency is often reduced significantly.[17]

2.1.5   Stability of rDNA, mRNA, and protein

The stability of rDNA, plasmid vector, mRNA, and protein is very important in enhancing the productivity of gene products. In general, the stability of mRNA and protein product depends a great deal on the growth rate of the host organism, and the rates of degradation of protein and mRNA increase under unfavorable environmental conditions, such as an increase in temperature.[21,22] The stability of protein also depends on the intracellular location of the product accumulation. In the case of insulin, it was found to be more stable when accumulated in the periplasm as compared with accumulation in the cytoplasm. The half-life of insulin outside the periplasmic membrane was estimated to be about 20 min, whereas it is only about 2 min in the cytoplasm.[23]

When multiple structural genes are put together or protein product accumulates in the form of inclusion bodies, the rate of protein degradation was found to decrease.[24]

2.1.6   Efficiencies of protein secretion and termination of transcription

Although the molecular mechanisms of protein secretion are not completely understood, some progress has been made with the signal peptide and efficient secretion of product using the host cells including *Bacillus*,[25] *Pseudomonas*,[26] and yeast.[27]

When a strong promoter is used, the transcription of the replicon contained in the same vector system is often affected adversely, and low productivity results. In this situation, a strong terminator could be used to prevent any interaction between the replicon and the vector system.[28]

2.2   Recombinant Fermentation Process Variables

In recombinant fermentation processes, basically the same process variables or parameters that are used in the traditional fermentation processes should be considered. The important fermentation process parameters are listed in Tables 2 and 3. These parameters are related to the bioreactor system design and operation, media design and optimization, and the state and control variables involved in the recombinant fermentation processes.

TABLE 2. Fermentation process design parameters.

1. Medium design
   ○ concentration of carbon source
   ○ concentration of nitrogen source
   ○ concentration of trace elements
   ○ other nutrients and growth factors
   ○ precursor and/or effector concentration

2. Bioreactor design
   ○ agitator speed
   ○ air flow rate
   ○ viscosity of culture fluid
   ○ power input
   ○ medium feed rate
   ○ cooling water flow rate and heat transfer

3. Other culture conditions
   ○ pH
   ○ temperature
   ○ dissolved oxygen (DO) level
   ○ dilution rate

For commercial-scale fermentation processes, one has to consider the "scale-up" of bioreactors (including microbial, animal, and plant-cell cultures), the mode of operation (including batch, semi-continuous, and continuous), and the strategy for optimal process control. When the recombinant fermentation processes are to be scaled up, one of the critical problems is the instability of recombinant organisms harboring unstable plasmids. When such unstable recombinants are used, the inoculum development protocol and medium design must be carefully examined.

Recombinants show far more sensitivity to the conditions of their cellular environment, and a more accurate control strategy and technique will be required for optimal process control, especially when the gene expression is turned on by a temperature shift or inducer addition.[29] Recombinants appear to have a higher specific oxygen uptake rate with the induction of gene expression, in large part due to their faster specific production rate.

TABLE 3. Parameters and other factors in rDNA fermentation.

1. Inoculum preparation
   ○ stable maintenance of recombinant cells during inoculum development
2. Media design
   ○ medium design for maintenance of recombinant cells
3. Oxygen transfer
   ○ increased oxygen demand
   ○ sensitivity to oxygen perturbation
4. Process control and optimization
   ○ fine control of process variables
      (pH, temperature, DO level, etc.)
   ○ optimization of induction time
5. Others
   ○ reduced shear rate when protein is secreted into medium
   ○ optimization of dilution rate in continuous fermentation
   ○ biosafety guidelines

Changes in the metabolic rate of recombinants due to cloned-gene expression require the fine control of recombinant fermentation processes more accurately.[30,31] In some cases when hazardous recombinants are used, post-treatment and/or sterilization of recombinants is required in accordance with the biosafety guidelines. Productivity can be improved by using immobilized cell cultures or high-density cell cultures in a continuous or fed-batch mode of operation.[32,33]

3. Optimization of Recombinant Fermentation Processes

Both the microbiological process variables and fermentation process variables affect productivity, and must be carefully studied and evaluated when the recombinant fermentation process is to be scaled up and optimized.

3.1   Microbiological Process Variables

Very often recombinants are challenged to overproduce gene products and problems associated with their metabolic strain or imbalance are very common. For this reason, we will have to find the optimal set of conditions with respect to DNA replication, transcription, and translation, as discussed in detail below.

3.1.1   Gene concentration

Aiba et al. found that a low copy-number plasmid, pSC101, yielded more tryptophan as compared with other high copy-number plasmids, pBR322 or RSF1010, because of their instability.[13] Their experimental results are shown in Table 4. They also showed that plasmid stability decreased significantly with an increase in the plasmid copy number, although the tryptophan synthetase (TSase) activity increased with the plasmid copy number. They suggested that, because of these opposing effects, there is an optimal gene or plasmid concentration corresponding to the maximum gene product productivity.

3.1.2   Transcription efficiency

One method of increasing transcription efficiency is to increase the inducer concentration. There are some evidences that recombinant organisms respond to the optimal inducer concentration with maximum productivity.[34-36] The well-known examples are production of haemaglutinin and immunoglobulin using the trp promoter and indoleacetic acid (IAA) as an inducer, and production of beta-interferon using the recA promoter and nalidixic acid as an inducer (see Fig. 3). The growth rate of cells is sometimes affected adversely by increasing the inducer concentration.

3.1.3   Translation efficiency

Among the many factors that affect translation efficiency, the optimal distance between the SD sequence and the initiation codon ATG is considered very important (see Table

TABLE 4. Effect of plasmid copy number on gene product and plasmid stability. (summary of experimental data from Aiba et al., Ref. 13)

Strain	TSase U. (mg protein)$^{-1}$	Copy number	Plasmid stability (%)	Tryptophan (g l^{-1})
Tna (RP4 trp I15)	36	1 $\sim$ 3	$\sim$ 100	1.7
Tna (pSC101 trp I15)	107	$\sim$ 5	$\sim$ 85	3.1
Tna (RSF1010 trp I15)	215	10 $\sim$ 50	$\sim$ 35	2.6
Tna (pBR322 trp I15)*	—	60 $\sim$ 80	—	—

* stable transformants difficult to obtain.

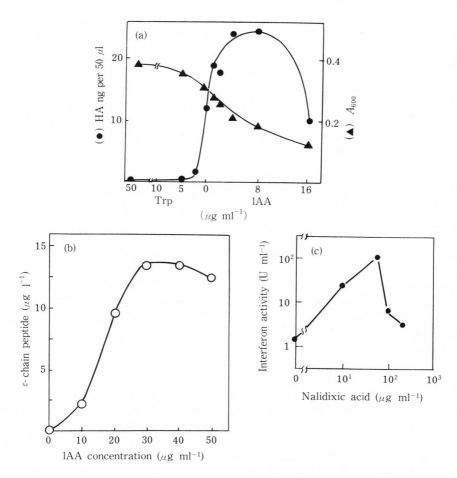

FIG. 3. Effect of inducer concentration on rDNA protein; (a) haemaglutinin (b) ε-chain peptide (c) β-interferon. (see Refs. 34–36)

TABLE 5. Effect of nucleotide spacing between SD and ATG on the production of ε chain peptides.

	Sequence between SD and ATG	Nucleotide residues	ε-chain peptides (μg l^{-1})
pGET*trp*302	*AAGGG*TATCG*ATG*	6	2.5
–a	*AAGGG*TATCTAGAATTCTAG*ATG*	16	0.1
–b	*AAGGG*TATCGAATTCTAG*ATG*	14	7.5
–c	*AAGGG*TATCGCTAG*ATG*	10	3.0
–d	*AAGGG*TATCGAATT*ATG*	10	17.5

IAA was added at a concentration of 30 μg ml.$^{-1}$
(see Ref. 35)

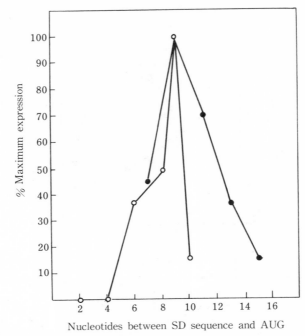

FIG. 4. Effect of spacing on efficiency of IFN synthesis. Data for IFN-β_1 ($\circ$) and IFN-αA ($\bullet$) are expressed as the percent of maximal synthesis observed for the various spacings between the Shine-Dalgarno and AUG sequences (see Ref. 37).

5). Figure 4 shows an example where the optimal space appears to be about 9 base pairs for the case of α- and β-interferon production using the *trp* promoter and *E. coli* host cells.[37] Another example shows that the optimal spacing between the SD sequence and ATG initiation codon is 10 base pairs for immunoglobulin when the *trp* promoter is used.[35]

3.2   Fermentation Process Variables

So far, very little is known about scaling up and optimizing recombinant fermentation processes, and a great deal of work needs to be done in this area in order to commercialize many more new gene products in the future.

3.2.1   Temperature

Temperature optimization is still an important area of investigation for the recombinant fermentation process development. While the optimal growth temperature for the host organism, *E. coli*, is 37°C, the optimal temperature for production of both interferon and insulin was found to be about 30°C.[38-40] In the case of insulin production, productivity was increased threefold when the cultivation temperature was maintained at 30°C as compared with the productivity obtained when the fermentation temperature was 37°C. This increase was attributable to a significant reduction in product degradation at lower temperature.

3.2.2   Time of induction and oxygen demand

Recent studies by Botterman et al.[30] showed that an early induction of gene expression during the logarithmic phase for production of the *Eco*RI restriction enzyme resulted in about a fivefold increase in productivity as compared to that of late induction during the stationary phase. In case of early induction the intracellular restriction enzyme reached as much as 22% of the dry cell weight. In both cases cited, a significant increase in the oxygen uptake rate coincided with the induction of gene expression and production of the enzyme. These results suggest that the selection of the best induction time for gene expression during the growth cycle and a sufficient oxygen supply to the recombinant cells, especially during the production phase immediately following gene expression, are important in optimizing the recombinant fermentation process.

3.2.3   Dilution rate

In a chemostat culture, the dilution rate corresponds to the specific growth rate, and in principle these parameters should have the same effect on productivity. However, one needs to distinguish these parameters for plasmid-harboring and plasmid-free cell populations when highly unstable recombinant organisms are used and there is a mixed population with different growth characteristics.

Usually the cellular content of chromosomal DNA tends to increase slightly with an increase in the specific growth rate.[41] However, the cellular content of certain plasmid DNA tends to decrease with an increase in the specific growth rate. Such examples include the R1, ColEl, and pBR322 plasmids.[42–44]

Figure 5 shows the relationship between plasmid content and dilution rate for the pPLc23*trpA1* plasmid. However, the pLP11 plasmid contained in *Bacillus stearother-*

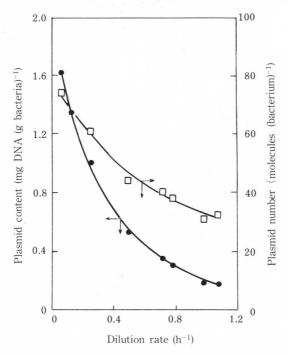

Fig. 5.   Effect of dilution rate on plasmid DNA content and plasmid copy number: (●) plasmid content and (☐) plasmid number (see Ref. 44).

mophilus increased its copy number with the increase in dilution rate, although it showed a constant value at a relatively lower temperature.[45]

In the case of pPLc23*trpA1* (Fig. 6), there appears to be an optimal dilution rate corresponding to maximum productivity.[46,47] In studies, by Lee, Siegel et al., they used a two-stage continuous culture system and the gene expression was switched on by shifting the cultivation temperature to 41°C in the second stage while maintaining the growth stage (or the first stage) under repressed conditions at 37°C.

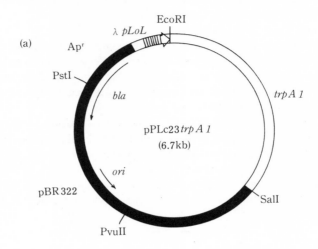

(a)

E.coli K12ΔH1 Δ *trp* [λ*N*am7 *N*am53 *cI*857]

(b)

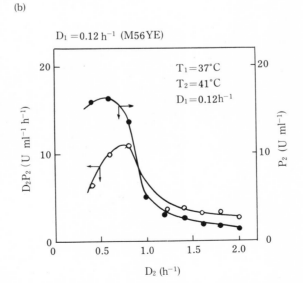

FIG. 6. Production of tryptophan synthetase using pPLc23*trpA1* in a two-stage continuous culture system (see Refs. 46, 47).

4. Instability of Recombinants

Despite many attempts to optimize recombinant fermentation processes, the instability of recombinant plasmids remains one of the most important problems in the commercial application of many recombinants. There is both the structural instability and segregational instability in rDNA. Structural instability is often caused by deletion, insertion, recombination, or other events, while segregational instability is caused by uneven distribution of plasmids during cell division.[48] Both types of instability result in a loss of productivity. Thus, it is also important to examine: (1) which parameters determine the degree of instability; (2) which environmental factors affect the instability of recombinants; and (3) how one can make best use of unstable recombinants.

4.1   Factors that Affect Instability of Recombinants

4.1.1   Parameters determining the degree of instability

The degree of instability of a recombinant can be determined by the growth ratio parameter and the plasmid-loss rate or segregation rate parameter. The growth ratio (μ^+/μ^-) is defined as the ratio of specific growth rate of plasmid-harboring cells (μ^+) to that of plasmid-free cells (μ^-). The segregation rate parameter (θ) is defined as the ratio of the specific rate of generation of plasmid-free cells from the plasmid-harboring cells (Θ;

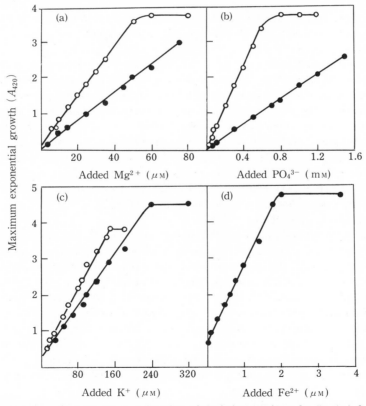

FIG. 7. Maximum exponential growth as a function of depleting nutrients for R^- ($\circ$) for R^+ ($\bullet$) cells (see Ref. 50).

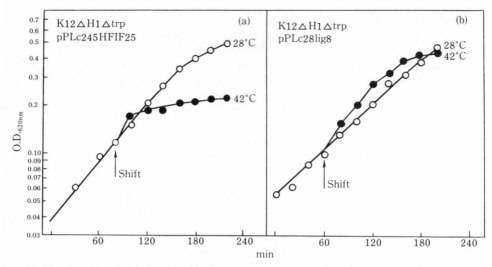

FIG. 8. Effect of fibroblast interferon synthesis on the growth of *E. coli* (see Ref. 51).

plasmid loss rate) to the specific growth rate of plasmid-harboring cells (μ^+).[46,47] The segregation rate parameter ($\theta = \Theta/\mu^+$) depends, in large part, on genetic characteristics and the segregation effect.

In general, the growth ratio ($\alpha = \mu^+/\mu^-$) is less than 1.0.[49] The growth rate of recombinants is adversely affected by the expression of recombinant DNA which causes stress on the metabolism of the host organism, although exceptions to this general rule have been found. The growth ratio is also influenced by the medium composition, mineral ion concentration, temperature, other environmental factors, and the characteristics of the gene product.[50,51] Figures 7 and 8 illustrate these observations. Plasmid-harboring cells and plasmid-free cells respond differently to mineral ions (Fig. 7). In the case of β-interferon production, the specific growth rate decreases shortly after the gene is expressed through a temperature shift (Fig. 8-a), while a slight increase in growth rate was observed after gene expression in the case of T4 DNA ligase production (Fig. 8-b).

4.1.2 Factors affecting instability

The instability of recombinants is affected by both genetic and environmental factors, including properties of plasmid vectors, genetic properties of host cell chromosomes and interaction with plasmids, the degree of gene expression, cultivation temperature, limiting nutrients, dilution rate, and the mode of bioreactor operation.

When pBR322 and pBR325 plasmids are compared, pBR325 is somewhat more unstable; *E. coli* K-12 GM31 strain is also slightly more unstable compared to *E. coli* strain K-12 Gy2354.[52] Another study with *Streptomyces lividans* harboring the pIJ2 plasmid showed that the recombinants became more unstable when continuously cultivated at a higher temperature and lower dilution rate. The degree of instability in terms of the plasmid loss rate also depends on the kind of limiting nutrient, temperature, and dilution rate[53] (see Fig. 9).

Although single-stage continuous culture does not improve the stability problem,[66] the two-stage continuous culture system was found to be very useful in overcoming it.

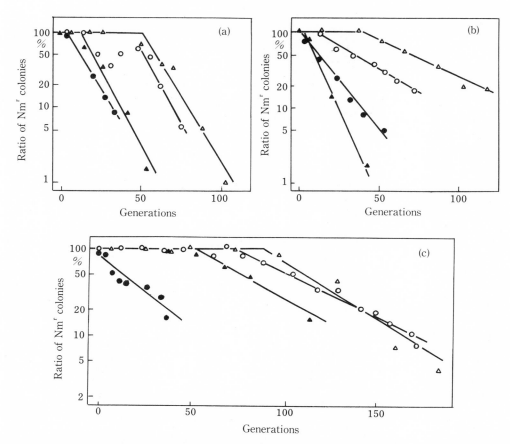

FIG. 9. Segregation kinetics of the recombinant plasmid pIJ2 from chemostat cultures of *S. lividans* 66-PM1. The presence of the plasmid was detected by plating samples withdrawn from continuous cultures on nonselective and selective medium containing neomycin. The resulting ratio of Nm^r colonies is plotted against generations. Regression lines were calculated for the decreasing part of the kinetics.

(a) glucose limitation: ○, dilution rate $D = 0.10$ h^{-1}; ●, $D = 0.11$ h^{-1}; △, $D = 0.19$ h^{-1}; ▲, $S = 0.20$ h^{-1};

(b) phosphate limitation: ○, $D = 0.10^{-1}$; ●, $D = 0.12$ h^{-1}; △, ▲, $D = 0.21$ h^{-1};

(c) ammonium limitation: ○, ●, $D = 0.11$ h^{-1}; △, $D = 0.23$ h^{-1}; ▲, $D = 0.22$ h^{-1};

Open symbols, continuous cultivation at 28°C; filled symbols, cultivation at 36°C (see Ref. 53).

Ryu and others[44,46,47,54] used the first stage as the growth stage under repressed conditions and the second stage as the expression stage or production stage where high productivity could be maintained (see Fig. 10-a and -b). They showed that a high level of the plasmid-harboring recombinant cell fraction (Φ) and a high level of gene product concentration could be maintained in the second stage of the two-stage continuous culture system.

4.2.   Strategies for Improving Recombinant Stability

Since recombinant instability depends on genetic and environmental factors, the in-

(a)

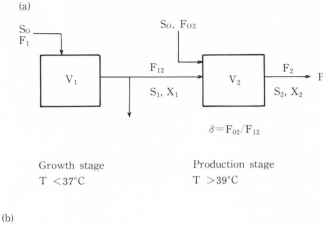

Growth stage
T $<37°$C

Production stage
T $>39°$C

$\delta = F_{02}/F_{12}$

(b)

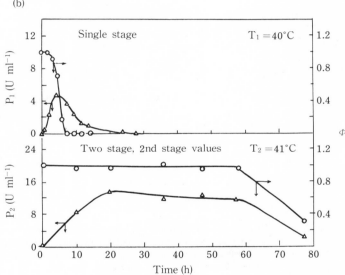

Fig. 10. (a) Two-stage continuous culture system. (b) Comparison of plasmid stability (Φ) and product formation (P). (see Refs. 44, 46, 47)

stability can be improved by preventing the growth of plasmid-free cells or altering the genetic stability of recombinants. A selected list of these approaches is presented in Table 6, and more details are discussed below.

4.2.1   Addition of antibiotics

Antibiotics are widely used to prevent the growth of plasmid-free cells because of their convenience, although this can be a problem in large-scale operations. The problems are: (1) a large amount of antibiotics is required; (2) the antibiotics must be separated out in a downstream purification stage; and (3) in some cases the plasmid-harboring cell population becomes unstable in the presence of antibiotics.[55] Antibiotics like novobiocin can also be used to control excessive DNA replication.[33]

TABLE 6. Strategies for improving or maintaining stability of recombinant cells.

1. Addition of antibiotics
   ○ ampicillin, tetracycline, etc.
   ○ novobiocin
2. Use of an auxotrophic mutant as host cell
   ○ amino acid auxotroph
3. Use of genes modulating stable maintenance of plasmid
   ○ *par* loci
   ○ *cer* gene
4. Use of genes involving natural fitness
   ○ λ *cos* site
5. Use of internal selection pressure
   ○ λ lysogens
   ○ streptomycin dependency
6. Use of recombination-deficient host strains
   ○ *recA*, etc.
7. Regulation of gene expression
   ○ proper induction/repression
8. Design of bioreactor operating mode
   ○ two-stage continuous culture

4.2.2   *Auxotrophic mutants*

Using auxotrophic mutants as the host organism in combination with the gene coded for the required amino acid, improves the stability of a recombinant cell population.[56] However, this method also presents some problems: (1) a defined medium must be used, often resulting in added expense; and (2) when the required amino acid is overproduced by the plasmid-harboring cells and leaked out into the medium, the plasmid-free cells can also grow competitively with the plasmid-harboring cells, resulting in an unstable recombinant population.[54,57]

4.2.3   *Genes modulating stable plasmid maintenance*

In nature, very stable plasmids exist and certain mechanisms by which the stability of plasmids are maintained have been studied. The *par* (partitioning) function of the pSC101 plasmid and the *cer* (ColE1 resolution) function of the ColE1 plasmid have been identified.[58,59] The *par* function is for equal partitioning of plasmids and *cer* makes monomer plasmids from multimers by multimer resolution at the time of cell division. The CloDF13 plasmid also has the *cer* function.[60] Certain plasmids containing the *trp* operon can be stabilized by inserting the *par* function, although some problems remain.[61]

4.2.4   *Control of gene expression*

One approach to control gene expression is to use a promoter in which induction and repression can be controlled. With such a promoter, it would be possible to separate the growth stage from the product stage, thus maximizing productivity. The two-stage continuous culture concept discussed earlier is an example of this approach.[44,46,70]

4.2.5   *Other approaches*

Insertion of the λ*cos* site into plasmids has been reported in another approach to plasmid stabilization,[62] although the mechanism is not clearly understood. Built-in intracellular selection pressure is another method to prevent the growth of plasmid-free cells. Use of λ lysogen forms suicidal cells when the host cells lose plasmids.[63] Streptomycin dependency is another intracellular selective pressure that has been used.[64] In some cases, *recA* deficient host cells can stabilize the plasmids to a certain degree.[65]

4.3   Scale-up Problems Associated with Recombinant Instability

Imanaka and Aiba addressed the instability problem in their studies.[49] Typically, commerical-scale inoculum development requires time for about 25 generations. They estimated the fraction of the plasmid-harboring recombinant cell population as a function of the growth ratio and the plasmid loss rate or segregation coefficient. The results of their analysis are shown in Fig. 11. Changes in the fraction of plasmid-harboring cell population is very sensitive to the growth ratio (α). When the inoculum development is not handled carefully,

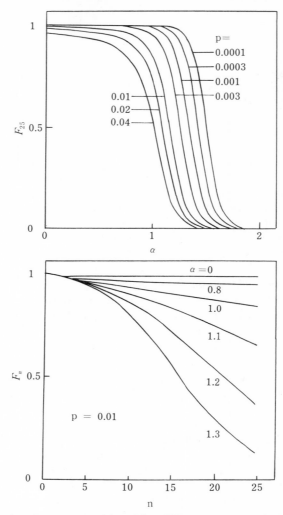

FIG. 11.   Effect of growth ratio parameter (α) and instability parameter (p) on the fraction of plasmid-harboring cells at the nth generation (F_n) (see Ref. 49).

and if the plasmid-harboring cell fraction in the inoculum of production stage is very low, it will be difficult to achieve high productivity.

Many have studied the plasmid instability problem theoretically and experimentally.[44,46,66-68] Miwa and others studied a recombinant containing pBR322 and the threonine operon for production of threonine. Their inoculum development required 17 generations and they found that the inoculum culture for the production-stage fermentor contained about 20% of plasmid-free cells.[69]

Other criteria for scale-up have also been studied.[70-72] Some problems are common to more traditional fermentation processes, and similar scale-up criteria must apply to both conventional and recombinant fermentation processes.

5. Bioprocess Kinetics of Recombinant Fermentation

When the recombinant fermentation process is optimized, we have to deal with many more genetic recombinant parameters as discussed earlier in this paper. Thus, a more in-depth understanding of the recombinant fermentation process and quantitative analyses of bioprocess kinetics are necessary for optimization.

5.1   Bioprocess Modeling and Simulation

Many recombinant fermentation processes have been developed to make gene products, and the related bioprocess kinetics involve transcription and translation processes. The kinetic expressions can be formulated for transcription and translation and are presented in the Appendix. These kinetic expressions (Eqs. (A1) and (A2) in the Appendix) show three terms: the rate of biosynthesis, the rate of degradation, and the rate of dilution by cell growth for the mRNA and gene products. These expressions include genetic and microbiological process design parameters and can be used to evaluate the relationships among these parameters through modeling and simulation.

Our bioprocess model is organized according to five levels of complexity: (1) the molecular level model, (2) the single-cell model, (3) the population model, (4) the bioreactor model, and (5) the bioplant model[7,73] (see Fig. 12). The first two levels are microscopic process kinetics and the next two levels are macroscopic process kinetics; a combination of all four levels gives the overall bioprocess kinetics. We have very little information about the first two levels of complexity in bioprocess systems.

5.1.1   Molecular model

Recent advances in molecular biology have enhanced our understanding of the regulatory mechanisms of microbial metabolism and have enabled us to analyze them more accurately than ever before. Analysis of the molecular model provides insight into metabolic regulations taking place at the molecular level. We are beginning to build up the molecular level model based on genetic control and regulatory mechanisms. For instance, the control mechanisms of the *lac* promoter-operator and the λdv plasmid replication, which are now well understood, have been studied in detail using molecular mechanism models.[74-77]

Our ultimate aim is to develop the strategy that gives the best gene expression efficiency. The *trp* or λP$_L$ promoter with the autorepressor control system have been found to yield high gene expression efficiency. Some model simulation studies on the effect of rDNA design on the transcription efficiency indicate that the best strategy is cloning the repressor

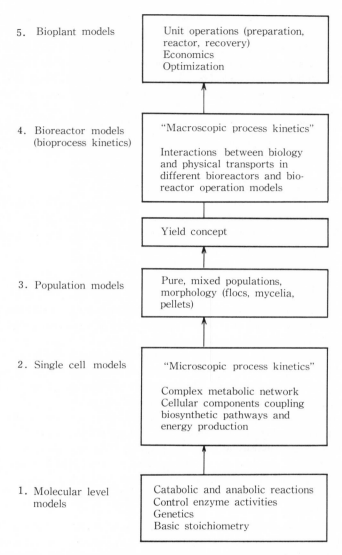

5. Bioplant models — Unit operations (preparation, reactor, recovery)
Economics
Optimization

4. Bioreactor models (bioprocess kinetics) — "Macroscopic process kinetics"

Interactions between biology and physical transports in different bioreactors and bioreactor operation models

Yield concept

3. Population models — Pure, mixed populations, morphology (flocs, mycelia, pellets)

2. Single cell models — "Microscopic process kinetics"

Complex metabolic network
Cellular components coupling biosynthetic pathways and energy production

1. Molecular level models — Catabolic and anabolic reactions
Control enzyme activities
Genetics
Basic stoichiometry

FIG. 12. An overview of research problem areas relevant to bioprocess modeling (Refs. 7,73).

gene with the target gene on the plasmid and putting the repressor formation under the autorepression control system.

5.1.2 Single-cell model

It is convenient to use the specific growth rate as a lumped parameter representing the overall metabolic activities of the whole cell system, although other representations can also be used. Some attempts have been made to develop a cellular model that relates the specific growth rate to the plasmid DNA and other biosynthetic activities of recombinant organisms.[78] Model simulation of λdv replication shows that the plasmid DNA content

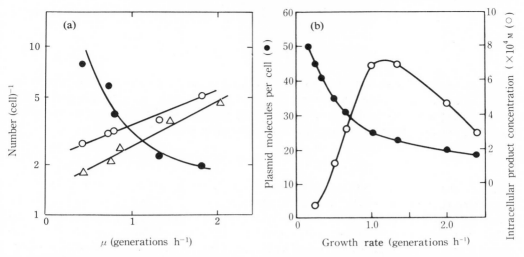

FIG. 13. Effect of growth rate on the plasmid copy number. (a) Experimental data for R1 plasmid; ($\bullet$) plasmid, ($\circ$, $\triangle$) chromosome (see Ref. 42). (b) Simulation results of λdv model (see Refs. 7,78).

decreases with an increase in the specific growth rate in accordance with the experimental results obtained with plasmids, R1, pBR322, and ColE1 [42-44] (see Fig. 13).

5.1.3   Population model and bioreactor model

The recombinant fermentation bioreactor system contains both plasmid-harboring and plasmid-free cells, and mixed cell population dynamics must be considered in the design and analysis of recombinant fermentation processes. From the model simulation it was found that there appears to be an optimal gene or plasmid concentration that corresponds to maximum productivity due to two opposite effects of gene concentration.[79] Although the rate of gene product formation is related to gene concentration, too high a gene concentration puts a severe metabolic strain on the primary metabolism of the host cells. When continuous culture is employed, the plasmid-harboring cell population decreases as the dilution rate decreases.

5.2   Kinetics of Gene Product Formation

Based on the molecular mechanism of transcription and translation, Lee et al. have derived a mathematical model for gene product formation by which the efficiency of gene expression can be evaluated.[46,47,74-79] Equation (1) comes from their earlier work and represents the specific production rate of gene product (q_p) as a function of the gene concentration (G_p), efficiency of gene expression (ϵ), the specific growth rate of plasmid-harboring cells (μ^+), the overall rate constant for gene product biosynthesis (k_0), and another lumped specific rate parameter implicit in contributions of nongrowth-associated parameters toward the product biosynthesis (b):

$$q_p/G_p = k_0 \, \epsilon \, (\mu^+ + b). \tag{1}$$

It was pointed out earlier that a two-stage continuous culture system was employed to evaluate these kinetic parameters accurately.[44,46] Under the apparent steady state conditions of a two-stage continuous culture system, q_p and μ^+ in Eq.(1) are given:

$$q_p \approx (D_2 p)/x_2 \tag{2}$$

$$\mu_2^+ \approx \mu_2^{app} = D_2 \left\{ 1 - \left(\frac{1}{1+\delta} \right) \left(\frac{x_1}{x_2} \right) \right\} \tag{3}$$

where D_2 and μ_2^{app} represent the dilution rate and apparent specific growth rate in the second stage, x_1 and x_2 the cell concentration in the first and second stages; and δ is the ratio of flow rates of fresh feed into the second stage to the feed coming from the first stage (see Fig. 10). The productivity equation is rearranged in Eq. (4)

$$q_p/G_p = k_0 \epsilon (\mu_2^{app} + b). \tag{4}$$

Based on experimental values the parameters q_p/G_p are correlated to μ_2^{app} as shown in Fig. 14 (see Refs. 46 and 47 for details). From correlation of the type shown in Fig. 14, the values of $k_0 \epsilon$ and b were determined as 41.8 mg protein gene product per mg DNA and 0.036 h^{-1}, respectively. The regression coefficient for this correlation was 0.973, and the plasmid used in the experiment was pPLc23$trpA1$ (Fig. 6). Since the value of b is about 0.036 h^{-1}, one may safely conclude that the specific production rate of the given gene product was strongly growth associated. The value of $k_0 \epsilon$ estimated above is equivalent to 5.7×10^3 molecules of protein gene product per molecule of DNA.

Under optimal conditions, the average value of k_0 for the E. coli K-12 strain is estimated to be 17.5 mg protein per mg DNA independent of growth rate.[46] Assuming that the host E. coli K-12 cell used in this experiment is approximately equal to the E. coli K-12 strain cited above in protein biosynthetic activity, the value of ϵ can be estimated to be 2.4. This suggests that the efficiency assessed in gene expression of TrpA protein is approximately twenty times higher than that for the total protein encoded in the host cell chromosomal DNA, since the $trpA$ gene which was under the control of the λP_L promoter is about one eighth of pPLc 23 $trpA1$ in its size.

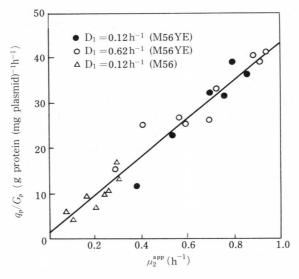

FIG. 14. Relationship between q_p/G_p and μ_2^{app}(M56YE; M56 medium + yeast extract) (see Refs. 46, 47).

The examples discussed here show how to develop a rationale for bioprocess kinetics based on molecular and cellular biology and how to apply the results to practical bio-technology process design and development.

Appendix. Kinetics of Gene Product Formation

The rate equations for transcription and translation for mRNA and protein gene product, respectively, are:

$$\frac{d\hat{m}}{dt} = k_m^\circ \, \eta \, \hat{G}_p - k_{-m} \, \hat{m} - \mu \hat{m}. \tag{A1}$$

$$\frac{d\hat{p}}{dt} = k_p^\circ \, \zeta \, \hat{m} - k_{-p} \, \hat{p} - \mu \hat{p}. \tag{A2}$$

We may obtain a quasi-steady state solution for $\hat{m}$ from Eq. (A1),

$$\hat{m} = \frac{k_m^\circ \, \eta \, \hat{G}_p}{k_{-m} + \mu} \tag{A3}$$

and solving Eqs. (A1) and (A2),

$$\frac{d\hat{p}}{dt} = \hat{G}_p \, \eta \, \zeta \left(\frac{k_m^\circ k_p^\circ}{k_{-m} + \mu} \right) - k_{-p} \hat{p} - \mu \hat{p}. \tag{A4}$$

Now the intracellular parameters can be replaced by the easily measurable bulk parameters.

$$p = \hat{p} x^+ / \rho_b \tag{A5}$$

$$\frac{dp}{dt} = \frac{1}{\rho_b} \left(x^+ \, \frac{d\hat{p}}{dt} + \hat{p} \, \frac{dx^+}{dt} \right)$$

$$= \frac{1}{\rho_b} \, \epsilon \, \hat{G}_p f(\mu) \, x^+ - k_{-p} \hat{p} \tag{A6}$$

where

$$\epsilon = \eta \cdot \zeta \tag{A7}$$

$$f(\mu) = \frac{k_m^\circ \, k_p^\circ}{k_{-m} + \mu} \cong k_0(\mu^+ + b). \tag{A8}$$

Assuming that k_{-p} is negligible,

$$\frac{dp}{dt} = \frac{1}{\rho_b} \cdot \epsilon \, \hat{G}_p f(\mu) \, x^+ = q_p x^+ \tag{A9}$$

$$q_p = \frac{1}{\rho_b} \cdot \epsilon \, \hat{G}_p f(\mu)$$

$$= \frac{1}{\rho_b} \cdot k_0 \, \epsilon \, \hat{G}_p (\mu^+ + b)$$

$$q_p = k_0 \, \epsilon \, G_p (\mu^+ + b). \tag{A10}$$

Now, the product formation kinetics may be expressed as:

$$\frac{dp}{dt} = k_0\, \epsilon\, G_p \cdot \frac{dx^+}{dt} + k_0\, \epsilon\, G_p b x^+ \tag{A11}$$

$$= A\, \frac{dx^+}{dt} + B x^+ \tag{A12}$$

$$A = k_0\, \epsilon\, G_p \text{ and } B = Ab. \tag{A13}$$

Equation (A12) is in the form of a Leudeking-Piret equation and the coefficients A and B, now show biological significance far more meaningfully than do the empirical constants proposed by Leudeking and Piret.[80]

Nomenclature

A	parameter of the Leudeking-Piret product formation model
B	parameter of the Leudeking-Piret product formation model
D	dilution rate
$\hat{G}_p$	intracellular cloning vector concentration
G_p	plasmid concentration
k_0	constant
k_m°	rate constant used in Eq. (A1) for mRNA synthesis
k_p°	rate constant used in Eq. (A2) for protein synthesis
k_{-m}	decay constant for mRNA used in Eq. (A1)
k_{-p}	decay constant of protein product (tryptophan synthetase)
$\hat{m}$	intracellular mRNA concentration
p	concentration of product
$\hat{p}$	intracellular protein concentration
q_p	specific production rate $[= k_0\, \epsilon\, G_p\, (\mu^+ + b)]$
s	concentration of limiting nutrient
t	culture time
x	concentration of cell
x_T	total cell concentration $(= x^+ + x^-)$

Greek Symbols

α	ratio of specific growth rate of plasmid-harboring cells to that of plasmid-free cells $(= \mu^+/\mu^-)$, or growth ratio
δ	ratio of the flow rate of fresh medium into second stage to that of fermentor broth coming from the first stage
η	transcription efficiency (relative to E. coli average value)
ϵ	combined efficiency for both transcription and translation $(= \eta \cdot \zeta)$
θ	ratio of the specific rate of generation of plasmid-free cell to the specific growth rate of the plasmid-harboring cell, or segregation rate
Θ	specific rate of generation of plasmid-free cells or plasmid loss rate $(= \mu^+\, \theta)$
μ	specific growth rate
μ^{app}	apparent specific growth rate
μ_m	maximum specific growth rate

ζ translation efficiency (relative to *E. coli* average value)

ρ_b cell density

Φ fraction of plasmid-harboring cells ($= x^+/x_T$)

Subscripts

j   culture stage (1, first stage or 2, second stage)

o   inlet feed concentration

T   total value of cell population

Superscripts

+   plasmid-harboring cells

−   plasmid-free cells

app   apparent value

°   used for intracellular rate constants

REFERENCES

1. Cohen, S.N., Chang, A.C.Y., Boyer, H.W., Helling, R.B., "Construction of biologically functional bacterial plasmids *in vitro*." *Proc. Nat. Acad. Sci. USA*: **70**, 3240–3244 (1974).

2. Webber, D., "Consolidation begins for biotechnology firms." *Chem. & Eng. News*: p. 25–60, November 18 (1985).

3. *Genetic Technology: A new Frontier*, Office of Technology Assessment, 1982.

4. *Commercial Biotechnology: An International Analysis*, Office of Technology Assessment, 1984.

5. Fiechter, A., "Physical and chemical parameters of microbial growth." *Adv. Biochem. Eng./Biotechnol.*: **30**, 7–60 (1984).

6. Aiba, S., Humphrey, A.E., Millis, N.F., "Instrumentation for environmental control." In *Biochemical Engineering 2nd ed.*, p. 317–345. Academic Press, 1973.

7. Bailey, J.E., Hjortso, M., Lee, S.B., Srienc, F., "Kinetics of product formation and plasmid segregation in recombinant microbial populations." *Ann. New York Acad. Sci.*: **413**, 71–87 (1983).

8. Bok, S.H., "Overproduction of recombinant DNA fermentation products: theoretical and practical considerations." *Dev. Ind. Microbiol.*: **24**, 255–270 (1983).

9. Thompson, R. "Plasmid and phage M13 cloning vectors." In *Genetic Engineering*. p. 1–52. Williamson, R. (ed.), Academic Press, New York, 1982.

10. Uhlin, B.E., Molin, S., Gustafsson, P., Nordström, K., "Plasmids with temperature-dependent copy number for amplification of cloned genes and their products." *Gene*: **6**, 91–106 (1979).

11. Yarranton, G.T., Wright, E., Robinson, M.K., Humphreys, G.O., "Dual-origin plasmid vectors whose origin of replication is controlled by the coliphage lambda promoter P_L." *Gene*: **28**, 293–300 (1984).

12. Seo, J.-H., Bailey, J.E., "Continuous cultivation of recombinant *Escherichia coli*: Existence of an optimum dilution rate for maximum plasmid and gene product concentration." *Biotechnol. Bioeng.*: **28**, 1590–1594 (1986).

13. Aiba, S., Tsunekawa, H., Imanaka, T., "New approach to tryptophan production by *Escherichia coli*: Genetic manipulation of composite plasmids In Vitro." *Appl. Environ. Microbiol.*: **43**, 289–297 (1982).

14. De Boer, H.A., Comstock, L.J., Vasser, M., "The *tac* promoter: A functional hybrid derived from the *trp* and *lac* promoters." *Proc. Nat. Acad. Sci. USA*: **80**, 21–25 (1983).

15. Shirakawa, M., Tsurimoto, T., Matsubara, K., "Plasmid vectors designed for high-efficiency expression controlled by the portable *recA* promoter-operator of *Escherichia coli*." *Gene*: **28**, 127–132 (1984).

16. Gold, L., Pribnow, D., Schneider, T., Shinedling, S., Singer, B.S., Stormo, G., "Translational initiation in prokaryotes." *Ann. Rev. Microbiol.*: **35**, 365–403 (1981).

17. Schottel, J.L., Sninsky, J.J., Cohen, S.N., "Effects of alterations in the translation control region on bacterial gene expression: use of *cat* gene constructs transcribed from the *lac* promoter as a model system." *Gene*: **28**, 177–193 (1984).

18. Grosjean, H., Fiers, W., "Preferential codon usage in prokaryotic genes: the optimal codon – anticodon interaction energy and the selective codon usage in efficiently expressed genes." *Gene*: **18**, 199–209 (1982).
19. Shine, J., Dalgarno, L., "The 3′-terminal sequence of *Escherichia coli* 16S ribosomal RNA: complementarity to nonsense triplets and ribosome binding sites." *Proc. Nat. Acad. Sci. USA*: **71**, 1342–1346 (1974).
20. Crowl, R., Seamans, C., Lomedico, R., McAndrew, S., "Versatile expression vectors for high-level synthesis of cloned gene products in *Escherichia coli*." *Gene*: **38**, 31–38 (1985).
21. Nilsson, G., Belasco, J.G., Cohen, S.N., von Gabain, A., "Growth-rate dependent regulation of mRNA stability in *Escherichia coli*." *Nature*: **312**, 75–77 (1984).
22. Fish, N.M., Lilly, M.D., "The interactions between fermentation and protein recovery." *Bio/Technol.*: **2**, 623–627 (1984).
23. Talmadge, K., Gilbert, W., "Cellular location affects protein stability in *Escherichia coli*." *Proc. Natl. Acad. Sci. USA*: **79**, 1830–1833 (1982).
24. Shen, S.-H., "Multiple joined genes prevent product degradation in *Escherichia coli*." *Proc. Natl. Acad. Sci. USA*: **81**, 4627–4631 (1984).
25. Palva, I., Lehtovaara, P., Kääriäinen, L., Sibakov, M., Cantell, K., Schein, C.H., Kashiwagi, K., Weissmann, C., "Secretion of interferon by *Bacillus subtilis*." *Gene*: **22**, 229–235 (1983).
26. Gray, G.L., McKeown, K.A., Jones, A.J.S., Seeburg, P.H., Heyneker, H.L., "*Pseudomonas aerugionosa* secretes and correctly processes human growth hormone." *Bio/Technol.*: **2**, 161–165 (1984).
27. Singh, A., Lugovoy, J.M., Kohr, W.J., Perry, L.J., "Synthesis, secretion and processing of α-factor-interferon fusion proteins in yeast." *Nucleic Acids Res.*: **12**, 8927–8938 (1984).
28. Gentz, R., Langner, A., Chang, A.C.Y., Cohen, S.N., Bujard, H. "Cloning and analysis of strong promoters is made possible by the downstream placement of a RNA termination signal." *Proc. Natl. Acad. Sci. USA*: **78**, 4936–4940 (1981).
29. Hopkins, D.J., Betenbaugh, M.J., Dhurjati, P., "Environmental effects on growth of a recombinant population of *Escherichia coli* containing plasmid pKN 401." Paper presented at the annual meeting of Am. Inst. Chem. Engrs. (1985).
30. Bottermann, J.H., De Buyser, D.R., Spriet, J.A., Zabeau, M., Vansteenkiste, G.C., "Fermentation and recovery of the *Eco*RI restriction enzyme with a genetically modified *Escherichia coli* strain." *Biotechnol. Bioeng.*: **27**, 1320–1327 (1985).
31. Betenbaugh, M.J., Dhurjati, P., "Growth kinetics of temperature-sensitive recombinant cells." Paper presented at the 190th ACS meeting (1985).
32. Allen, B.R., Luli, G.W., "A gradient-feed process for obtaining high cell densities for recombinant *Escherichia coli* and *Bacillus subtilis*." Paper presented at the 190th ACS meeting (1985).
33. Mosbach, K., Birnbaum, S., Hardy, K., Davies, J., Bülow, L., "Formation of proinsulin by immobilized *Bacillus subtilis*." *Nature*: **302**, 543–545 (1983).
34. Emtage, J.S., Tacon, W.C.A., Catlin, G.H., Jenkins, B., Porter, A.G., Carey, N.H. "Influenza antigenic determinants are expressed from haemagglutinin genes cloned in *Escherichia coli*." *Nature*: **283**, 171–174 (1980).
35. Kurokawa, T., Seno, M., Sasada, R., Ono, Y., Onda, M., Igarashi, K., Kikuchi, M., Sugino, Y., Honjo, T., "Expression of human immunoglobulin Eε chain cDNA in *E. coli*." *Nucleic Acids Res.*: **11**, 3077–3085 (1983).
36. Feinstein, S.I., Chernajovsky, Y., Chen, L., Maroteaux, L., Mory, Y., "Expression of human interferon genes using the *recA* promoter of *Escherichia coli*." *Nucleic Acids Res.*: **11**, 2927–2941 (1983).
37. Shepard, H.M., Yelverton, E., Goeddel, D.V., "Increased synthesis in *E. coli* of fibroblast and leukocyte interferons through alterations in ribosome binding sites." *DNA*: **1**, 125–131 (1982).
38. Backman, K., Ptashne, M., "Maximizing gene expression on a plasmid using recombination in vitro." *Cell*: **13**, 65–71 (1978).
39. Meyer, H.-P., Kuhn, H.-J., Brown, S.W., Fiechter, A., "Production of human leucocyte interferon by *E. coli*." *Proc. 3rd Eur. Cong. Biotechnol.*: **1**, 499–505 (1984).
40. Emerick, A.W., Bertolani, B.L., Ben-Bassat, A., White, T.J., Konrad, M.W., "Expression of a β-lactamase preproinsulin fusion protein in *Escherichia coli*." *Bio/Technol.*: **2**, 165–168 (1984).
41. Cooper, S., Helmstetter, C.E., "Chromosome replication and the division cycle of *Escherichia coli* B/r." *J. Mol. Biol.*: **31**, 519–540 (1968).

42. Engberg, B., Nordström, K., "Replication of R-factor R1 in *Escherichia coli* K-12 at different growth rates." *J. Bacteriol.*: **123**, 179–186 (1975).

43. Stueber, D., Bujard, H., "Transcription from efficient promoters can interfere with plasmid replication and diminish expression of plasmid specified genes." *EMBO J.*: **1**, 1399–1404 (1982).

44. Siegel, R., Ryu, D.D.Y., "Kinetic study of instability of recombinant plasmid pPLc23*trpA1* in *E. coli* using two-stage continuous culture system." *Biotechnol. Bioeng.*: **27**, 28–33 (1985).

45. Koizumi, J.-I., Monden, Y., Aiba, S., "Effects of temperature and dilution rate on the copy number of recombinant plasmid in continuous culture of *Bacillus stearothermophilus* (pLP11)." *Biotechnol. Bioeng.*: **27**, 721–728 (1985).

46. Lee, S.B., Ryu, D.D.Y., Siegel, R., Park, S.H., "Performance of recombinant fermentation and evaluation of gene expression efficiency for gene product in a two-stage continuous culture system." *Biotechnol. Bioeng.* in press (1987).

47. Ryu, D.D.Y., Siegel, R., Lee, S.B., "Kinetics of gene expression in recombinant *E. coli* K-12 *ΔH1 Δtrp*/pPLc 23*trpA1*." Paper presented at the annual meeting of Am. Inst. Chem. Engrs. (1985).

48. Primrose, S.B., Ehrlich, S.D., "Isolation of plasmid deletion mutants and study of their instability." *Plasmid*: **6**, 193–201 (1981).

49. Imanaka, T., Aiba, S., "A perspective on the application of genetic engineering: stability of recombinant plasmid." *Ann. N.Y. Acad. Sci.*: **369**, 1–14 (1981).

50. Klemperer, R.M.M., Ismail, N.T.A.J., Brown, M.R.W., "Effect of R plasmid RP1 on the nutritional requirements of *Escherichia coli* in batch culture." *J. Gen. Microbiol.*: **115**, 325–331 (1979).

51. Remaut, E., Stanssens, P., Fiers, W., "Inducible high level synthesis of mature human fibroblast interferon in *Escherichia coli*." *Nucleic Acids Res.*: **11**, 4677–4688 (1983).

52. Noack, D., Roth, M., Geuther, R., Müller, G., Undisz, K., Hoffmeier, C., Gaspar, S., "Maintenance and genetic stability of vector plasmids pBR322 and pBR325 in *Escherichia coli* K12 strains grown in a chemostat." *Mol. Gen. Genet.*: **184**, 121–124 (1981).

53. Roth, M., Noack, D., Geuther, R., "Maintenance of the recombinant plasmid pIJ2 in chemostat cultures of *Streptomyces lividans* 66 (pIJ2)." *J. Basic Microbiol.*: **25**, 265–271 (1985).

54. Kim, J.H., personal communication.

55. Caulcott, C.A., "Competition between plasmid-positive and plasmid-negative cells." *Biochem. Soc. Trans.*: **12**, 1140–1142 (1984).

56. Anderson, D.M., Herrman, K.M., Somerville, R.L., "*E. coli* bacteria carrying recombinant plasmids and their use in the fermentative production of L-tryptophan." U.S. Pat. No. 4,371,614 (1983).

57. Dwivedi, C.P., Imanaka, T., Aiba, S., "Instability of plasmid-harboring strain of *E. coli* in continuous culture." *Biotechnol. Bioeng.*: **24**, 1465–1468 (1982).

58. Meacock, P.A., Cohen, S.N., "Partitioning of bacterial plasmids during cell division: a cis-acting locus that accomplishes stable plasmid inheritance." *Cell*: **20**, 529–542 (1980).

59. Summers, D.K., Sherratt, D.J., "Multimerization of high copy number plasmids causes instability: ColE1 encodes a determinant essential for plasmid monomerization and stability." *Cell*: **36**, 1097–1103 (1984).

60. Hakkaart, M.J.J., Veltkamp, E., Nijkamp, H.J.J., "Maintenance of the bacteriocinogenic plasmid CloDF13 in *Escherichia coli* cells. II. Specific recombination functions involved in plasmid maintenance." *Mol. Gen. Genet.*: **188**, 338–344 (1982).

61. Skogman, G., Nilsson, J., Gustafsson, P., "The use of a partition locus to increase stability of tryptophan-operon-bearing plasmids in *Escherichia coli*." *Gene*: **23**, 105–115 (1983).

62. Edlin, G., Tait, R.C., Rodriguez, R.L., "A bacteriophage λ cohesive ends (*cos*) DNA fragment enhances the fitness of plasmid-containing bacteria growing in energy-limited chemostats." *Bio/Technol.*: **2**, 251–254 (1984).

63. Rosteck, P.R.Jr., Hershberger, C.L., "Selective retention of recombinant plasmids coding for human insulin." *Gene*: **25**, 29–38 (1983).

64. Miwa, K., Nakamori, S., Sano, K., Momose, H., "Novel host-vector system for selection and maintenance of plasmid-bearing, streptomycin-dependent *Escherichia coli* cells in antibiotic-free media." *Gene*: **31**, 275–277 (1984).

65. Ream, L.W., Crisona, N.J., Clark, A.J., "ColE1 plasmid stability in ExoI⁻ ExoV⁻ strain of *Escherichia coli* K-12." In *Microbiology-1978*, p. 78–80. Schlessinger, D. (ed.), ASM, Washington, D.C., 1978.

66. Olyis, D.F., "Industrial fermentations with (unstable) recombinant cultures." *Phi. Trans. Roy. Soc.*

Lond.: **B297**, 617–629 (1982).

67. Kim, S.H., Ryu, D.D.Y., "Instability kinetics of *trp* operon plasmid ColEl-*trp* in recombinant *Escherichia coli* MV12 (pVH5) and MV12*trp*R (pHV5)." *Biotechnol. Bioeng.*: **26**, 497–502 (1984).
68. Seo, J.-H., Bailey, J.E., "A segregated model for plasmid content and product synthesis in unstable binary fission recombinant organisms." *Biotechnol. Bioeng.*: **27**, 156–165 (1985).
69. Miwa, K., Nakamori, S., Sano, K., Momose, H., "Stability of recombinant plasmids carrying the threonine operon in *Escherichia coli.*" *Agric. Biol. Chem.*: **48**, 2233–2237 (1984).
70. Ryu, D.D.Y., "Bioprocess scale-up using genetically engineered microbes." *Proc. BioExpo*: **85**, p. 351–367 (1985).
71. Naveh, D., "Scale-up of fermentation for recombinant DNA products." *Food Technol.*: **39**, 102–109, October (1985).
72. Muth, W.L., "Scale-up biotechnology safely." *CHEMTECH*: **15**, 356–361 (1985).
73. Moser, A., "General strategy in bioprocessing." In *Biotechnology* Vol. 2, p. 173–197. Brauer, H. (ed.), VCH Verlagsgesellschaft, Weinheim, 1985.
74. Lee, S.B., Bailey, J.E., "Genetically structured models for *lac* promoter-operator function in the *Escherichia coli* chromosome and in multicopy plasmids: *lac* operator function." *Biotechnol. Bioeng.*: **26**, 1372–1382 (1984).
75. Lee, S.B., Bailey, J.E., "Genetically structured models for *lac* promoter-operator function in the chromosome and in multicopy plasmids: *lac* promoter function." *Biotechnol. Bioeng.*: **26**, 1383–1389 (1984).
76. Lee, S.B., Bailey, J.E., "A mathematical model for λdv plasmid replication: analysis of wild-type plasmid." *Plasmid*: **11**, 151–165 (1984).
77. Lee, S.B., Bailey, J.E., "A mathematical model for λdv plasmid replication: analysis of copy number mutants." *Plasmid*: **11**, 166–177 (1984).
78. Lee, S.B., Bailey, J.E., "Analysis of growth rate effects on productivity of recombinant *Escherichia coli* populations using molecular mechanism models." *Biotechnol. Bioeng.*: **26**, 66–73 (1984).
79. Lee, S.B., Seressiotis, A., Bailey, J.E., "A kinetic model for product formation in unstable recombinant populations." *Biotechnol. Bioeng.*: **27**, 1699–1709 (1985).
80. Leudeking, R., Piret, E.L., "A kinetic study of the lactic acid fermentation. Batch process at controlled pH." *J. Biochem. Microbiol. Technol. Eng.*: **1**, 393–412 (1959).

Fermentor Production of HBsAg Via Expression in the Periplasmic Space of *E. coli*

Kisay Lee and Cha Yong Choi

Laboratory of Biotechnology and Bioengineering, Department of Chemical Technology, College of Engineering, Seoul National University, Seoul, Korea

A truncated HBsAg gene lacking the sequence encoding the NH_2-terminal hydrophobic region was put behind the *E. coli lpp - lac* double promoters and successful expression was achieved. The transformant harboring the newly constructed recombinant plasmid, pKSB292, did not inhibit growth and much of the produced antigen was present in the osmotic shock fluids of the *E. coli* host cells, indicating the transport of the gene product to the periplasmic space.

1. A Mini Review

Professor Aiba and Dr. Imanaka were the first biochemical engineers to master recombinant DNA techniques, apply the techniques to conventional biotechnology, and establish gene manipulation as a branch of biotechnology.

Their first paper on the cloning and expression of the *trp* operon in *E. coli* in the *Journal of General Microbiology* appeared in 1980, and was followed by their work on the theoretical background and experimental observation of plasmid stability in bioreactors, general plasmid behavior upon induction for gene expression in bioreactors, and the discovery and development of the high-temperature vector in *Bacillus stearothermophilus*. Subsequent publications concerned the cloning and expression of various genes in *Bacillus* and *Streptomyces*, the elucidation of processing mechanisms in the signal region for secreted gene products, and protein engineering via site-directed mutagenesis.

A few years later reports by Bailey and Ryu on the theoretical treatment of hypothetical model systems and observation on model experiments were published. Similar work on recombinant DNA in yeast, *E. coli*, and *Streptomyces* as well as culture studies with anchorage-dependent cells on micro-carrier and self-made hybridoma cells began in our laboratory in 1981. Among the structural genes we worked on were those for alpha interferon and hepatitis B virus surface antigen (HBsAg). This report on the occasion of Professor Aiba's retirement concerns our work with the HBsAg gene.

The final level of yield and productivity in bioreactor production of gene product protein can be influenced by many factors, including regulations at the cellular and molecular levels, transport effects, and kinetic effects. Among the molecular biological factors of importance are the regulatory sequences of DNA, the regulatory events at the level of replication, transcription, and translation, and host-vector interaction. Some of the problems more specifically related to plasmid vectors include plasmid stability, gene expression efficiency, and relative efficacy of various modes of induction for gene expression. The use of the stabilizing DNA sequences and the optimization of reactor configuration and operation mode are some of the measures that have been proposed to cope with these problems.

Among the promoters utilized in our laboratory for the last several years are: (1) yeast acid phosphatase promoter controlled by phosphate level and wild-type or mutated repressor; (2) the *tac* promoter; (3) *lpp* + *lac* double promoters; and (4) the λP_L promoter appropriately modified. Inducing agents used were heat, IPTG, nalidixic acid, and phosphate level for gene expression or replication runaway.

2. Introduction

The production of a vaccine against viral hepatitis B has drawn much attention due to the worldwide spread of the disease. The commercial production of the vaccine using recombinant DNA technology is now a reality.

The structural gene coding for hepatitis B virus surface antigen (HBsAg) has been cloned into *E. coli*,[1-4] *B. subtilis*,[5] *S. cerevisiae*,[6-9] and animal cell lines[10,11] using various kinds of vectors. Many have tried to express the HBsAg gene in *E. coli* and have obtained expressions as fusion proteins.[1,4,12] However, neither the direct expression of the HBsAg gene in *E. coli* nor synthesis at high levels has been reported because the synthesis of a complete HBsAg peptide sequence in bacteria is impossible either due to the high toxicity of the N-terminal part of the protein or due to rapid proteolytic degradation in the cytoplasm. *E. coli* cells carrying the complete HBsAg gene have been shown to exhibit both growth inhibition and low-level antigen production mainly because viral hydrophobic protein inhibits the growth of *E. coli*.[4,13]

A truncated HBsAg gene lacking the sequence that encodes the N-terminal hydrophobic region was prepared in order to avoid the growth inhibition. A substantial portion of the HBsAg produced in *E. coli* under the influence of the *lpp* (lipoprotein) and *lac* (lactose) double promoters was found to have been translocated to the periplasmic space, helping prevent cytoplasmic degradation and facilitating product recovery.

3. Materials and Methods

3.1   Strains, Plasmids, and Microbial Techniques

E. coli strains HB101,[14] JM109,[15] and JE5505[16] and plasmids pHBV130[17] from Professor Murray at Edinburgh, and pUC18[18] and pINIII-B1[19] from Professor Inouye at Stonybrook were used.

For HBsAg expression, cells were grown overnight at 37°C in a modified L medium (LB) supplemented with 50 μg of amplicillin (Amp) ml⁻¹ and 1mM isopropyl thiogalactoside (IPTG).

3.2   DNA Manipulation

Plasmids were isolated as described by Kieser[20] and DNA fragments were recovered from electrophoresis agarose gel by the freeze squeeze method[21] with some modifications.

Restriction enzymes *Hin*dIII and *Eco*RI from Bethesda Research Laboratories, Inc., Rockville, Md., *Pst*I from New England Biolabs (NEB), Beverly, Mass., and *Xba*I from Takara Shuzo Co., Kyoto, Japan were used. T4 DNA ligase was purchased from NEB.

The procedure of Cohen et al.[22] with some modifications was used to transform *E. coli* strains with ligation mixture.

3.3 Preparation of Crude Cell Extract

Forty ml of culture was grown in LB and harvested by centrifugation. Cells were washed in 4 ml of washing buffer (50 mM Tris-HCl, pH 8.0, and 30 mM NaCl) 2–3 times, and resuspended in 4 ml of lytic solution consisting of 50 mM Tris-HCl (pH 8.0), 50 mM EDTA (ethylenediaminetetraacetic acid), 15% sucrose, 2 mg ml^{-1} of lysozyme, and 20 mg ml^{-1} of PMSF (phenylmethylsulfonyl fluoride), followed by alternate rapid freezing ($-70°C$) and rapid thawing ($40°C$) 5–6 times. After centrifugation at 200,000 × g for 2 h, the supernatant was taken as the extract sample and assayed using an AUSRIA II RIA kit (Abbott Laboratories, North Chicago, IL) (RIA).

3.4 Purification of Periplasmic Proteins

In order to ascertain the cellular localization of the produced HBsAg, culture medium separated from the cell mass was assayed as the secreted fraction. Osmotic shock fluid was collected as the periplasmic fraction[23] and the crude extract of *E. coli* cells was taken as the sum of the periplasmic fraction and cytoplasmic fraction.

Osmotic shock fluid was prepared as follows: 4 ml of culture were harvested in a microtube and suspended in 600 μl of 100 mM Tris-HCl (pH 7.5) and 20% sucrose solution. All procedures were carried out at $0°C$. After adding 20 μl of 0.5 M EDTA (pH 8.0) to the tube, the mixture was incubated in ice for 10 min and centrifuged for 5 min. The supernatant was discarded and 400 μl of cold water was added. After vigorous vortexing for 1 min, the tube was put on ice for 10 min and centrifuged. The supernatant thus produced was considered to be osmotic shock fluid containing periplasmic protein.

3.5 Induction and HBsAg Assay

Plates for screening the β-galactosidase nonproducing *E. coli* JM109 (pHX9) were overlaid with soft agar containing 1 mM IPTG and 40 μg ml^{-1} X-gal. White colonies were selected as the recombinants.

For induction of HBsAg from JE5505 (pKSB292), 1 mM IPTG was added to the culture broth and the presence of HBsAg was detected with RIA.

3.6 Fermentor Operation

A 5-l fermentor (Marubishi MD300, Tokyo, Japan) was used for batch production in a bioreactor with operating conditions of pH 7.5, $37°C$, 200 rpm, working volume of 2 l LB, and 1 vvm aeration rate.

Twenty μg ml^{-1} of ampicillin was added to minimize segregants and 1 mM of IPTG was used for induction. pH was maintained at 7.5 with 2N NaOH and 2N H_2SO_4. Fermentation was started by inoculating 40 ml of stationary-phase culture into fresh medium (2 l).

4. Results

4.1 Construction of Expression Vector (Fig. 1)

The recombinant plasmid pHBV130 harboring the HBV/adyw gene was digested with

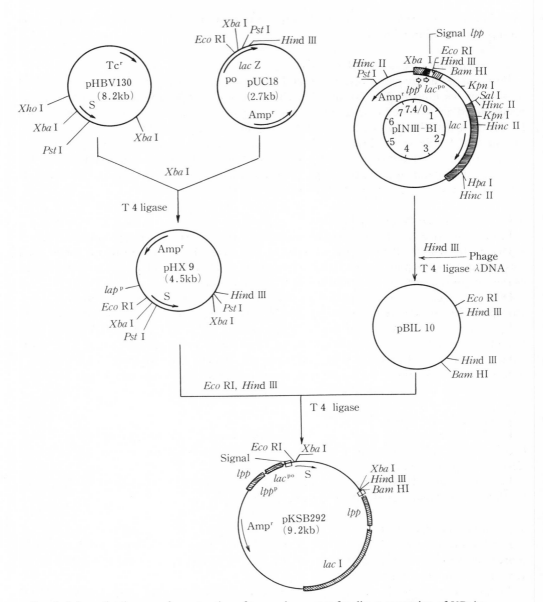

FIG. 1. Schematic diagram of construction of expression vector for direct expression of HBsAg.

*Xba*I and the 1.8 kb fragment with HBsAg gene lacking part of the N-terminal nucleotides was isolated. One of the *Xba*I sites on pHBV130 is located 80 nucleotides downstream from the ATG initiation codon of the 678bp HBsAg gene.

The 1.8 kb band recovered from agarose gel was ligated with the *Xba*I-digested pUC18 and the ligation mixture was used to transform *E. coli* JM109. The white colony with the inserted DNA was screened using the soft agar overlay containing IPTG and X-gal. Insert

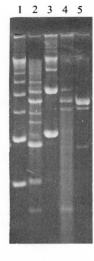

FIG. 2. Agarose gel electrophoresis of pKSB292
and relevant plasmids in Fig. 1 (from left
to right).
lane 1: pHX9;
lane 2: pHX9/*Xba*I;
lane 3: pKSB292;
lane 4: pKSB292/*Eco*RI-*Hin*dIII;
lane 5: pINIII-B1/*Hin*dIII.

orientation was checked by *Pst* I digestion and the recombinant plasmid with the correctly oriented insert was designated pHX9.

Another vector, pINIII-B1, has a *lpp* promoter of *E. coli* lipoprotein, a *lac* promoter, and a *lpp* signal sequence. The sites for *Eco*RI, *Hin*dIII, and *Bam*HI exist at the end of the *lpp* signal sequence. When a foreign gene is inserted into one of these sites, the gene product can be secreted into the periplasmic space.[24,25] In addition, there is a *lac*I regulatory gene downstream from the *lpp* signal region to avoid the dilution of the *lac* promoter by the host repressor.

The distance between the closely positioned *Eco*RI and *Hin*dIII sites of pINIII-B1 was enlarged by inserting a *Hin*dIII fragment of bacteriophage λDNA into the *Hin*dIII site of pINIII-B1 and the resulting recombinant plasmid was called pBIL10. The *Eco*RI-*Hin*dIII fragment of pHX9 (1.8 kb) containing the HBsAg gene was inserted into the large fragment of pBIL10 *Eco*RI-*Hin*dIII digest. The ligation mixture was used to transform *E. coli* JE5505 and the transformants were selected on the agar plate containing 50 μg ml^{-1} of ampicillin. These selected transformants are presumed to harbor the desired recombinant plasmid, pKSB292 (Fig. 2), in which the truncated HBsAg gene was in line with the *lpp* promoter, the *lac* promoter, and the *lpp* signal sequence.

4.2 Effect of pH on the Production of HBsAg

The effect of induction pH on HBsAg production is shown in Fig. 3. The optimum pH range was 7.5–8.0. All the subsequent experiments were routinely conducted at pH 7.5.

4.3 Effect of IPTG Concentration

As indicated in Table 1, HBsAg synthesis increased in rough proportion to the concentration of IPTG. However, the extent of production enhancement was not that great and 1 mM IPTG was used in later experiments for induction.

4.4 Growth Pattern of JE5505 (pKSB292)

The time course of growth and HBsAg formation of transformants carrying the re-

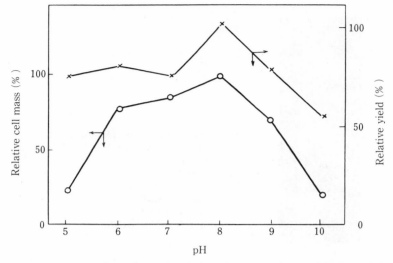

FIG. 3. Effect of induction pH on the production of HBsAg(×) and cell growth (○).

TABLE 1. Effect of concentration of inducer IPTG.

IPTG (mM)	cpm in 1 ml of extract	cpm in 1 ml of supernatant
0.1	2208	1010
0.5	2606	1421
1.0	2945	1530
1.5	2950	1545
2.0	2995	1551
2.5	3310	1583

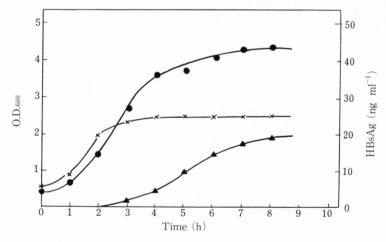

FIG. 4(a). Time course of HBsAg formation by *trp* promoter in *E. coli* C600 and growth.
● : C600,   × : C600 (*Bom*HI-*Hpo*I fragment containing HBsAg DNA)
▲ : HBsAg activity of C600 (*Bom*HI-*Hpo*I fragment containing HBsAg DNA).

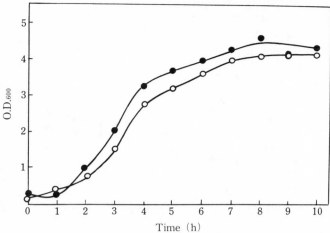

Fɪɢ. 4(b). Growth curves of JE5505 (●) and JE5505 (pKSB 292) (○) upon early induction and 1 mM IPTG.

combinant plasmid pKSB292 was investigated. Overnight test-tube culture of the transformants was inoculated into 50 ml of LB in a 250-ml flask.

As shown in Fig. 4(a), *E. coli* C600 harboring a plasmid with the *trp* promoter and a complete HBsAg gene at 37°C, pH 7.5, displayed both a severe growth inhibition and a low level of HBsAg production.[26] However, JE5505 (pKSB292) showed nearly the same

Tᴀʙʟᴇ 2. Specific growth rate of JE5505 and JE5505 (pKSB292).

Culture medium	Host JE5505	JE5505 (kPSB292)			
	LB	LB	LB + 1 mM IPTG	LB + 50 μg ml^{-1}Amp	LB + 1 mM IPTG + 50 μg ml^{-1}Amp
μ_m(h^{-1})	1.10	0.84	0.81	0.79	0.73

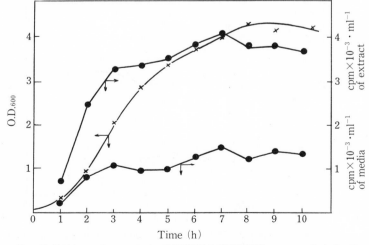

Fɪɢ. 5. Fermantation pattern of JE5505 (pKSB 292) upon early induction.

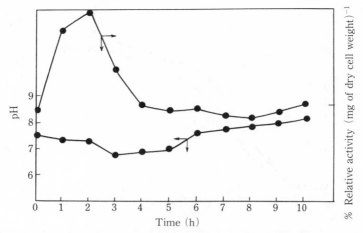

FIG. 6. Relative value of produced antigen per unit mass of dry cell weight upon early induction and pH
        profile.

growth rate and final cell density as the host JE5505 when it was cultured in LB + IPTG
medium (Fig. 4(b) and Table 2).

4.5 Production Pattern at Early Induction

IPTG was added to a 5-l Marubishi fermentor with 2 l of medium at the time of inoculation and the time course of HBsAg production under the otherwise optimal condition
was investigated.

As illustrated in Fig. 5, about 4,000 cpm of HBsAg was detected in 1 ml of extract after
7 h of fermentation, and the cell density was 2.3 g of dry cell weight per l of medium. The
maximum relative value of HBsAg activity in the extract was obtained 2 h after the start
of culture (Fig. 6).

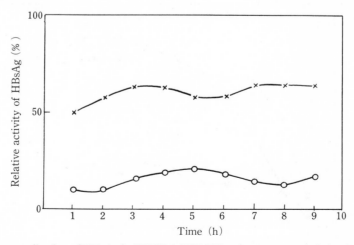

FIG. 7. Compartmentalization of HBsAg in JE5505 (pKSB 292) culture upon early induction.
        o-o-o : Activity in supernatant, x-x-x : Activity in osmotic shock fluid (periplasmic protein).

4.6 Translocation of Produced HBsAg

The compartmentalization of the produced HBsAg in JE5505 (pKSB292) culture upon early induction is shown in Fig. 7. Fractions from the cytoplasm, periplasmic space, and the supernatant showed relative activity distributions of 15–30%, 50–65%, and 10–20%, respectively, although each fraction exhibited oscillation-like variation.

5. Discussion

E. coli JE5505 transformed with a newly constructed recombinant plasmid pKSB292 that carries the viral HBsAg gene but lacks the region for the amino terminal hydrophobic domain did not inhibit growth and produced a considerable amount of HBsAg. Although it is hard to compare directly the productivity of HBsAg between Figs. 5 and 4(a), the antigen production in this work (Fig. 5) would have exceeded that in the previous work (Fig. 4(a)), judging from the least growth inhibition of *E. coli* JE5505 (pKSB292) (Fig. 4(b)). In addition, more than 50% of produced antigen was secreted into the periplasmic space with the help of the *lpp* signal peptide, thus facilitating the isolation of the product.[27,28]

It was found that the optimal pH was 7.5–8.0, and the maximum production at a level of more than 4,000 cpm ml^{-1} of extract was attained at an early stationary growth phase, i.e., 7 h after the culture started in a jar fermentor.

The effect of removing the hydrophobic terminal region on the immunological character of HBsAg and also on the protein structure remains to be studied. However, the antigen produced in this work showed a normal antigen-antibody interaction in RIA.

E. coli JE5505 (pKSB292) has clearly manifested the secretory apparatus of *E. coli* functions in the secretion of viral HBsAg protein when coupled with the *lpp* signal sequence.

REFERENCES

1. Pasek, M., Goto, T., Gilbert, W., Zink, B., Schaller, H., Mackay, P., Leadbetter, G., Murray, K., "Hepatitis B virus genes and their expression in *E. coli*." *Nature*: **282**, 575–579 (1979).
2. Burrell, C.J., Mackay, P., Greenaway, P.J., Hofschneider, P.H., Murray, K., "Expression in *Escherichia coli* of hepatitis B virus DNA sequences cloned in plasmid pBR322." *Nature*: **279**, 43–47 (1979).
3. Stahl, S. Mackay, P., Magazin, M., Bruce, S.A., Murray, K., "Hepatitis B virus core antigen: Synthesis in *Escherichia coli* and application in diagnosis." *Proc. Natl. Acad. Sci. USA*: **79**, 1606–1610 (1982).
4. Edman, J.C., Hallewell, R.A., Valenzuela, P., Goodman, H.M., Rutter, W.J., "Synthesis of hepatitis B surface and core antigens in *E. coli*." *Nature*: **291**, 503–506 (1981).
5. Hardy, K., Stahl, S., Küpper, H., "Production in *B. subtilis* of hepatitis B core antigen and of major antigen of foot and mouth disease virus." *Nature*: **293**, 481–483 (1981).
6. Hitzeman, R.A., Chen, C.Y., Hagie, F.E., Patzer, E.J., Liu, C., Estell, D.A., Miller, J.V., Yaffe, A., Kleid, D.G., Levinson, A.D., Oppermann, H., "Expression of hepatitis B virus surface antigen in yeast." *Nucleic Acids Res.*: **11**, 2745–2763 (1983).
7. Valenzuela, P., Medina, A., Rutter, W.J., Ammerer, G., Hall, B.D. "Synthesis and assembly of hepatitis B virus surface antigen particles in yeast." *Nature*: **298**, 347–350 (1982).
8. Miyanohara, A., Toh-E, A., Nozaki, C., Hamada, F., Ohtomo, N., Matsubara, K., "Expression of hepatitis B surface antigen gene in yeast." *Proc. Natl. Acad. Sci. USA*: **80**, 1–5 (1983).
9. Lee, H.C., Choi, C.Y., "Production of HBsAg via recombinant DNA technology in yeast." *Proc. 1st Int. Biotech. Symp.*, Seoul, Korea, 41–52 (1985).
10. Laub, O., Rall, L.B., Truett, M., Shaul, Y., Standring, D.N., Valenzuela, P., Rutter, W.J., "Synthesis of Hepatitis B Surface Antigen in Mammalian Cells: Expression of the Entire Gene and the Coding Region." *J. Virol.*: **48**, 271–280 (1983).

11. Moriarty, A.M., Hoyer, B.H., Shih, J.W., Gerin, J.L., Hamer, D.H., "Expression of the hepatitis B virus surface antigen gene in cell culture by using a simian virus 40 vector." *Proc. Natl. Acad. Sci. USA*: **78**, 2606–2610 (1981).

12. Charnay, P., Gervais, M., Louise, A., Galibert, F., Tiollais, P., "Biosynthesis of hepatitis B virus surface antigen in *Escherichia coli*." *Nature*: **286**, 893–895 (1980).

13. Rose, J.K., Shafferman, A., "Conditional expression of the vesicular stomatitis virus glycoprotein gene in *Escherichia coli*." *Proc. Natl. Acad. Sci. USA*: **78**, 6670–6674 (1981).

14. Boyer, H.W., Rolland-Dussoix, D., "A complementation Analysis of the Restriction and Modification of DNA in *Escherichia coli*." *J. Mol. Biol.*: **41**, 459–472 (1969).

15. Yanisch-Perron, C., Vieira, J., Messing, J., "Improved M13 phage cloning vectors and host strains: nucleotide sequences of the M13mp18 and pUC19 vectors." *Gene*: **33**, 103–119 (1985).

16. Hirota, Y., Suzuki, H., Nishimura, Y., Yasuda, S., "On the process of cellular division in *Escherichia coli*: A mutant of *E. coli* lacking a murein-lipoprotein." *Proc. Natl. Acad. Sci. USA*: **74**, 1417–1420 (1977).

17. Gough, N.M., Murray, K., "Expression of the Hepatitis B Virus Surface, Core and E Antigen Genes by Stable Rat and Mouse Cell Lines." *J. Mol. Biol.*: **162**, 43–67 (1982).

18. Norrander, J., Kempe, T., Messing, J., "Construction of improved M13 vectors using oligodeoxy-nucleotide-directed mutagenesis." *Gene*: **26**, 101–106 (1983).

19. Inouye, M., "Multipurpose Expression Cloning Vehicles in *Escherichia coli*." In *Experimental Manipulation of Gene Expression*, p. 15–31. Academic Press, New York (1983).

20. Kieser, T., "Factors Affecting the Isolation of cccDNA from *Streptomyces lividans* and *Escherichia coli*." *Plasmid*: **12**, 19–36 (1984).

21. Tautz, D., Renz, M., "An Optimized Freeze-Squeeze Method for the Recovery of DNA Fragments from Agarose Gels." *Anal. Biochem.*: **132**, 14–19 (1983).

22. Lederberg, E.M., Cohen, S.N., "Transformation of *Salmonella typhimurium* by Plasmid Deoxyribonucleic Acid." *J. Bacteriol.*: **119**, 1072–1074 (1974).

23. Neu, C.H., Heppel, L.A., "The Release of Enzymes from *Escherichia coli* by Osmotic Shock and during the Formation of Spheroplasts." *J. Biol. Chem.*: **240**, 3685–3692 (1965).

24. Inouye, S., Soberon, X., Franceschini, T., Nakamura, K., Itakura, K., Inouye, M., "Role of positive charge on the amino-terminal region of the signal peptide in protein secretion across the membrane." *Proc. Natl. Acad. Sci. USA*: **79**, 3438–3441 (1982).

25. Nakamura, K., Inouye, M., "DNA sequence of the Gene for the Outer Membrane Lipoprotein of *E. coli*: An Extremely AT-Rich Promoter." *Cell*: **18**, 1109–1117 (1979).

26. Fujisawa, Y., Ito, Y., Sasada, R., Ono, Y., Igarashi, K., Marumoto, R., Kikuchi, M., Sugino, Y., "Direct expression of hepatitis B surface antigen gene in *E. coli*." *Nucleic Acids Res.*: **11**, 3581–3591 (1983).

27. Oka, T., Sakamoto, S., Miyoshi, K., Fuwa, T., Yoda, K., Yamasaki, M., Tamura, G., Miyake, T., "Synthesis and secretion of human epidermal growth factor by *Escherichia coli*." *Proc. Natl. Acad. Sci. USA*: **82**, 7212–7216 (1985).

28. Gray, G.L., Baldridge, J.S., McKeown, K.S., Heyneker, H.L., Chang, C.N., "Periplasmic production of correctly processed human growth hormone in *Escherichia coli*: natural and bacterial signal sequences are interchangeable." *Gene*: **39**, 247–254 (1985).

II-4
Enhancement of Thermostability of Enzymes

Tadayuki Imanaka

Department of Fermentation Technology, Faculty of Engineering, Osaka University, Suita-shi, Osaka 565, Japan

1. Introduction

A number of enzymes from thermophiles are more thermostable and tolerant to chemical denaturing agents than their psychrophile and mesophile counterparts.[1] Thermostable enzymes are utilized extensively in industry.[2] If the nature of a useful enzyme could be altered to a more thermostable one, this alteration *per se* would be advantageous not only for better understanding of basic research but also for wider practical application.

The purpose of this paper is to present a brief review of procedures for obtaining thermostable enzymes, to propose some criteria to improve protein thermostability, and to demonstrate the effectiveness of the criteria for enhancing thermostability of the neutral protease from *Bacillus stearothermophilus*.

2. Obtaining Thermostable Enzymes

Thermostable enzymes can be obtained by the following methods:

I. Screening of useful thermophiles from a natural environment, and isolation of thermostable enzymes from them.

II. Physical/chemical modifications of thermolabile enzymes.

Enzyme thermostability can be induced by interactions with substrate, solvent, and salt as well as by physical/chemical modifications.[3]

III. Mutation of enzyme gene.

III - A. Random mutation and screening of thermostable enzymes.

Which mutants are subjected to screening for thermostable enzymes is a crucial step in this method. To cite an example, thermostable mutants of the T4 lysozyme were isolated by assay on lawns of infected *Escherichia coli* that were incubated at a temperature prohibitive to the wild-type enzyme.[4] However, this screening system is not as effective as direct selection of thermostable variants.

Recently, *in vivo* screening systems for thermostable mutant enzymes have been developed. Matsumura and Aiba[5] used a transformation system of the thermophile *B. stearothermophilus*.[6,7] A structural gene of kanamycin nucleotidyltransferase cloned into bacteriophage M13 was subjected to mutagenesis with hydroxylamine. After recloning the mutagenized gene of the enzyme in a vector plasmid, the recombinant plasmid was used to transform *B. stearothermophilus*. The transformants carrying more thermostable enzyme than the wild type were isolated by shifting from a permissive (55°C) to a non-permissive (61°C) temperature in the presence of kanamycin.

Another method developed by Liao et al.[8] consists of cloning a gene encoding an enzyme (kanamycin nucleotidyltransferase) from a mesophile, transfer of the gene into a thermophile (*B. stearothermophilus*), and isolation of the desired mutant via enzymatic activity at higher growth temperatures of the host bacterium.

III - B. Site-directed mutagenesis.

Enzyme thermostability can be enhanced by single-amino-acid substitutions.[9,10] This offers the merits of site-directed mutagenesis. This mutagenesis is supported by recent advances both in chemical synthesis of oligonucleotides and in genetic engineering, and has made it possible to create "novel" proteins in a predictable manner, as long as their structures are known.[11] This aspect of "protein engineering" has already been proved effective for enhancing thermostabilities of the T4 lysozyme[12] and subtilisin[13] by the formation of engineered disulfide bonds. However, no general means for deciding which amino acid at which position in the protein structure should be replaced by which other amino acid has been established; in other words, a guiding principle for "designing protein" for a specific purpose must be set in future.

3. Criteria in Protein Design for the Enhancement of Thermostability

In view of the status of protein engineering described above, some criteria for enhancing enzyme thermostability without loss of enzymatic activity are proposed below:

1) Compare the primary structure (amino acid sequence) among enzymes that are of the same kind but differ in origin. This comparison is useful for finding regions that are essential not only for activity but also for substrate specificity, because active and/or substrate-binding sites are highly conserved in homologous regions.[14] Therefore, amino acids in the highly conserved regions, whenever confirmed, might well be left untouched in engineering protein.

2) Refer to statistical data[15] on comparison of various enzymes between mesophilic and thermophilic origins. Amino acid substitutions (Gly to Ala, Ser to Ala, Ser to Thr, Lys to Arg, Asp to Glu, and so on, in order of significance)[15] should be checked.

3) Refrain from substituting an amino acid that would drastically change the secondary structure (e.g., Pro in the α-helix). The secondary structure of an enzyme can be partly predicted from its primary structure by the method of Chou and Fasman,[16] although this prediction is not widely accepted yet. Information on the secondary structure of a protein (or a closely related protein) is indispensable.

4) Pay attention to amino acids lying on the protein surface in between hydrophobic and hydrophilic cores, because additional and electrostatic interaction (salt bridge) with a neighboring charged amino acid that is effective for thermostability enhancement without causing any significant change in tertiary structure can be acquired in this particular area. The hydropathic character of protein can be assessed by the method of Kyte and Doolittle.[17]

5) Specify an amino acid that can increase internal hydrophobicity and stabilize helices for strong internal packing. This specification can be made by consulting the secondary and tertiary protein structures.

6) Remember previous examples that were successful in enhancing protein thermo-

stability, for example, disulfide bonds engineered by the introduction of a new Cys near the pre-existing Cys in the wild-type enzyme.[12,13]

7) Delete amino acid sequences not needed for normal functioning of the protein.

4. Comparison of Amino Acid Sequences of Four Different Proteases

From now on, the above Criteria 1, 2, and 5 will be followed to demonstrate the alteration of the thermostability of the neutral protease of *B. stearothermophilus*. The gene had

```
                .                    .       -   .                  .        -+   .              C
B.st.  VAGASTVGVGRGVLGDQKYINTTYSSYYGYYYLQDNTR   GSGIFTYDGRNRT VLPGSLWTD 60
B.th.  ITGTSTVGVGRGVLGDQKNINTTYSTYY    YLQDNTR   GDGIFTYDAKYRT TLPGSLWAD
B.su.  AAA T GSGTTLKGATVPLN   ISYEGGKYVLRDLSKPTGTQIITYDLQNRQSRLPGTLVSS
B.am.  AAT T GTGTTLKGKTVSLN   ISSESGKYVLRDLSKPTGTQIITYDLQNREYNLPGTLVSS

       A(M2)
        ↑    S(M3)
       ⌐C
B.st.  GDNQFTASYDA AAVDAHY YAGVVYDYYKNVHGRLSYDGSNAA IRSTVHYG RGYNNAFWNG 120
B.th.  ADNQFFASYDA PAVDAHY YAGVTYDYYKNVHNRLSYDGNNAA IRSSVHYS QGYNNAFWNG
B.su.  TTKTFTSSSQR AAVDAHY NLGKVYDYFYSNFKRNSYDNKGSK IVSSVHYGT QYNNAAWTG
B.am.  TTNQFTTSSQR AAVDAHY NLGKVYDYFYQKFNRNSYDNKGGK IVSSVHYGS RYNNAAWIG

                                   A(M1)
                                    ↑
              .             .C  ⌐Z  Z .           .          Z.            C
B.st.  SQMVYGDGDGQ TFLPFSGGI DVVGHELTH AVTDYTAGLVYQNESGAINEAMSDIFGTLVE 180
B.th.  SEMVYGDGDGQ TFIPLSGGI DVVAHELTH AVTDYTAGLIYQNESGAINEAISDIFGTLVE
B.su.  DQMIYGDGDGS FFSPLSGSL DVTAHEMTH GVTQETANLIYENQPGALNESFSDVFG
B.am.  DQMIYGDGDGS FFSPLSGSM DVTAHEMTH GVTQETANLNYENQPGALNESFSDVFG

             C   .  CC    .+  C+        .               .                    .
B.st.  FYANRNP DWEIGEDI YTPGVAGD ALRSMSDPAKY GDPDHYSK RYT     GTQ DNGGVHTNSGII 240
B.th.  FYANKNP DWEIGEDV YTPGISGD SLRSMSDPAKY GDPDHYSK RYT     GTQ DNGGV◉INSGII
B.su.  YFNDTE DWDIGEDI T   VSQP ALRSLSNPTKY NQPDNYANYRNLPNTDEG DYGGVHTNSGIP
B.am.  YFNDTE DWDIGEDI T   VSQP ALRSLSNPTKY GQPDNFKNYKNLPNTDAG DYGGVHTNSGIP

             .                .              .                .      +    .
B.st.  NKAAY LLSQGGVHYGVSVNGIGRDKMGK IFYRALVYYLTPTS NFSQLRAACV QAAADLYG 300
B.th.  NKAAY LISQGGTHYGVSVVGIGRDKLGK IFYRALTQYLTPTS NFSQLRAAAV QSAYDLYG
B.su.  NKAAY NTITK      LGVSKSQQ IYYRALTTYLTPSS TFKDAKAALI QSARDLYG
B.am.  NKAAY NTITK      IGVNKAEQ IYYRALTVYLTPSS TFKDAKAALI QSARDLYG

            .
B.st.  ST SQEVNSVKQAFNAVGVY 319
B.th.  ST SQEVASVKQAFDAVGVK
B.su.  ST D AAKVEAAWNAVGL
B.am.  SQ D AASVEAAWNAVGL
```

FIG. 1. Comparison of amino acid sequences of various extracellular neutral proteases[23]. Amino acid residues are shown by single letters as follows: A, Ala; C, Cys; D, Asp; E, Glu; F, Phe; G, Gly; H, His; I, Ile; K, Lys; L, Leu; M, Met; N, Asn; P, Pro; Q, Gln; R, Arg; S, Ser; T, Thr; V, Val; W, Trp; Y, Tyr. Blank indicates the absence of a corresponding amino acid. Enzyme sources are abbreviated as: B.st., *Bacillus stearothermophilus*[19]; B.th., *Bacillus thermoproteolyticus*[20]; B.su., *Bacillus subtilis*[21]; B.am., *Bacillus amyloliquefaciens*.[22] Homologous sequence regions are surrounded by rectangles. Active and substrate-binding sites of thermolysin are indicated by ○ and □, respectively. Protein ligands for Zn and Ca ions for thermolysin are indicated above the sequence by Z and C, respectively. Amino acid substitutions that are expected to enhance or reduce the thermostability of *B. stearothermophilus* protease in comparison with that of thermolysin from *B. thermoproteolyticus* are indicated above the sequence by + or −, respectively. Vertical arrows indicate amino acid substitutions in *B. stearothermophilus* protease.

previously been cloned and sequenced.[18,19] According to Criterion 1, amino acid sequences of the extracellular portion for four neutral proteases from *B. stearothermophilus*,[19] *Bacillus thermoproteolyticus*,[20] *Bacillus subtilis*,[21] and *Bacillus amyloliquefaciens*[22] were aligned to facilitate the search for homologous regions when compared (Fig. 1).[23] As a result, the amino acid sequence of thermostable neutral protease from *B. stearothermophilus* was homologous (85%) with that of thermolysin from *B. thermoproteolyticus*. Similarly, the sequences of thermolabile neutral proteases from *B. subtilis* and *B. amyloliquefaciens* were also homologous (89%).

In contrast, the homology between the thermostable and thermolabile enzymes was lower (about 45%) (Fig. 1). Nevertheless, nine highly homologous regions were found

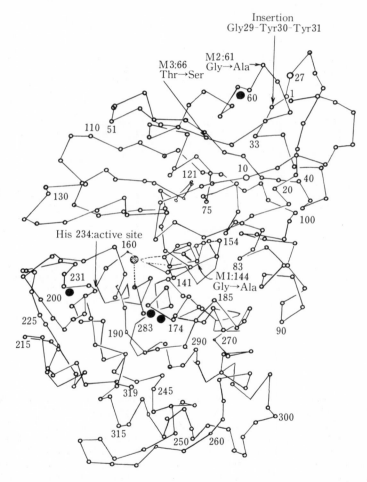

Fig. 2. Three-dimensional structure of thermolysin.[23] Open circles correspond to the α-carbon positions. Zinc atom is drawn stippled with its three protein ligands shown diagramatically as dotted lines. Four calcium atoms are shown as solid circles. The amino acid number from the NH_2-terminus is for *B. stearothermophilus* neutral protease, i.e., originally, Gly29-Tyr30-Tyr31 were absent in thermolysin.

in the amino acid sequences of all the enzymes despite their different origins. These highly homologous regions are most likely to be essential for enzyme activity. In fact, it has been reported that His234 and Arg206 are the active and substrate-binding sites of neutral protease (thermolysin), respectively.[24] Zn ion is essential for the enzyme activity of thermolysin, and the three protein ligands for Zn (His145, His149, and Glu169)[24] are conserved in all four of the enzymes (Fig. 1). Glu146, reported to promote the attack of water molecule on carbonyl carbon of the substrate,[24] is also conserved in these enzymes (Fig. 1).

The comparison of *B. stearothermophilus* protease and thermolysin revealed that the amino acid frames were completely matched except for three additional amino acids (Gly29-Tyr30-Tyr31) in the enzyme from *B. stearothermophilus* (Fig. 1). Therefore, the three-dimensional structure of *B. stearothermophilus* protease was assumed to be basically similar to the known tertiary structure of thermolysin[25] shown in Fig. 2. Although His234 (active site), Arg206 (substrate-binding site), Leu205, Glu193, and Asn162 are located separately in the primary structure of the enzyme (Fig. 1), these amino acid residues are in the vicinity of the active site in the three-dimensional structure (Fig. 2), and they are completely conserved in the four neutral proteases. In accordance with Criterion 1, alteration of amino acids in the highly conserved sequences was avoided.

5. Protein Design to Enhance Thermostability of Neutral Protease

According to the three criteria (1, 2, and 5) above, the position and species of amino acid to be replaced for the enhancement of enzyme thermostability were decided as follows. Amino acid substitutions that were pointed out by Argos et al.[15] were discovered in both *B. stearothermophilus* protease and thermolysin (Fig. 1). Thermolysin and *B. stearothermophilus* protease contain some amino acids effective for higher thermostability in the first and second halves, respectively. Therefore, the area where the substitution of amino acid was to be attempted was confined to the first half. Promising substitutions (Gly to Ala) are available at positions 47, 61, and 144. Since Gly144 exists within the α-helix which combines the two domains (Fig. 2), the amino acid substitution (Gly144 to Ala144, mutation M1) would increase the internal hydrophobicity and stabilize the α-helix, and hence, the M1 mutant is expected to enhance the thermostability of protease. This substitution (i.e. addition of the methyl group) is least likely to interrupt the function or internal residue packing arrangement. Another mutation M3 (Thr66 to Ser66) which is the reverse replacement in Criterion 2 would decrease thermostability of the enzyme.

6. Site-directed Mutagenesis

Amino acid substitutions were performed by site-directed mutagenesis (Fig. 3). *B. stearothermophilus* neutral protease gene *nprT*+ was cloned in pNP22-1.[19] pNP22-2 is a deletion derivative of pNP22-1. Since the plasmid contains three *Sal*I sites, one site contiguous to the *Eco*RI site was eliminated through the following series of steps: partial digestion of pNP22-2 with *Sal*I, digestion with Bal31, and ligation. The plasmid thus obtained was designated pNP22-3. The 888 bp *Sal*I fragment containing most of the extracellular portion of protease was isolated from pNP22-3 (Npr+), and the resultant plasmid (Npr−) was designated pDMP10. Following the method of Norris et al.,[26] the 888 bp *Sal*I fragment was cloned in M13mp11, and single-stranded (ss) DNA was isolated. The ss DNA was annealed with a synthetic oligonucleotide whose 5′-end had been phosphorylated with polynucleotide kinase. After DNA extension using the primer and liga-

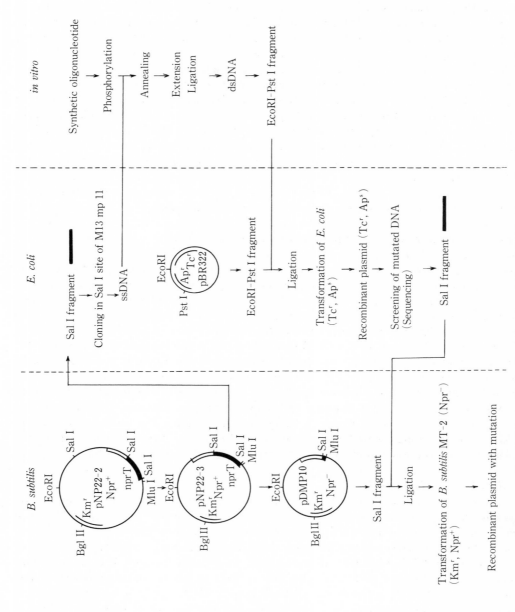

Fig. 3. Scheme of site-directed mutagenesis and DNA construction.[23]

tion, the double-stranded DNA was digested with both *Eco*RI and *Pst*I. The *Eco*RI-*Pst*I fragment was ligated with a larger *Eco*RI-*Pst*I fragment of pBR322. The ligation mixture was used to transform *E. coli*. Recombinant plasmids were isolated from the transformants (Tc^r Ap^s). Mutated DNA was screened by either restriction enzyme treatment (mutated primer 5'-CGTCGTGG*C̊*GCATGAGT-3', *Hha*I for mutation M1, and 5'-GACCG*Å*TGCCGACAACC-3', *Sfa*NI for mutation M2, asterisks show the location of mismatches and the sequences in italics show the new restriction site) or colony hybridization (M3) as described by Carter et al.,[27] and the mutation was confirmed by DNA sequencing. The 888 bp *Sal*I fragment with mutation was isolated from the recombinant plasmid, and ligated with *Sal*I-treated pDMP10. The ligation mixture was used to transform *B. subtilis* MT-2 (Npr^-)[18] to Km^r Npr^+. *B. substilis* MT-2 carrying the mutant plasmid was used to produce extracellular neutral protease.

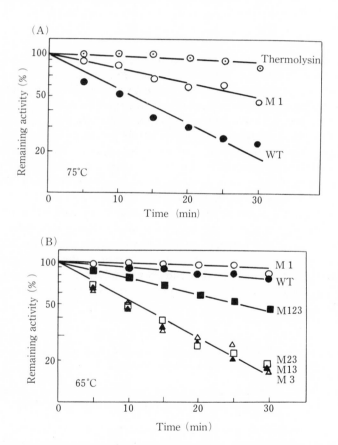

FIG. 4. Thermostability of neutral proteases.[23] Extracellular protease produced by *B. subtilis* carrying the *nprT*^+ gene was purified by a method described previously.[18] Remaining activity after heating at 75°C (A) or 65°C (B) was assayed as described earlier[18] and was expressed as percentage of original activity. ⊙, thermolysin; ●, wild-type enzyme (WT) from *B. stearothermophilus*; ○, mutant M1; △, mutant M3; ▲, double-mutant M13; □, double-mutant M23; ■, triple-mutant M123.

7. *Thermostability of Wild-type Enzyme and Its Variants*

Thermostabilities of the wild-type (WT) neutral protease of *B. stearothermophilus* and thermolysin were tested at 75°C (Fig. 4A). Thermolysin was more thermostable than the WT protease from *B. stearothermophilus*. When the M1 mutation (Gly144 to Ala144, GGG to GCG) was introduced into the WT strain, the M1 enzyme became more thermostable than the WT enzyme (Fig. 4A, 4B). The M3 mutation (Thr66 to Ser66, ACC to TCC) was also introduced into the WT strain, and the M3 enzyme was fairly thermolabile (Fig. 4B). To restore the thermostability of M3 enzyme, a double-mutant M13 was constructed. However, M13 enzyme exhibited the same thermostability as M3. When another mutation M2 (Gly61 to Ala61, GGC to GCC) was added into M3, the double-mutant enzyme M23 showed the same stability as M3. However, when both mutations M1 and M2 were introduced simultaneously into M3, thermostability of the triple mutant enzyme M123 was recovered to some extent compared to that of M3 (Fig. 4B). This result indicates the synergistic effect of enhancing thermostability by the two mutations.

8. *Discussion*

Enzyme thermostability is influenced by many factors, e.g., amino acid sequence, three-dimensional structure, cofactors, pH, etc. The *Bacillus* neutral proteases require Ca ion as the most important stabilizing factor of the enzymes. Four Ca-binding sites, Asp141, Asp188, Glu193, and Asp194, are conserved for all proteases examined (Fig. 1). However, another Ca-binding site, Glu180, of thermostable proteases from *B. stearothermophilus* and *B. thermoproteolyticus*, is deleted for thermolabile proteases of *B. subtilis* and *B. amyloliquefaciens* (Fig. 1). This may account partly for the fact that the latter neutral proteases are rather thermolabile. If a particular amino acid sequence from Thr177 to Phe181 were inserted into the corresponding vacant positions of *B. subtilis* protease, the latter enzyme would be expected to become more thermostable.

The criteria used in this example for the enhancement of thermostability could be applied to other kinds of proteins if the secondary and tertiary structures were available. The same technique may also help envisage the three-dimensional structure of a protein from its primary structure. To achieve this goal, further studies on point mutations and theoretical discussion are needed. In this category, analysis of the folding and unfolding processes of polypeptides in their denaturing phenomena would be rewarding.

REFERENCES

1. Amelunxen, R.E., Murdock, A.L., "Microbial life at high temperatures: Mechanisms and molecular aspects." In *Microbial Life in Extreme Environments*. Kushner, D.J. (ed.), p. 217–278. Academic Press, New York, 1978.
2. Ng, T.K., Kenealy, W.R., "Industrial applications of thermostable enzymes." In *Thermophiles: General, Molecular, and Applied Microbiology*. Brock, T.D. (ed.), p. 197–215. John Wiley & Sons, New York, 1986.
3. Schmid, R.D., "Stabilized soluble enzymes." *Adv. Biochem. Eng.*: **12**, 41–118 (1979).
4. Alber, T., Wozniak, J.A., "A genetic screen for mutations that increase the thermal stability of phage T4 lysozyme." *Proc. Natl. Acad. Sci. USA*: **82**, 747–750 (1985).
5. Matsumura, M., Aiba, S., "Screening for thermostable mutant of kanamycin nucleotidyltransferase by the use of a transformation system for a thermophile, *Bacillus stearothermophilus*." *J. Biol. Chem.*: **260**, 15298–15303 (1985).
6. Imanaka, T., Fujii, M., Aramori, I., Aiba, S., "Transformation of *Bacillus stearothermophilus* with

plasmid DNA and characterization of shuttle vector plasmids between *Bacillus stearothermophilus* and *Bacillus subtilis*." *J. Bacteriol.*: **149**, 824–830 (1982).

7. Imanaka, T., "Host-vector systems in thermophilic bacilli and their applications." *Trends Biotechnol.*: **1**, 139–144 (1983).

8. Liao, H., McKenzie, T., Hageman, R., "Isolation of a thermostable enzyme variant by cloning and selection in a thermophile." *Proc. Natl. Acad. Sci. USA*: **83**, 576–580 (1986).

9. Yutani, K., Ogasahara, K., Sugino, Y., Matsushiro, A., "Effect of a single amino acid substitution on stability of conformation of a protein." *Nature*: **267**, 274–275 (1977).

10. Matsumura, M., Katakura, Y., Imanaka, T., Aiba, S., "Enzymatic and nucleotide sequence studies of a kanamycin-inactivating enzyme encoded by a plasmid from thermophilic bacilli in comparison with that encoded by plasmid pUB110." *J. Bacteriol.*: **160**, 413–420 (1984).

11. Ulmer, K.M., "Protein engineering." *Science*: **219**, 666–671 (1983).

12. Perry, L.J., Wetzel, R., "Disulfide bond engineered into T4 lysozyme: Stabilization of the protein toward thermal inactivation." *Science*: **226**, 555–557 (1984).

13. Wells, J.A., Powers, D.B., "*In vitro* formation and stability of engineered disulfide bonds in subtilisin." *J. Biol. Chem.*: **261**, 6564–6570 (1986).

14. Nakajima, R., Imanaka, T., Aiba, S., "Comparison of amino acid sequences of eleven different α-amylases." *Appl. Microbiol. Biotechnol.*: **23**, 355–360 (1986).

15. Argos, P., Rossmann, M.G., Grau, U.M., Zuber, H., Frank, G., Tratschin, J.D., "Thermal stability and protein structure." *Biochem.*: **18**, 5698–5703 (1979).

16. Chou, P.Y., Fasman, G.D., "Empirical predictions of protein conformation." *Ann. Rev. Biochem.*: **47**, 251–276 (1978).

17. Kyte, J., Doolittle, R.F., "A simple method for displaying the hydropathic character of a protein." *J. Mol. Biol.*: **157**, 105–132 (1982).

18. Fujii, M., Takagi, M., Imanaka, T., Aiba, S., "Molecular cloning of a thermostable neutral protease gene from *Bacillus stearothermophilus* in a vector plasmid and its expression in *Bacillus stearothermophilus* and *Bacillus subtilis*." *J. Bacteriol.*: **154**, 831–837 (1983).

19. Takagi, M., Imanaka, T., Aiba, S., "Nucleotide sequence and promoter region for the neutral protease gene from *Bacillus stearothermophilus*." *J. Bacteriol.*: **163**, 824–831 (1985).

20. Titani, K., Hermodson, M.A., Ericsson, L.H., Walsh, K.A., Neurath, H., "Amino-acid sequence of thermolysin." *Nature New Biol.*: **238**, 35–37 (1972).

21. Yang, M.Y., Ferrari, E., Henner, D.J., "Cloning of the neutral protease gene of *Bacillus subtilis* and the use of the cloned gene to create an in vitro-derived deletion mutation." *J. Bacteriol.*: **160**, 15–21 (1984).

22. Vasantha, N., Thompson, L.D., Rhodes, C., Banner, C., Nagle, J., Filpula, D., "Genes for alkaline protease and neutral protease from *Bacillus amyloliquefaciens* contain a large open reading frame between the regions coding for signal sequence and mature protein." *J. Bacteriol.*: **159**, 811–819 (1984).

23. Imanaka, T., Shibazaki, M., Takagi, M., "A new way of enhancing the thermostability of proteases." *Nature*: **324**, 695–697 (1986).

24. Kester, W.R., Matthews, B.W., "Comparison of the structures of carboxypeptidase A and thermolysin." *J. Biol. Chem.*: **252**, 7704–7710 (1977).

25. Matthews, B.W., Jansonius, J.N., Colman, P.M., Schoenborn, B.P., Dupourque, D., "Three-dimensional structure of thermolysin." *Nature New Biol.*: **238**, 37–41 (1972).

26. Norris, K., Norris, F., Christiansen, L., Fiil, N., "Efficient site-directed mutagenesis by simultaneous use of two primers." *Nucleic Acids Res.*: **11**, 5103–5112 (1983).

27. Carter, P.J., Winter, G., Wilkinson, A.J., Fersht, A.R., "The use of double mutants to detect structural changes in the active site of the tyrosyl-tRNA synthetase (*Bacillus stearothermophilus*)." *Cell*: **38**, 835–840 (1984).

28. Matthews, B.W., Colman, P.M., Jansonius, J.N., Titani, K., Walsh, K.A., Neurath, H., "Structure of thermolysin." *Nature New Biol.*: **238**, 41–43 (1972).

III. Metabolites

III-1
The *In Vivo* Longevity of Antibiotic Synthetases

S. N. Agathos* and A. L. Demain[2*]

* *Department of Chemical and Biochemical Engineering and the Waksman Institute of Microbiology, Rutgers-The State University of New Jersey, Piscataway, New Jersey 08854, U.S.A.*
[2*] *Department of Applied Biological Sciences, Massachusetts Institute of Technology, Cambridge, Massachusetts 02139, U.S.A.*

1. In Vivo Instability of Synthetases of Secondary Metabolites

Increases in yields and productivities of microbial processes for the production of antibiotics and other secondary metabolites are attained through a number of well-established and currently evolving environmental and genetic approaches including the control of the environmental conditions within the bioreactor (Küenzi, 1978; DeTilly et al., 1983; Calam, 1986), mutational programs (Demain, 1973; Weinberg, 1970), strain selection and media design based on by-passing or deregulation of metabolic controls (Demain, 1981; Martin and Demain, 1980; Demain et al., 1983), and modern genetic engineering methodology (Hopwood and Chater, 1980; Kurth and Demain, 1984; Elander and Vournakis, 1986).

Due to the complexity of the biosynthesis of most secondary metabolites and the frequent lack of adequate information on the genetics and physiology of the majority of industrial microorganisms, the above approaches are often used in combination until an optimized process is developed. Despite the individual differences among the various production systems, the overall efficiency of every microbial process depends on (a) the supply of precursor molecules, (b) the amount of relevant biosynthetic enzymes (level of gene expression), (c) the degree of catalytic activity of the biosynthetic enzymes, and (d) the operational stability of these biosynthetic enzymes *in vivo*.

In the field of secondary metabolite production, the involvement of precursors and the management of their availability has been studied to some extent (Zähner and Kurth, 1982; Šupek et al., 1985; Aharonowitz et al., 1986), although there exists a clear need for a better understanding and control of these potentially rate-limiting factors. Enzyme levels in secondary metabolite-producing microorganisms are known to be regulated through the control of the expression of the corresponding structural genes (induction and repression), whereas enzyme activity is controlled by noncovalent binding of various ligands, including regulatory molecules and precursors from intermediary metabolism (activation and inhibition). These control modes, which affect factors (b) and (c) above, have dominated the literature and have been the almost exclusive targets of process development through environmental and genetic approaches.

In contrast, the *in vivo* stability of biosynthetic enzymes has been all but ignored, despite the fact that the continued functioning of such enzymes is often the limiting constraint in the long-term production of secondary metabolites. This article specifically addresses factor (d) above, i.e., the *in vivo* stability of biosynthetic enzymes (synthetases) in secondary metabolism.

2. Decline and Cessation of Secondary Metabolism

Although the duration of the production phase (idiophase) may differ between different species or vary for the same species under different environmental conditions (Weinberg, 1970), the rate of secondary metabolite formation gradually decreases, and, eventually, the process comes to a halt in batch fermentations (Weinberg, 1981; Demain et al., 1983). The cessation of the idiophase has been attributed to a variety of cellular events, including cell lysis, depletion of precursors, feedback control of the synthetases by the product, and irreversible decay (*in vivo* inactivation) of the synthetases (Abraham et al., 1965; Bu'Lock, 1967; Weinberg, 1970; Malik, 1982; Demain et al., 1983). While there is a dearth of studies addressing the reasons for the decline of antibiotic production, the experimental evidence points in the direction of (i) end-product inhibition or repression of several antibiotic synthetases, and (ii) synthetase instability *in vivo*. Because *in vivo* enzyme inactivation in secondary metabolism has remained relatively neglected, the mechanisms responsible for the disappearance of such enzyme activities remain obscure. Increased understanding of the process would make it possible to achieve prolonged periods of synthesis of desired metabolites in industrial practice or easy harvesting of cells with adequate synthetase activity for cell-free synthesis applications.

The *in vivo* inactivation of secondary metabolite synthetases is consistent with the notion of control of enzymatic activity via selective inactivation *in vivo* (Thurston, 1972; Holzer et al., 1975; Switzer, 1977; Holzer, 1981). Inactivation is defined as the irreversible loss *in vivo* of an enzyme's catalytic activity as distinct from inhibition of catalytic activity through noncovalent, reversible binding of inhibitor molecules (Switzer, 1977). Inactivation in viable microbial cells includes both (i) modification inactivation, where covalent binding or a change in physical state destroys enzyme activity despite the enzyme protein being left intact, and (ii) degradation inactivation, where at least one peptide bond of the enzyme protein is cleaved. The latter is a process that is distinct from protein turnover, in that it does not result in the cleavage of the protein into its constituent amino acids, but it may be the first step of such a process. Since the disappearance of enzyme activity seems to be a controlled aspect of metabolism under specific physiological conditions, it qualifies as an additional mode of enzyme regulation aimed at phasing out a particular enzyme when a metabolic shift renders it wasteful or directly harmful to the cell, e.g., when repression of further synthesis is not sufficient for efficient regulation. These generalizations have been derived from studies performed exclusively on microbial enzymes of primary metabolism.

Similar physiological rationales appear plausible in the *in vivo* disappearance of synthetase activity in secondary metabolic pathways, as the process seems to be associated with a specific physiological condition, such as the exhaustion of a nutrient, the onset of the stationary growth phase, or a stage of differentiation. Several instances of *in vivo* inactivation of antibiotic synthetases have been documented that reveal the transient presence of such synthetases in batch culture.

Among peptide antibiotics produced by bacilli, rapid inactivation of their synthetases has been well established: e.g., gramicidin S (Tomino et al., 1967; Otani et al., 1969; Laland and Zimmer, 1973; Matteo et al., 1975), tyrocidin (Fujikawa et al., 1968), and bacitracin (Frøyshov, 1977). The *in vivo* activity of these synthetases is lost within 1–2 h after attaining a peak value toward the end of the exponential growth phase. No concrete mechanism has been proposed for such inactivation prior to our own studies.

Enzymes participating in the biosynthesis of various antibiotics by streptomycetes also exhibit a loss of activity *in vivo*, which usually extends over longer periods of time as compared with the peptide antibiotic synthetases. In the case of *Streptomyces* sp. 3022, the producer of chloramphenicol, Jones and Westlake (1974) found that the activity of arylamine synthetase, which catalyzes the initial step in the chloramphenicol pathway, starts dropping after reaching a maximum level by the mid-exponential growth phase. These workers further indicated that arylamine synthetase is also repressed by the end product, chloramphenicol. However, it is not clear whether endogenously produced chloramphenicol is directly responsible for the gradual fall in enzyme activity observed during the production phase.

Enzymes following the same general pattern after reaching a peak in the course of the fermentation include arginine: X amidinotransferase in the streptomycin producing organisms, *Streptomyces bikiniensis* and *Streptomyces griseocarneus* (Walker and Hnilica, 1964), NADPH-dependent dTDP-dihydrostreptose synthetase (Ortmann et al., 1974), and streptomycin phosphorylase (Nimi et al., 1971) in the streptomycin producer, *Streptomyces griseus*, anhydrotetracycline hydratase (oxygenase) and tetracycline dehydrogenase in the tetracycline producer, *Streptomyces aureofaciens* (Behal, 1986), and the cyclase, epimerase, and expandase enzymes of the cephalosporin pathway in *Cephalosporium acremonium* (Heim et al., 1984), *Streptomyces clavuligerus* (Braña et al., 1985), and *Nocardia lactamdurans* (Cortés et al., 1986).

From among secondary metabolic pathways leading to the formation of products other than antibiotics, enzymes exhibiting similar *in vivo* instabilities can be cited, such as dimethylallyltryptophan synthetase (Heinstein et al., 1971; Maier and Gröger, 1976) and chanoclavine-1 cyclase (Erge et al., 1973) of ergot alkaloid biosynthesis by the fungus *Claviceps*; cyclopenin *m*-hydroxylase (Richter and Luckner, 1976) and cyclopeptine dehydrogenase (Aboutabl and Luckner, 1975) of benzodiazepene alkaloid biosynthesis by *Penicillium cyclopium*; and rhodotorulic acid synthetase (Anke and Dickmann, 1972), which catalyzes the biosynthesis of the fungal sideramine, rhodotorulic acid in the yeast *Rhodotorula glutinis*. In all of the above-mentioned cases, the mechanisms responsible for the *in vivo* loss of enzyme activity remains unclear.

The only instances of even a partial understanding of inactivation of antibiotic synthetases *in vivo* have come from studies on biosynthesis of patulin and gramicidin S. Studies on patulin biosynthesis have been published in a series of papers by Gaucher and co-workers (Grootwassink and Gaucher, 1980; Gaucher et al., 1981; Neway and Gaucher, 1981; Scott et al., 1984, 1986). The synthesis of this polyketide antibiotic and mycotoxin by *Penicillium urticae* has been used by this Canadian group as a comprehensive model of secondary metabolism in order to answer fundamental questions concerning regulation, including the dynamics and control of appearance and disappearance of pathway-specific enzymes. Grootwassink and Gaucher (1980) achieved a clear temporal separation of growing cells devoid of secondary metabolism-specific enzymes from nongrowing cells that rapidly produce these enzymes. These workers were able to show the sequential induction of the first (6-methylsalicyclic acid synthetase) and the fourth (*m*-hydroxybenzyl alcohol dehydrogenase) enzymes at approximately 19 and 22 h, respectively, after inoculation. The appearance of the fourth enzyme was determined by the nutritional status of the culture (nitrogen source level) rather than its age. The appearance of both enzymes was arrested by protein synthesis inhibitors and therefore, the operation of the patulin pathway was not a result of activation of preformed proteins. These studies were extended to include the dynamics of appearance and disappearance of the seventh enzyme (iso-epoxydon dehydrogenase), which, like the first and the fourth enzymes, was transient.

An examination of the half-lives of the first, fourth, and seventh enzyme of the pathway revealed that the earliest pathway enzymes were also the most labile (half-lives of 7, 17, and 19 h, respectively). It was ascertained by Neway and Gaucher (1981) that, after the removal of all extrinsic limitations to continued production of patulin, the only remaining intrinsic limitation was the short lifetime of the first enzyme of the pathway.

Addition of 6-MSA to fermentor cultures of *P. urticae* brought about the partial re-tardation of the *in vivo* inactivation of the fourth and seventh but not the first enzyme, whereas the most dramatic increase in the half-life of overall patulin production (by about threefold) was achieved when cells of *P. urticae* were immobilized in carrageenan beads. No mechanistic explanation was given for the increased longevity of the idiophase upon cell immobilization.

The correlation between *in vivo* stability and enzyme position in the pathway seems to apply to many secondary metabolic pathways. The first committed step in the biosynthesis of many antibiotics is often catalyzed by a large, multifunctional enzyme of rather limited stability. For example, the first unique step in the biosynthesis of chloramphenicol in *Streptomyces* sp. is catalyzed by arylamine synthetase, an enzyme of approximately 248,000 daltons, whose *in vivo* half-life is between 7 and 13 h (Jones and Westlake, 1974). In contrast, a later enzyme, *p*-aminophenylalanine aminotransferase has a half-life of about 90 h. The large synthetases which initiate many antibiotic biosyntheses may be more susceptible to degradation because of their size (Dice and Goldberg, 1975). Such degradation may be preceded or initiated by an oxygen-dependent or product-dependent destabilization that may be operative in cases such as the *in vivo* inactivation of the large multienzyme complex of gramicidin S synthetase (see below).

3. In Vivo Inactivation of Enzymes of Gramicidin S Biosynthesis

3.1 Introduction

Studies of *in vivo* inactivation of antibiotic synthetases have used as a model system the production of the cyclic decapeptide antibiotic, gramicidin S (GS), by *Bacillus brevis*. The structure of this antibiotic is shown in Fig. 1. This system was chosen in view of (1) the large amount of literature available on GS biosynthesis, (2) the existence of only two synthetases (a "light" enzyme and a "heavy" enzyme) catalyzing the formation of the antibiotic by functioning together in a polyenzyme complex collectively known as GS synthetase, and (3) the advantage of working with a rapidly growing prokaryote that produces a single antibiotic.

Our present understanding of the enzymatic mechanism of GS synthesis is that the antibiotic is formed in a nonribosomal manner by (1) sequential activation of the constitu-ent amino acids on specific functional domains of the synthetase complex, (2) initiation of dipeptide formation, (3) peptide elongation, and (4) cyclization (Kleinkauf, 1979; Kura-hashi, 1981; Zimmer et al., 1979). The transient appearance and disappearance of the GS

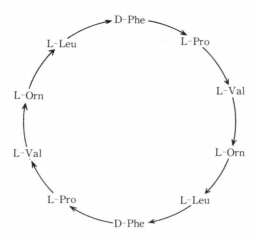

Fig. 1. Structure of gramicidin S.

synthetase complex *in vivo* has been well established. The first specific efforts to understand and control the *in vivo* inactivation of GS synthetase were undertaken by Friebel and Demain (1977a, b). They found that in growing cells of *B. brevis*, the disappearance of the synthetase, which occurs in the mid- or late-log growth phase (half-life: 0.5–1 h), is oxygen dependent and can be arrested by nitrogen sparging. In similar experiments on frozen-thawed, nongrowing cells incubated in buffer, they found that agitation and aeration led to a similar synthetase inactivation as had been observed in growing cells. The exclusion of air in this resting cell system prevented the loss of activity. It was also found that the reducing agent, dithiothreitol (DTT), was effective in partially preserving the activity of the synthetase when added to whole cell suspensions. However, protease and protein synthesis inhibitors were incapable of preventing or retarding inactivation. These initial observations suggested the likely involvement of molecular oxygen in a mechanism of direct oxidative inactivation of the synthetase's sensitive thiol groups.

3.2   *Redox Modulation of GS Synthetase Stability In Vivo; The Carbon Source Effect*

A more comprehensive analysis of O_2-dependent inactivation of GS synthetases was performed by us after confirming the above findings (Demain et al., 1981; Agathos, 1983; Agathos and Demain, 1983, 1984; Demain and Agathos, 1986). The inactivation system used frozen-thawed cells of *B. brevis* exhibiting inactivation in short-term experiments simulating the aeration and agitation conditions of a batch fermentation while circumventing the fermentation itself. We found that inhibitors of energy metabolism and of protein synthesis did not protect the synthetase against inactivation; hence the process does not depend on energy-yielding metabolism or on *de novo* protein synthesis. Organic thiols including DTT, L-cysteine, and reduced glutathione added to partially anaerobic cell suspensions provided long-term retardation of the inactivation, while in short-term incubations of previously aerated cells they partially restored activity. When total intracellular titratable SH groups were assayed during shaking, loss of synthetase activity was accompanied by a parallel drop in total SH concentration. These results indicated that oxidation of enzyme SH groups is a mechanism of *in vivo* inactivation of the synthetase.

The catalytic involvement of SH groups in carrying out the 20 or 30 reactions of GS biosynthesis has been indicated previously (Lipmann, 1973; Laland and Zimmer, 1973). The direct demonstration of SH groups was reported recently in both the light and heavy enzymes of the synthetase complex (Kanda et al., 1981, 1984; Schlumbohm et al., 1985). These recent studies have proven that accessible SH groups, six on the light enzyme and 24 on the heavy one, are essential in several distinct catalytic functions of GS synthetase.

The above-mentioned beneficial effect of organic thiols (either retardation or partial reversal of inactivation) upon the *in vivo* stability of the synthetase can be explained by a thiol-disulfide exchange (Torchinsky, 1981) that tends to protect accessible SH groups on the enzyme from autooxidation and maintain them in the reduced, and hence catalytically active, form. Such a mechanism of partial protection of GS synthetase against oxidative inactivation could be operative under physiological conditions intracellularly. Our further investigations provided good evidence for this possibility.

Retardation of inactivation was obtained upon addition of utilizable carbon sources (glycerol, fructose, inositol) to aerated cell suspensions. The magnitude of the stabilizing effect correlated with the degree of utilization of these carbon sources and their concentrations. The carbon source-mediated retardation of synthetase inactivation approximately doubled the half-life of the enzyme and the effect was abolished by the uncoupler dinitrophenol (Demain and Agathos, 1986). These results are compatible with a direct oxidative mechanism of enzyme inactivation. A mechanistic scheme proposed to explain the carbon source effect takes into account the fact that O_2-labile SH groups on the synthetase may be preserved by the catabolism of the carbon source (Demain and Agathos, 1986).

Catabolism under aerobic conditions would lead to a low intracellular redox potential conducive to maintaining intracellular cysteine residues in the SH (reduced) state. Under these conditions, reduced pyridine nucleotides (NADH, NADPH) generated by aerobic carbon source catabolism may lead to regeneration of enzyme SH groups via thiol interchange with intracellular thiols such as glutathione, dihydrolipoamide, thioredoxin, or glutaredoxin. Support for such a mechanism comes from experimental evidence that the time-course of the decrease in titratable intracellular SH concentrations (a measure of the intracellular redox state) in both carbon source-supplemented and -unsupplemented aerated cell suspensions was paralleled by corresponding drops in synthetase activity, while, at the same time, the SH concentration in the cells supplemented with a carbon source remained higher than in the unsupplemented controls (Agathos and Demain, 1984; Demain and Agathos, 1986). Thus, it appears that the oxygen-dependent inactivation of GS synthetase *in vivo* is modulated by the intracellular redox state of the cells.

Recent data from the literature include several examples of thiol-disulfide interchange reactions *in vivo*, which appear to be of general regulatory significance. For instance, such redox modulation of enzyme activity has been reported for glyceraldehyde-3-phosphate dehydrogenase and glucose-6-phosphate dehydrogenase in cyanobacteria (Udvardy et al., 1982, 1984) and glutathione reductase in *Escherichia coli* (Mata et al., 1984, 1985). The activation and inactivation of latent collagenase in human leukocytes is another example (Tschesche and Macartney, 1981). Gilbert (1982, 1984) suggested that many enzymes are controlled *in vivo* through the operation of exchange reactions between regulatory SH groups on the enzyme and widely available intracellular disulphides such as oxidized glutathione, oxidized coenzyme A, and mixed disulfides between the two. The same author suggested that the modulation of the thiol-disulfide ratio *in vivo* regulates key enzymes of glycolysis-gluconeogenesis.

In view of our own results, it is probable that enzymes of secondary metabolism may be regulated in terms of both activity and stability by the intracellular redox state, as set by thiol/disulfide ratios fluctuating in response to the cultural/nutritional environment. Indirect evidence of *in vivo* regulation of aflatoxin biosynthesis by the redox state has been recently reported in *Aspergillus parasiticus* (Bhatnagar et al., 1986). Furthermore, recent data on the fourth enzyme of patulin biosynthesis suggests a possible link between its *in vivo* instability and its *in vitro* stabilization by DTT (Scott et al., 1986). In this regard it is of note that *in vitro* inactivation of the expandase enzyme of cephalosporin biosynthesis in *C. acremonium* can be prevented or reversed by DTT (Lübbe et al., 1985). A system bearing similarities to the GS synthetase system is the secondary metabolic process of cyanogenesis by *Pseudomonas* sp. (Castric, 1981; Castric et al., 1981). The enzymatic apparatus for HCN formation (cyanide synthase) is inactivated towards the end of exponential growth at the point of exhaustion of the carbon/energy source, L-glutamate. By increasing the initial concentration of glutamate, these workers were able to prolong cyanide production. The inactivation is oxygen dependent, is not slowed down by protein synthesis inhibitors, and can be prevented by anaerobiosis. Significant protection of this enzyme system was provided either by inclusion of a carbon source or by DTT under aerobic conditions, and the authors concluded that the inactivation is triggered by a state of C-source starvation accompanied by high cultural dissolved oxygen levels. It was also shown that progressively anaerobic conditions brought about a regeneration of cyanogenic activity that could be blocked by chloramphenicol; thus it was suggested that cyanogenesis is regulated by a combination of enzyme repression and inactivation.

3.3   *Substrate Amino Acid-Mediated Stabilization of GS Synthetase In Vivo*

We have also studied the influence of the amino acid substrates of GS synthetase on its *in vivo* stability (Agathos and Demain, 1984, 1986). In aerated suspensions of *B. brevis* cells, the addition of a mixture of the five constituent amino acids of GS in the L-form (phenylalanine, proline, leucine, ornithine, and valine) provided significant stabilization of the synthetase, increasing the half-life of the enzyme complex from a little over 1 h to approximately 9 h in the absence of growth or *de novo* protein synthesis. L-leucine and L-ornithine stabilized the synthetase synergistically almost as well as the five amino acids combined. The amino acid effect is compatible with substrate-mediated protection of SH groups on the active sites by thioesterified amino acid substrates (Otani et al., 1969; Kristensen et al., 1973) and is corroborated by partial protection of the synthetase *in vitro* by the presence of saturating concentrations of substrates (Kleinkauf and von Döhren, 1981). Ligand-induced *in vitro* enzyme stabilization against a variety of inactivating agents and actions such as denaturants, covalent effectors, and proteolysis is well established (Grisolia, 1964; Wiseman, 1978; Goldberg and Dice, 1974). Partial *in vitro* protection of the antibiotic synthetase from inactivation by phenylglyoxal can be provided by substrate ligands, including ATP and constituent amino acids (Kanda et al., 1982a, b). Also, protection of the heavy enzyme of GS synthetase against inactivation by N-ethylmaleimide, a thiol-specific reagent, has been achieved by incubation with the constituent amino acids (Schlumbohm et al., 1985). Our findings on the special character of L-leucine and L-ornithine as *in vivo* stabilizing agents of the synthetase have been explained (Agathos and

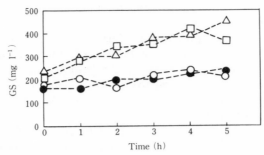

FIG. 2. Time-course of GS production in aerated cell suspensions supplemented with carbon source and/or amino acid substrates. ○, buffer only; ●, 80 g l⁻¹ fructose; △6 mM of the five constituent amino acids of GS; □, 6 mM of five amino acids + 80 g l⁻¹ of fructose.

Demain, 1986) on the basis of the particular importance that their corresponding activating units (enzyme functional domains) have for the structural integrity and the catalytic efficiency of the synthetase complex (Kleinkauf and Koischwitz, 1980; Kanda et al., 1982b; Schlumbohn et al., 1985).

3.4 GS Production Under Conditions of Partial Synthetase Stabilization

The early studies by Friebel and Demain (1977b) on growing cells of *B. brevis* showed that nitrogen sparging during the late exponential growth phase resulted in stabilization of GS synthetase but arrested antibiotic synthesis, presumably by blocking ATP synthesis. The cessation of GS production under anaerobiosis provided us with the impetus to investigate whether we could demonstrate net production of antibiotic under the conditions established above that are conducive to stabilization or retardation of synthetase inactivation *in vivo*. Addition of the utilizable carbon source, fructose, did not result in any antibiotic production (Fig. 2). However, it can be seen that the use of the five amino acid substrates of the synthetase (at 6 mM each) resulted in sustained production of GS throughout the length of the incubation.

An additional objective of our studies was to determine whether compounds found to retard *in vivo* inactivation of GS synthetase (i.e., carbon sources and the five constituent amino acids) were capable of prolonging the active life of the synthetase and improving GS production when added to actively growing (late exponential) cultures of *B. brevis* (Agathos and Demain, unpublished data).

Portions of whole broth taken from the main fermentation at four different times after the point of mid-exponential growth (Table 1) were supplemented with fructose or the five amino acids or both, and a series of four "satellite" fermentations in complex (YP)

TABLE 1. Time course of GS synthetase activity in the main fermentation.

Time (h)	Sample	Specific Activity (μg GS mg⁻¹h⁻¹)
3.5	I, early exponential growth phase	3.3
4.5	II, mid-exponential growth phase	1.5
5.5	III, later exponential growth phase	1.1
7.5	IV, early stationary growth phase	0.6

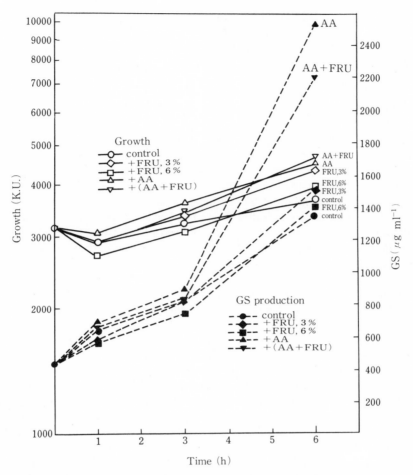

FIG. 3. Cell growth and GS production in satellite fermentation III (cells taken during the late exponential growth phase).

medium were carried out. For brevity, only one such satellite fermentation will be discussed in detail. The time profile of satellite fermentation III (cells taken from the initial fermentation harvested in late exponential phase) appears in Fig. 3. It is apparent that the culture supplemented with the five amino acids (or the amino acids plus fructose) exhibited a marked increase in GS production as compared to control (unsupplemented) cultures or cultures supplemented with fructose alone.

Table 2 shows GS production data, appropriately normalized per unit biomass and unit time, that reflect the GS biosynthetic capacity of the different stage cells under different conditions of supplementation. For comparing the different conditions within each satellite fermentation, we normalized the productivities with respect to that observed in each control (i.e., unsupplemented broth).

Inspection of the data presented in Table 2 shows that, in general, the culture supplemented with the five amino acids (or the five amino acids plus fructose) had up to 50%

TABLE 2. Relative stability of GS biosynthetic capacity.

| Additive | Growing Cells Satellite Fermentations | | | |
	I	II	III	IV
None	1.00	1.00	1.00	1.00
5 Amino Acids	1.10	1.26	1.50	1.48
5 Amino Acids + Fructose (3%)	0.76	1.70	1.27	1.49
Fructose (3%)	0.75	0.88	0.92	0.93
Fructose (6%)	0.60	0.75	0.94	0.95

higher biosynthetic capacity than the unsupplemented controls. On the other hand, fructose alone did not enhance the inherent biosynthetic capacity of the growing cells of *B. brevis*. Thus, both with cells incubated in buffer (Fig. 2) and cells incubated in complex YP medium (Fig. 3), the addition of the five amino acids allowed production of the antibiotic but incubation with fructose did not. We do not know why the carbon source-mediated retardation of GS synthetase inactivation does not result in the production of GS whereas the amino acid-mediated stabilization does. The only clue we have, at this point, is that the amino acids have a much greater effect than the carbon source, i.e., the amino acids increased the half-life of GS synthetase ninefold whereas fructose elicited only a doubling of half-life.

4. Final Comments

While a detailed understanding of the *in vivo* stability of GS synthetase has still not been attained, our studies on this system have shed some light on the process(es) of its *in vivo* inactivation and reveal some possible mechanistic bases for this apparently physiologically controlled phenomenon. Specifically, the intracellular redox state appears to play at least a partial role in the protection of the synthetase against oxygen-mediated inactivation. By adding utilizable carbon sources, the intracellular environment becomes more reduced and a retardation of inactivation ensues. This metabolic effect indicates that the physiological requirements for synthetase inactivation are reflected by the nutritional status of the culture *in vivo* at the time of synthetase disappearance. Specifically, it appears as if the inactivation requires aerobic starvation for carbon source and/or a deficiency in processes associated with aerobic carbon source catabolism, namely, generation of reducing power in the form of NADH and NADPH. This is in agreement with the generalizations provided by Switzer (1977), i.e., that *in vivo* inactivation most often requires a drastic metabolic shift in carbon or nitrogen utilization. In the case of the GS synthetase it seems that at the end of exponential growth the continued operation of this enzyme for continued production of the peptide antibiotic from amino acids would become wasteful of protein precursors and of ATP. Thus, it makes physiological/regulatory sense that the inactivation of the synthetase is necessary to spare metabolites and energy and may be brought about by signals of depletion in C/energy source and/or ATP pools and/or reduced pyridine nucleotide pools.

In addition, the natural substrates of the enzyme complex, i.e., the five amino acids, provide an even more dramatic (ninefold as measured by synthetase half-life) stabilizing effect, resulting in prolonged antibiotic production.

Some of these factors affecting *in vivo* stability of GS synthetase may reveal fundamental parameters of secondary metabolite regulation that could be extended to a number of other secondary metabolic processes. Indeed, they appear to suggest strategies for prolonged and, ultimately, continuous production systems for commercially important antibiotics.

5. Summary

Synthetases of secondary metabolism appear in many cases to be unstable *in vivo*, losing their activity in a matter of hours after reaching their peaks. As a result, production of secondary metabolites soon ceases. Thus, *in vivo* inactivation of such enzymes appears to be a limiting factor in the prolonged production of antibiotics and other secondary metabolites. Study of this phenomenon has been meager compared to the effort devoted to the formation and catalytic activities of these synthetases.

Work on patulin by the Gaucher's group has shown the half-life of the first enzyme, 6-methylsalicylic acid synthetase, to be the most crucial factor in the production of this antibiotic-mycotoxin.

The oxygen-dependent *in vivo* inactivation of the enzyme complex gramicidin S (GS) synthetase was investigated in *Bacillus brevis* ATCC 9999, the producer of the peptide antibiotic gramicidin S. The inactivation, which had been previously found to be preventable by anaerobiosis, was further characterized biochemically and specific additives and conditions were found leading to retardation of *in vivo* inactivation.

Inhibitors of energy metabolism and of protein synthesis added to aerated cell suspensions in buffer did not provide any protection against inactivation, thus indicating that the process does not depend on energy-yielding metabolism nor on *de novo* protein synthesis.

Organic thiols added to anaerobic long-term incubations of *B. brevis* cells provided retardation of synthetase inactivation by several hours, whereas in short-term incubations of previously air-exposed cells they resulted in partial restoration of activity. The *in vivo* inactivation of the enzyme was found to be accompanied by a parallel drop in intracellular thiols. These results implicate enzyme SH autooxidation as a mechanism of *in vivo* inactivation.

Retardation of inactivation was achieved upon addition of utilizable carbon sources (glycerol, fructose, inositol) to aerated cell suspensions in buffer, the degree of stabilization being proportional to the ease of uptake and to the concentration of C-source. The magnitude of the carbon source effect was a doubling of synthetase half-life. Carbon source-supplemented cell suspenions incubated under air had a decreased intracellular redox state, as revealed by intracellular SH concentrations.

The five substrate amino acids of GS synthetase protected the enzyme against O_2 inactivation *in vivo*, the effect not being due to cell growth or protein synthesis. The amino acid addition increased GS synthetase half-life ninefold. Omission of L-Leu or L-Orn from the five amino acids resulted in diminished protection of the synthetase, whereas the two amino acids used together were capable of stabilizing almost as well as the combination of the five amino acids.

Experiments were also carried out with growing *B. brevis* cells, which were supplemented with the five amino acids and/or fructose at various points in a batch cultivation cycle. The biosynthetic capacity of these cells increased significantly in the case of exogenously added amino acids. This behavior is compatible with the data from nongrowing cells and supports the notion of a substrate-mediated stabilization of GS synthetase *in vivo*.

Acknowledgments

We wish to express our appreciation to AiQi Fang for competent technical assistance in some of the studies done at M.I.T. The Greek Ministry of National Economy is ackowledged for a North Atlantic Treaty Organization (NATO) Science Fellowship to SNA. We thank Alexander M. Klibanov and George M. Whitesides for useful discussions and encouragement.

REFERENCES

1. Aboutabl, E.A., Luckner, M., "Cyclopeptine dehydrogenase in *Penicilium cyclopium.*" *Phytochemistry*: **14**, 2573–2577 (1975).
2. Abraham, E.P., Newton, G.G.F., Warren, S.C., "Problems relating to the biosynthesis of peptide antibiotics." In *Biogenesis of antibiotic substances*. Vanek, Z., Hostalek, Z. (eds.), p. 169–194. Academic Press, New York, 1965.
3. Agathos, S.N., "Studies on the *in vivo* inactivation of Gramicidin S synthetase." Ph.D. Thesis, Massachusetts Institute of Technology, Cambridge, Mass. 286 pp. (1983).
4. Agathos, S.N., Demain, A.L., "Optimization of fermentation processes through the control of In Vivo inactivation of microbial biosynthetic enzymes." *ACS Symp. Ser.*: **207**, 53–67 (1983).
5. Agathos, S N., Demain, A.L., "Gramicidin S synthetase stabilization *in vivo.*" *Ann. N.Y. Acad. Sci.*: **434**, 44–47 (1984).
6. Agathos, S.N., Demain, A.L., "Substrate amino acid-mediated stabilization of gramicidin S systhetase activity against inactivation *in vivo.*" *Enzyme Microb. Technol.*: **8**, 465–468 (1986).
7. Aharonowitz, Y., Mendelovitz, S., Ben-Artzi, H., Magal, N.M., "Implication of lysine metabolism on regulation of cephamycin C biosynthesis in *Streptomyces clavuligerus.*" In *Regulation of Secondary Metabolite Formation*. Kleinkauf, H., von Döhren, H., Dornauer, H., Nesemann, G. (eds.), p. 89–101. VCH Publishers, Weinheim, 1986.
8. Anke, T., Diekmann, H., "Biosynthesis of sideramines in fungi. Rhodotorulic acid synthetase from extracts of *Rhodotorula glutinis.*" *FEBS Lett.*, **27**, 259–262 (1972).
9. Behal, V., "Regulation of tetracycline biosynthesis." In *Regulation of Secondary Metabolite Formation*. Kleinkauf, H., von Döhren, H., Dornauer, H., Nesemann, G. (eds.), p. 265–281. VCH Publishers, Weinheim/Deerfield Beach, 1986.
10. Bhatnagar, R.K., Ahmad, S., Mukerji, K.G., Subramanian, T.A.V., "Pyridine nucleotides and redox state regulation of aflatoxin biosyntheses in *Aspergillus parasiticus* NRRL 3240." *J. Appl. Bacteriol.*: **60**, 135–141 (1986).
11. Braña, A.F., Wolfe, S., Demain, A.L., "Ammonium repression of cephalosporin production by *Streptomyces clavuligerus.*" *Can. J. Microbiol.*: **31**, 736–743 (1985).
12. Bu'Lock, J.D., "The regulation of secondary biosynthesis." In *Essays in Biosynthesis and Microbial Development*. p. 42–67. John Wiley and Sons, New York, 1967.
13. Calam, C.T., "Physiology of the overproduction of secondary metabolites." In *Overproduction of Microbial Metabolites*. Vanek, Z., Hostalek, Z. (eds.), p. 27–50. Butterworths, Boston, 1986.
14. Castric, P.A., "The metabolism of hydrogen cyanide by bacteria." In *Cyanide in Biology*. Vennelsand, B., Conn, E.J., Knowles, C.J., Wissing, F. (eds.), p. 233–262. Academic Press, New York, 1981.
15. Castric, P.A., Castric, K.F., Meganathan, R., "Factors influencing the termination of cyanogenesis in *Pseudomonas aeruginosa.*" In *Cyanide in Biology*. Vennesland, B., Conn, E.J., Knowles, C.J., Wissing, F. (eds.), p. 263–274. Academic Press, New York, 1981.
16. Cockburn, M.A., Hodgson, B., Walker, J.S., "Properties of an unfractionated tyrocidine synthesizing system from *Bacillus brevis* ATCC 10068." *Microbios*: **8**, 215–239 (1973).
17. Cortés, J., Liras, P., Castro, J.M., Martin, J.F., "Glucose regulation of cephamycin biosynthesis in

Streptomyces lactamdurans is exerted on the formation of α-aminoadiply-cysteinyl-valine and deacetoxycephalosporin C synthase." *J. Gen. Microbiol.*: **132**, 1805–1814 (1986).

18. Demain, A.L., "Mutation and the production of secondary metabolites." *Adv. Appl. Microbiol.*: **16**, 177–202 (1973).
19. Demain, A.L., "Industrial microbiology." *Science*: **214**, 987–995 (1981).
20. Demain, A.L., Poirier, A., Agathos, S., Nimi, O., "Appearance and disappearance of gramicidin S synthetases." *Devel. Ind. Microbiol.*: **22**, 233–244 (1981).
21. Demain, A.L., Aharonowitz, Y., Martin, J.F., "Metabolic control of secondary biosynthetic pathways." In *Biochemistry and genetic regulation of commercially important antibiotics*. Vining, L.C. (ed.), p. 49–72. Addison-Wesley, Reading, Mass., 1983.
22. Demain, A.L., Agathos, S.N., "Studies on *in vivo* inactivation of gramicidin S synthetase and its retardation." *Can. J. Microbiol.*: **32**, 208–214 (1986).
23. DeTilly, G., Mou, D.G., Cooney, C.L., "Optimization and economics of antibiotics production." In *The Filamentous Fungi*, Vol. IV (*Fungal Technology*). Smith, J.E., Berry, D.R., Kristiansen, B. (eds.), p. 190–209. Edward Arnold, London, 1983.
24. Dice, J.F., Goldberg, A.L., "A statistical analysis of the relationship between degradative rates and molecular weights of proteins." *Arch. Biochem. Biophys.*: **170**, 231–219 (1975).
25. Elander, R.P., Vournakis, J.N., "Genetic aspects of overproduction of antibiotics and other secondary metabolites." In *Overproduction of Microbial Metabolites*. Vanek, Z., Hostalek, Z. (eds.), p. 63–79. Butterworth, Boston, 1986.
26. Erge, D., Maier, W., Gröger, D., "Untersuchungen uber die enzymatische Umwandlung von Chanoclavin I." *Biochem. Physiol. Pflanzen.*: **164**, 234–247 (1973).
27. Friebel, T.E., Demain, A.L., "Oxygen-dependent inactivation of gramicidin S synthetase in *Bacillus brevis*." *J. Bacteriol.*: **130**, 1010–1016 (1977a).
28. Friebel, T.E., Demain, A.L., "Stabilization by nitrogen of the gramicidin S synthetase complex during fermentation." *FEMS Microbiol. Lett.*: **1**, 215–218 (1977b).
29. Frøyshov, Ø., "The production of bacitracin synthetase by *Bacillus licheniformis* ATCC 10716." *FEBS Lett.*: **81**, 315–318 (1977).
30. Fujikawa, K., Suzuki, T., Kurahashi, K., "Biosynthesis of tyrocidine by a cell-free enzyme system of *Bacillus brevis* ATCC 8185. I. Preparation of partially purified enzyme system and its properties." *Biochim. Biophys. Acta*: **161**, 232–246 (1968).
31. Gaucher, G.M., Lam, K.S., Grootwassink, J.W.D., Neway, J., Deo, Y.M., "The initiation and longevity of patulin biosynthesis." *Devel. Ind. Microbiol.*: **22**, 219–232 (1981).
32. Gilbert, H.F., "Biological disulfides: The third messenger ?" *J. Biol. Chem.*: **257**, 12086–12091 (1982).
33. Gilbert, H.F., "Redox control of enzyme activities by thiol/disulfide exchange." *Methods Enzymol.*: **107**, 330–351 (1984).
34. Goldberg, A.L., Dice, J.F., "Intracellular protein degradation in mammalian and bacterial cells." *Annual Rev. Biochem.*: **43**, 835–869 (1974).
35. Grisolia, S., "The catalytic environment and its biological implications." *Physiol. Rev.*: **44**, 657–712 (1964).
36. Grootwassink, J.W.D., Gaucher, G.M., "De novo biosynthesis of secondary metabolism enzymes in homogeneous cultures of *Penicillium urticae*." *J. Bacteriol.*: **141**, 443–455 (1980).
37. Heim, J., Shen, Y.-Q., Wolfe, S., Demain, A.L., "Regulation of isopenicillin N synthetase and deacetoxycephalosporin C synthetase by carbon source during the fermentation of *Cephalosporium acremonium*." *Appl. Microbiol. Biotechnol.*: **19**, 232–236 (1984).
38. Heinstein, P.F., Lee, S.-I., Floss, H.G., "Isolation of dimethylallylpyrophosphate: Tryptophan dimethylallyl transferase from the Ergot fungus (*Claviceps spec.*)." *Biochem. Biophys. Res. Commun.*: **44**, 1244–1251 (1971).
39. Holzer, H., Betz, H., Ebner, E., "Intracellular proteinases in microorganisms." *Curr. Top. Cell Reg.*: **9**, 103–156 (1975).
40. Holzer, H. (ed.)., *Metabolic Interconversion of Enzymes 1980*. Springer Verlag, Berlin, 1981.
41. Hopwood, D.A., Chater, K.F., "Fresh approaches to antibiotic production." *Phil. Trans. R. Soc. Lond.*: **B290**, 313–328 (1980).
42. Jaklitsch, W.M., Hampel, W., Röhr, M., Kubicek, C.P., Gamerich, G., "α-aminoadipate pool concen-

tration and penicillin biosynthesis in strains of *Penicillium chrysogenum.*" *Can. J. Microbiol.*: **32**, 473–480 (1986).

43. Jones, A., Westlake, D.W.S., "Regulation of chloramphenicol synthesis in *Streptomyces* sp. 3022*a*. Properties of arylamine synthetase, an enzyme involved in antibiotic biosynthesis." *Can. J. Microbiol.*: **20**, 1599–1611 (1974).

44. Kanda, M., Hori, K., Kurotsu, T., Miura, S., Yamada, Y., Saito, Y., "Sulfhydryl groups related to the catalytic activity of gramicidin S synthetase 1 of *Bacillus brevis.*" *J. Biochem.*: **90**, 765–771 (1981).

45. Kanda, M., Hori, K., Kurotsu, T., Yamada, Y., Miura, S., Saito, Y., "Essential arginine residue in gramicidin S synthetase 1 of *Bacillus brevis.*" *J. Biochem.*: **91**, 939–943 (1982a).

46. Kanda, M., Hori, K., Miura, S., Yamada, Y., Saito, Y., "A comparative study of essential arginine residues in gramicidin S synthetase 2 and isoleucyl tRNA synthetase." *J. Biochem.*: **92**, 1951–1957 (1982b).

47. Kanda, M., Hori, K., Kurotsu, T., Miura, S., Saito, Y., "A comparative study of sulfhydryl groups required for the catalytic activity of gramicidin S synthetase and isoleucyl tRNA synthetase." *J. Biochem.*: **96**, 701–711 (1984).

48. Kleinkauf, H., "Antibiotic polypeptides—biosynthesis on multifunctional protein templates." *Planta Medica*: **35**, 1–18 (1979).

49. Kleinkauf, H., Koischwitz, H., "Gramicidin S-synthetase." In *Multifunctional Proteins.* Bisswanger, H., Schmincke-Ott, E. (eds.), p. 217–233. John Wiley and Sons, New York, 1980.

50. Kleinkauf, H., Döhren, H.v., "Cell-free biosynthesis of peptide antibiotics." In *Adv. Biotechnol.* Vol. 3. Vezina, C., Singh, K. (eds.), p. 83–88. Pergamon Press, Toronto, 1981.

51. Kristensen, T., Gilhuus-Moe, C.C., Zimmer, T.-L., Laland, S.G., "The inhibitory effect of AMP on the activation reactions of the amino acids involved in gramicidin S biosynthesis." *Eur. J. Biochem.*: **34**, 548–550 (1973).

52. Küenzi, M.T., "Process design and control in antibiotic fermentations." In *Antibiotics and Other Secondary Metabolites.* Hutter, R., Leisinger, T., Nuesch, J., Wehril, W. (eds.), p. 39–56. Academic Press, New York, 1978.

53. Kurahashi, K., "Biosynthesis of peptide antibiotics." In *Antibiotics IV-Biosynthesis.* Corcoran, J.W. (ed.), p. 325–352. Springer-Verlag, Berlin/New York, 1981.

54. Kurth, R., Demain, A.L., "The impact of the new genetics on antibiotic production." In *Biotechnology of Industrial Antibiotics.* Vandamme, E.J. (ed.), p. 781–789. Marcel Dekker, New York, 1984.

55. Laland, S.G., Zimmer, T.-L., "The protein thiotemplate mechanism of synthesis for the peptide antibiotics produced by *Bacillus brevis.*" *Essays Biochem.*: **9**, 31–57 (1973).

56. Lipmann, F., "Nonribosomal polypeptide synthesis on polyenzyme templates." *Acc. Chem. Res.*: **6**, 361–367 (1973).

57. Lübbe, C., Wolfe, S., Demain, A.L., "Dithiothreitol reactivates desacetoxy-cephalosporin C synthetase after inactivation." *Enzyme Microb. Technol.*: **7**, 353–356 (1985).

58. Maier, W., Gröger, D., "Dimethylallyltryptophan-synthetase in *Claviceps.*" *Biochem. Physiol. Pflanzen*, **170**, 9–15 (1976).

59. Malik, V.S., "Genetics and biochemistry of secondary metabolism." *Adv. Appl. Microbiol.*: **28**, 27–115 (1982).

60. Martin, J.F., Demain, A.L., "Control of antibiotic biosynthesis." *Microbiol. Rev.*: **44**, 230–251 (1980).

61. Mata, A.M., Pinto, M.C., López-Barea, J., "Purification by affinity chromatography of glutathione reductase (EC 1.6.4.2) from *Escherichia coli* and characterization of such enzyme." *Z. Naturforsch.*: **39C**, 908–915 (1984).

62. Mata, A.M., Pinto, M.C., López-Barea, J., "Redox interconversion of glutathione reductase from *Escherichia coli.* A study with pure enzyme and cell-free extracts." *Mol. Cell Biochem.*: **67**, 65–76 (1985).

63. Matteo, C.C., Glade, M., Tanaka, A., Piret, J., Demain, A.L., "Microbiological studies on the formation of gramicidin S synthetases." *Biotechnol. Bioeng.*: **17**, 129–142 (1975).

64. Neway, J., Gaucher, G.M., "Intrinsic limitations on the continued production of the antibiotic patulin by *Penicillium urticae.*" *Can. J. Microbiol.*: **27**, 206–215 (1981).

65. Nimi, O., Ito, G., Ohata, Y., Funayama, S., Nomi, R., "Streptomycin-phosphorylating enzyme produced by *Streptomyces griseus.*" *Agr. Biol. Chem.*: **35**, 856–861 (1971).

66. Ortmann, R., Matern, U., Grisebach, H., Stadler, P., Sinnwell, V., Paulsen, H., "NADPH-dependent formation of thymidine diphosphodihydrostreptose from thymidine diphospho-D-glucose in a cell-free

system from *Streptomyces griseus* and its correlation with streptomycin biosynthesis." *Eur. J. Biochem*: **43**, 265–271 (1974).

67. Otani, S., Yamanoi, T., Saito, Y., "Biosynthesis of gramicidin S; ornithine activating enzyme." *J. Biochem.*: **66**, 445–453 (1969).
68. Richter, I., Luckner, M., "Cyclopenin *M*-hydroxylase—an enzyme of alkaloid metabolism in *Penicillium cyclopium.*" *Phytochemistry*: **15**, 67–70 (1976).
69. Schlumbohm, W., Vater, J., Kleinkauf, H., "Reactive sulfhydryl groups involved in the aminoacyl adenylate activation reactions of the gramicidin S synthetase 2." *Biol. Chem. Hoppe-Seyler*, **366**, 925–930 (1985).
70. Scott, R.E., Jones, A., Gaucher, G.M., "A manganese requirement for patulin biosynthesis by cultures of *Penicillium urticae.*" *Biotechnol. Lett.*: **6**, 231–236 (1984).
71. Scott, R.E., Lam, K.S., Gaucher, G.M., "Stabilization and purification of the secondary metabolism specific enzyme, *m*-hydroxybenzylalcohol dehydrogenase." *Can. J. Microbiol.*: **32**, 167–175 (1986).
72. Šupek, V., Gamulin, S., Delić, V., "Enhancement of bacitracin biosynthesis by branched-chain amino acids in a regulatory mutant of *Bacillus licheniformis.*" *Folia Microbiol.*: **30**, 342–348 (1985).
73. Switzer, R.L., "The inactivation of microbial enzymes *in vivo.*" *Ann. Rev. Microbiol.*: **31**, 135–157 (1977).
74. Thurston, C.F., "Disappearing enzymes." *Process Biochem.*: **7**, 18–20, August (1972).
75. Tomino, S., Yamada, M., Itoh, H., Kurahashi, K., "Cell-free synthesis of gramicidin S." *Biochemistry*: **6**, 2552–2560 (1967).
76. Torchinsky, Yu.M., In *Sulphur in Proteins* (Transl. from the Russian). Pergamon Press, Oxford, New York, 1981.
77. Tschesche, H., Macartney, H.W., "A new principle of regulation of enzymic activity. Activation and regulation of human polymorphonuclear leukocyte collagenase via disulfide-thiol exchange as catalysed by the gluthathione cycle in a peroxidase-coupled reaction to glucose metabolism." *Eur. J. Biochem.*: **120**, 183–190 (1981).
78. Udvardy, J., Balogh, Á., Farkas, G.L., "Modulation of glyceraldehyde-3-phosphate dehydrogenase in *Anacystis nidulans* by glutathione." *Arch. Microbiol.*: **132**, 2–5 (1982).
79. Udvardy, J., Borbély, G., Juhász, A., Farkas, G.L., "Thioredoxins and the redox modulation of glucose-6-phosphate dehydrogenase in *Anabaena* sp. strain PCC 7120 vegetative cell and heterocysts." *J. Bacteriol.*: **157**, 681–683 (1984).
80. Vandamme, E.J., "Productie, Biosynthese en Werkingsmechanisme van *Bacillus* Oligopeptide Antibiotica." D.Sc. Thesis (in Flemish), Rijksuniversiteit Gent (Gent State University), Belgium, 1976.
81. Walker, J.B., Hnilica, V.S., "Developmental changes in arginine: X amidinotransferase activity in streptomycin-producing strains of Streptomyces." *Biochim. Biophys. Acta*, **89**, 473–482 (1964).
82. Weinberg, E.D., "Biosynthesis of secondary metabolites: roles of trace metals." *Adv. Microb. Physiol.*: **4**, 1–44 (1970).
83. Weinberg, E.D., "The concept of secondary metabolism." In *Microbiology 1981*. Schlesinger, D. (ed.), p. 356–359. American Society for Microbiology, Washington, D.C., 1981.
84. Wiseman, A., "Stabilisation of enzymes." In *Topics in Enzyme and Fermentation Biotechnology* 1978. Wiseman, A. (ed.), p. 280–303. Ellis Horwood, Chichester, 1978.
85. Zähner, H., Kurth, R., "Overproduction of microbial metabolites—the supply of precursors from the intermediary metabolism." In *Overproduction of Microbial Products*, Krumphanzl, V., Sikyta, B., Vanek, Z. (eds.), p. 167–179. Academic Press, New York, 1982.
86. Zimmer, T.-L., Froyshov, Ø, Laland, S.G., "Peptide antibiotics." In *Economic Microbiology*. Vol.3. Rose, A.H. (ed.), p. 123–150. Academic Press, New York, 1979.

III-2
From Petroleum to Muscone and Related Compounds

Jui-Shen Chiao

Shanghai Institute of Plant Physiology, Academia Sinica Shanghai, People's Republic of China

Thirty-five years ago, I first studied the science of fermentation at the University of Wisconsin, U.S.A., under the late Dr. W.H. Peterson, Dr. M.J. Johnson, and Dr. E. McCoy. After returning to China, I continued to work in this field. One of the microbiological projects in which I have been involved recently is petroleum fermentation for the production of long-chain dicarboxylic acids as the starting material for the synthesis of muscone. Efforts in this research focused on three areas: 1) developing a process for converting individual petroleum component to the corresponding saturated long-chain dicarboxylic acid (DC) by mutants of the polyploid *Candida tropicalis*; 2) studying the regulatory effect of urea, its structural analogs and several amino acids on alkane oxidation; and 3) extending microbial oxidation to the preparation of unsaturated long-chain dicarboxylic acid.

1. Muscone, a Traditional Chinese Medicine

Traditional Chinese medicines are widely used in China today, especially in the rural areas. The Compendium of *Materia Medica*, compiled by Lee Shi-zhen during the Ming Dynasty (1368–1644), listed 1,982 pharmaceutical ingredients, of which 461 were of animal origin. One of the most important of the animal-served ingredients is *she-xiang*, a product of the scent gland of the male musk deer (*Moschus moschiterus*).

In the traditional Chinese pharmacopoeia, *she-xiang* is a resuscitative, and is used in various patent medicines to treat diverse conditions including coronary heart disease (angina pectoris) and inflammation of the joints. The active ingredient in *she-xiang* is muscone (0.5–2%), for which the structure and synthesis were elucidated in the early 1920s by the famous Swiss organic chemist Leopold Ruzicka.

She-xiang was traditionally obtained by hunting the adult male musk deer and removing the entire scent gland. Gradually, the demand increased and musk deer farms were established to ensure a stable supply obtained by humane methods. However, natural *she-xiang* was still not available in sufficient quantities and many laboratories in China set to work to synthesize muscone. This synthesis requires the building up of a long-chain dicarboxylic acid with fifteen carbon atoms. Our laboratory decided to explore the possibility of producing muscone through microbial oxidation of petroleum.

164 Jui-Shen Chiao

2. Long-chain Dicarboxylic Acid Fermentation

Kester and Foster[1] were the first to observe that *Corynebacterium* sp. could oxidize *n*-alkanes to fatty dicarboxylic acids through diterminal oxidation. Although the recent literature on diterminal oxidation will not be reviewed here, our efforts in this area are summarized below.[2-5]

First, short-chain DC producers (*C. lipolytica* 197, *C. tropicalis* 79, 896, and 1336) were isolated from sludge and soil. Long-chain DC producers were screened by isolating mutants incapable of growing or growing only slightly on plates containing pentadecane and tridecane-1, 13-dicarboxylic acid as the only carbon sources.

Secondly, long-chain DC producers were induced to polyploid strains by camphor and colchicine treatments. The yields of DC from two polyploid strains, NPco22 from colchicine and NPco18 from camphor, were appreciably enhanced. It was also observed that these polyploid strains were much more sensitive to mutagens, probably due to their thin cell wall, and their yields of DC were further increased by treatment with NTG.

Third, the fermentation variables such as aeration, pH, carbon and nitrogen sources for growing the cells, effects of phosphate, phenobarbital, etc. were evaluated. Under optimal conditions, the yield of tridecane-1, 13-DC reached 72–75 g l⁻¹, and the efficiency of conversion was about 60%.

It was noted that the conversion of *n*-alkane to DC could be easily achieved by using resting cells, suspended in phosphate buffer, and a yield as high as 109 g l⁻¹ was obtained with pentadecane. The DCs produced by resting cells were of high purity (98%).

One feature of our mutant polyploid strains was their ability to convert *n*-alkanes from octane to octadecane to the corresponding DCs without pronounced preference for either even or odd carbon chains (Fig.1).

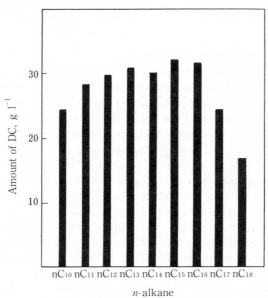

FIG. 1. Oxidation of *n*-alkanes to the corresponding dicarboxylic acids by polyploid mutant NPcoN22 of *C. tropicalis*.
Fermentation period: 144 h; alkane added: 6%; pH adjusted to 7.5 daily.

3. Regulatory Effect of Urea, Structual Analogs of Urea, Arginine, Alanine, and Aspartic Acid

During fermentation studies, it was found that as a nitrogen source the optimum level of urea for accumulation of DC was 0.1%, but at a level of 0.2% and above, the amount

TABLE 1. Effect of urea and ammonia on the production of DC-15.

	Addition time (h)	Nitrogen source added (%)	Nitrogen content (%)	Total nitrogen content (%)	DC-15 $A_{660nm} \times 10$ (144 h)
Urea	0	0.1	0.046	0.046	1.45
Urea	0	0.2	0.092	0.092	0.18
Urea	0 48	0.1 0.1	0.046 } 0.046 }	0.092	0.53
Urea Ammonia	0 48	0.1	0.046 } 0.046 }	0.092	1.20
Urea Ammonia	0 0	0.1	0.046 } 0.046 }	0.092	1.10
Ammonium tartarate	0	0.3	0.046	0.046	1.50
Ammonium tartarate	0	0.6	0.092	0.092	1.30
Ammonium sulfate	0	0.22	0.046	0.046	1.51
Ammonium sulfate	0	0.44	0.092	0.092	1.50

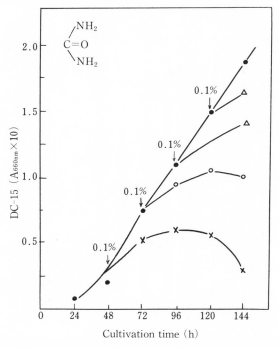

FIG. 2. Influence of supplementing urea at different fermentation periods on the production of long-chain dicarboxylic acids.

166   Jui-Shen Chiao

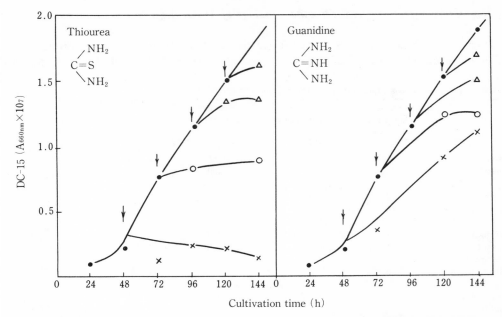

FIG. 3. Inhibitory effect of thiourea and guanidine on the biosynthesis of long-chain dicarboxylic acid. Arrows indicate the time at which thiourea or guanidine of equal molar concentration as 0.1% urea was added.

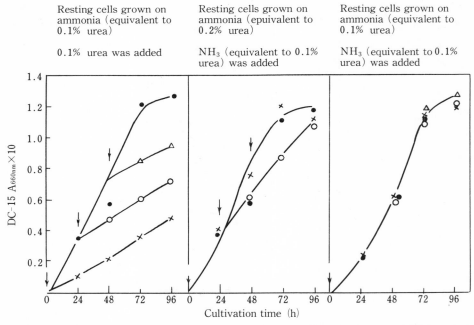

FIG. 4. Production of long-chain dicarboxylic acid with resting cells grown on appropriate amounts of ammonia (equivalent to 0.1% urea) and excessive amounts of ammonia (equivalent to 0.2% urea).

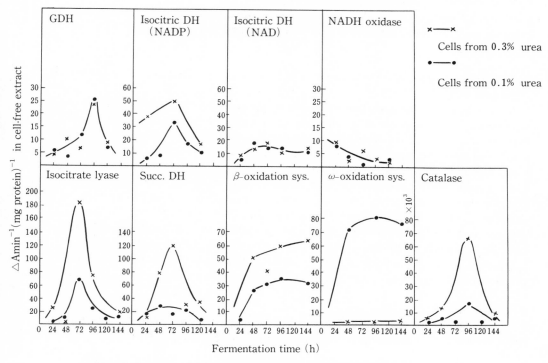

FIG. 5. Enzyme activities of cells cultivated with different amounts of urea.

of DC was greatly reduced. The replacement of urea with ammonium salts containing an equivalent amount of nitrogen did not affect the yield of DC appreciably (Table 1). Apparently the effect of excess urea is not due to its nitrogen content, and possibly urea serves as a modulator in the degradation of DC. The fact that thiourea and guanidine show similar inhibitory effects as urea (Figs. 2–4) supports this hypothesis. Further studies showed that urea and its structural analogs have both inhibitory and repressive effects on the accumulation of DC. In addition, excessive urea enhanced the specific activities of the main enzymes of the TCA cycle, glyoxylate pathway, catalase, and especially the β-oxidation system, but depressed that of the ω-oxidation and cytochrome P-450. (Fig. 5)

It is possible that our mutant polyploid yeast strains are partially blocked in the β-oxidation, TCA, and glyoxylate pathways, and that urea and its analogs function to switch on some of the partially blocked enzymatic steps along the β-oxidation, TCA, and glyoxylate pathways, and switch off others, such as ω-oxidation. These considerations are presented in a scheme (Fig. 6) that may serve as a working model for in-depth studies.[8-10]

Recently, Zhou[9] extended the regulation of DC production through the findings that L-aspartic acid, L-alanine, and L-arginine acted like urea, i.e., excessive amounts of these amino acids enhanced the degradation of DC. The most significant result was the accumulation of large amounts of L-arginine in the intracellular amino acid pool whenever excessive amounts of urea, alanine, or arginine were added. These observations brought arginine into the forefront of our search for the regulatory mechanism of alkane oxidation.

4. Extension of Microbial Oxidation to the Preparation of Unsaturated Long-chain Dicarboxylic Acids

In view of the strong ω-oxidation system of our polyploid mutants of *Candida tropicalis*, we extended their application to the preparation of dicarboxylic acids with double bond because of the difficulties inevitably encountered in chemical synthesis of these unsaturated acids. For example, an induced polyploid mutant was used for the co-oxidation of oleic acid. The oxidation product obtained was identified by gas chromotography, infrared spectroscopy, NMR, and reaction with ozone and the results agreed completely with those of the authentic sample of 1, 18-octadecen-9-dioic acid. The yield was 30 g l^{-1}, much higher than that (1.66 g l^{-1}) reported in a Japanese patent.[11]

Preliminary tests showed that the polyploid mutant of *C. tropicalis* could also oxidize both linoleic acid and linolenic acid to the corresponding dicarboxylic acids.

Obviously, the availability of the unsaturated dicarboxylic acids opens a route for the preparation of large alicylic compounds with one or more double bonds.

5. Conclusion

This essay summarizes our endeavors during the past several years. To our satisfaction, these results have been adopted for the production of muscone-T in large tonnage, and

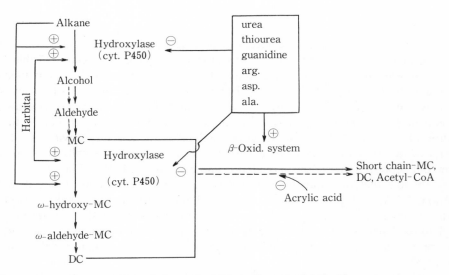

FIG. 6. A scheme for the regulation of alkane metablism in *C. tropicalis* NP$_{Co}$N$_{22}$.
$\oplus$ Induction or activation; $\ominus$ repression; $\cdots\cdots\rightarrow$ pathway for fatty alcohol; MC, monocarboxylic acid; DC, dicarboxylic acid.

cyclopentadecanone and muscone in small quantities. Further utilization of these long-chain dicarboxylic acids, especially the unsaturated dicarboxylic acid, is in sight. But our understanding of the regulatory aspect is still preliminary and I hope we may keep on pursuing this interesting problem in order to get a much clearer picture.

Acknowledgments

The laboratory work was carried out by my colleagues Dr. Shen Yong-qiang's group, Ms. Bao Huizhong and Mr. Wang Hao. I would like to express my appreciation for their cooperation all these years.

REFERENCES

1. Kester, A.S., Foster, J.W., "Diterminal Oxidation of Long-Chain Alkanes by Bacteria." *J. Bacteriol.*: **85**, 859–867 (1963).
2. Shen, Y,-Q., Lou, C.-J., Yao, P.-H., Xia, J., Xia, K.-R., Jiao, R.-S. (Chiao Jui-Shen), "Studies on Microbial Production of Long-Chain Dicarboxylic Acids from n-Alkanes (I): Screening of Polyploid Strains of *Candida tropicalis* capable of Producing High Yield of Long-Chain Dicarboxylic Acids from n-Alkanes." *Acta Phytophysiol. Sinica*: **5**, 161–170 (1979).
3. Shen, Y.-Q., Lou, C.-J., Yao, P.-H., Xia, J., Xia, K.-R., Jiao, R.-S. (Chiao Jui-Shen), "Studies on Microbial Production of Long-Chain Dicarboxylic Acids from n-Alkanes (II): Breeding of Polyploid Strains of *Candida tropicalis*." *Acta Phytophysiol. Sinica*: **5**, 171–179 (1979).
4. Shen, Y.-Q., Lou, C.-J., Xu, K.-R., Xia, J., Jiao, R.-S. (Chiao Jui-Shen), "Studies on Microbial Production of Long-Chain Dicarboxylic Acids from n-Alkanes (III): Some Factors Affecting the High Yield of Tridecane-1, 13-Dicarboxylic Acids with a Polyploid Mutant of *Candida tropicalis*." *Acta Phytophysiol. Sinica*: **5**, 385–393 (1979).
5. Shen, Y.-Q., Xia, G.-X., Lou, C.-J., Xu, K.-R., Jiao, R.-S. (Chiao Jui-Shen), "Studies on Microbial Production of Long-Chain Dicarboxylic Acids from n-Alkanes (IV): Production of Long-Chain Dicarboxylic Acids by Conversion with Resting *Candida* Yeast Cells." *Acta Phytophysiol. Sinica*: **6**, 29–35 (1980).
6. Chiao, J.-S., Shen, Y.-Q., Lou, C.-J., Xia, J., Xia, G.-X., "Regulatory Effect of Urea on the Diterminal Oxidation of n-Alkanes with *Candida tropicalis* to Produce Dicarboxylic Acids (I): Inhibition and Repression." *Acta Phytophysiol. Sinica*: **7**, 175–183 (1981).
7. Lou, C.-J., Xia, G.-X., Shen, X.-Q., Chiao, J.-S., "Regulatory Effect of Urea on the Diterminal Oxidation of n-Alkanes with *Candida tropicalis* to Produce Dicarboxylic Acids (VI): Increase and Decrease of Activities of Some Key Enzymes." *Acta Biochim. Biophys. Sinica*: **15**, 465–468 (1983).
8. Bao, H.-Z., Jiao, R.-S. (Chiao Jui-Shen), "Involvement of Cytochrome P-450 and Urea in the Alkane Metabolism by *Candida tropicalis*." *Acta Mycol. Sinica*: **3**, 45–53 (1984).
9. Zhou, J.-L., "Regulatory Effects of Amino Acids on the Production of Dicarboxylic Acid from n-Alkane by *Candida tropicalis* NPco22." *Thesis of the Institute of Plant Physiology, Academia Sinica*.
10. Chiao, J.-S., Xu, K.-R., Yao, P.-H., Xia, G.-X., Bai, Y.-T., Shen, Y.-Q., "Microbial Transformation of 1, 18-Octadecen-9-Dioic Acid from Oleic Acid by a Polyploid Mutant of *Candida tropicalis*." *Abstracts of VII International Biotechnology Symposium, New Delhi*, p. 333 (1984).
11. Morinaga, T., "Microbial production of saturated dicarboxylic acid." Jap. Pat. (A)57–65194 (1982).

Batch and Fed-Batch Culture Studies on a Hyperproducing Mutant of *E. coli* for L-Phenylalanine Production

N. H. Park and P. L. Rogers

Department of Biotechnology, University of N.S.W., Sydney, Australia

A detailed study has been carried out on the L-phenylalanine hyperproducing strain of *Escherichia coli* NST 74 to determine the factors likely to influence its growth kinetics. From batch culture data it was established that the increased initial glucose levels and increased L-phenylalanine product concentrations caused a decrease in the specific rates of growth and of product formation as well as reducing the yield (based on glucose utilized). Nutrient limitation of phosphate ions in batch culture resulted in increased L-phenylalanine yields, a result also supported by chemostat data. In fed-batch culture a relatively high L-phenylalanine concentration of 22.4 g l^{-1} was achieved at a productivity of 0.72 g l^{-1}h^{-1}. In comparison with other microorganisms available for the direct fermentation production of L-phenylalanine (*viz.*, *Corynebacterium glutamicum* and *Brevibacterium* spp.), hyperproducing mutants of *E. coli* offer the advantages of faster rates and simpler media.

1. Introduction

Commercial interest in "Aspartame" as a sweetening agent has stimulated research into the production of L-phenylalanine, one of the essential raw materials in its synthesis. Several possible methods exist for the production of L-phenylalanine, including direct fermentation using strains of *Corynebacterium glutamicum* or *Brevibacterium* spp. Other methods include the conversion of cinnamic acid or its derivatives to L-phenylalanine using an immobilized cell process (Klausner, 1985) and the chemical conversion of hydantoin into phenylpyruvic acid followed by an enzymatic conversion of phenylpyruvic acid to L-phenylalanine (Asai et al., 1959; Kitai et al., 1962). Although this hydantoin-to-phenylpyruvic acid-to-phenylalanine pathway may be the cheapest at present because of its direct route, high yields, and efficient conversion, the future cost of hydantoin will be an important factor in determining the total manufacturing cost.

Combinations of auxotrophy and metabolic regulation control have been investigated by many workers to improve L-phenylalanine producing strains. Nakayama et al. (1973) reported that tyrosine auxotrophs of *Corynebacterium glutamicum* resistant to various analogs of L-phenylalanine and tyrosine produced large amounts of L-phenylalanine. Hagino and Nakayama (1974) isolated a tyrosine auxotroph of *Corynebacterium glutamicum* resistant to both *p*-fluorophenylalanine, which excreted 9.5 g l^{-1} L-phenylalanine in molasses medium containing 10% sugar after 96 h of cultivation. Using a regulatory mutant of *Brevibacterium lactofermentum* requiring tyrosine and methionine for growth, Akashi et al. (1979) achieved improved L-phenylalanine production of 17.5 g l^{-1} L-phenylalanine after 62 h batch culture.

More recently, various auxotrophic regulatory mutant strains of *E. coli* have been reported. These include a tyrosine auxotroph resistant to β-thienylalanine (Park et al., 1984) and a tyrosine auxotroph resistant to both β-thienylalanine and *p*-fluorophenylalanine (Gil et al., 1985; Hwang et al., 1985).

Since Im et al. (1971) isolated mutant strains of *E. coli* K-12 in which the synthesis of 3-deoxy-D-arabino-heptulosonic acid 7-phosphate (DAHP) synthase phe was derepressed, systematic research has been carried out to identify specific regulatory mechanisms in the aromatic amino acid biosynthetic pathway and to improve the productivities of these amino acids by overcoming the limiting effects of these regulations (Dayan and Sprinson, 1971; Im and Pittard, 1973; Camakaris and Pittard, 1973; Tribe et al., 1976). With background information on the genetics of *E. coli*, several L-phenylalanine overproducing mutants have been constructed (Tsuchida and Sano, 1980; Choi and Tribe, 1982; Tsuchida and Sano, 1983).

From a survey of the patent literature on the direct production of L-phenylalanine by fermentation, the highest final concentration of L-phenylalanine reported so far is 24.8 g l^{-1} by a tyrosine and methionine requiring auxotroph of *Brevibacterium lactofermentum*, which is sensitive to decoynine and resistant to 5-methyltryptophan and *p*-fluorophenylalanine (Goto et al., 1981). Another patent (Tsuchida et al., 1975) describes the process in which a final concentration of 23.1 g l^{-1} phenylalanine was produced by a tyrosine requiring auxotroph, 5-methyltryptophan plus *p*-fluorophenylalanine-resistant mutant of *Brevibacterium lactofermentum*. Although the strains achieved relatively high concentrations of L-phenylalanine, such strains may not be suitable for commercial processes due to their requirements for relatively expensive and complex media. On the other hand, a process using *E. coli* offers the advantages of fast growth rates together with the use of defined media which are cheaper and may be more suitable for efficient product recovery.

2. Materials and Methods

2.1   Organism
A strain of *E. coli* K12 (NST 74) was used in the present study. Its genotype is *aro F* (feedback inhibition^{R}), *aro G* (feedback inhibition^{R}), *aro H*$^{-}$, *phe A* (feedback inhibition^{R}), *phe O* (constitutive), *tyr R* (constitutive) as described by Choi and Tribe (1982) and shown diagrammatically in Fig. 1. The metabolic controls of the strain have been attenuated to facilitate hyperproduction of L-phenylalanine.

2.2   Media Composition
The composition of media for the various experiments is given in Table 1.

2.3   Experimental Procedures
The batch and fed-batch culture experiments were carried out at pH 6.5, 37°C and under dissolved oxygen tension (DOT) maintained at 40–50% air saturation. The batch culture experiments were carried out in a laboratory fermentor, while for fed-batch cultures a Model MSc 10 fermentor was used with a maximum capacity of 7.5 l (MBR Bio Reactor

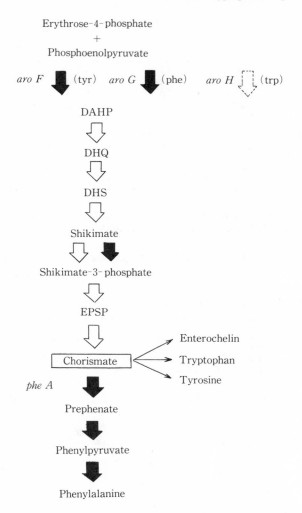

FIG. 1. An outline of reactions and controls involved in L-phenylalanine biosynthesis in *E. coli* K-12. The solid and empty arrows represent repressible and constitutive enzymes, respectively. The dotted arrow represents a deleted enzyme in strain NST 74.

AG, Switzerland). For the cell recycle studies, a hollow fiber concentration/dialysis system (Amicon Co., Model DC10) was used. The membrane used was an Amicon H10P5 unit with an effective surface area of 0.9 m². The nominal molecular weight "cut-off" of this ultrafiltration membrane was 5,000.

3. Batch Culture Studies

In an investigation of the batch culture kinetics of the hyperproducing strain *E. coli* NST 74, several factors were studied for their possible influence on L-phenylalanine production. These included: 1) the initial glucose concentration (substrate inhibition effect); 2) the L-phenylalanine concentration (product inhibition effect); and 3) limitation of growth

TABLE 1. Media compositions.

Component (g l⁻¹)	Batch culture			Fed-batch culture (Initial medium)
Glucose	25.0	65.0	107	20
NH₄Cl	3.03	8.03	9.63	7.0
K₂SO₄	0.1	0.3	0.6	2.3
K₂HPO₄	0.1	0.3	0.6	2.5
MgCl₂·6H₂O	0.1	0.3	0.6	2.6
Thiamine HCl	0.001	0.001	0.002	0.002
a) Iron citrate	0.5 ml	0.5 ml	1.0 ml	–
b) Micronutrients	1 ml	1 ml	2 ml	4 ml

a) Iron-citrate solution was prepared by dissolving 81 g $FeCl_3$ in 1 l of 0.5 M citrate buffer. For the fed-batch culture, the iron-citrate solution was replaced by 1.0 g l⁻¹ Na.citrate · $2H_2O$ and 0.14 g l⁻¹ $FeCl_3·6H_2O$.

b) The composition of the micronutrient solution as recommended by Neidhardt et al. (1974) was as follows in μM: $(NH_4)_6 Mo_7O_{24}$ 3; H_3BO_3 400; $CoCl_2$ 30; $CuSO_4$ 10; $MnCl_2$ 80 and $ZnSO_4$ 10.

by specific nutrients including carbon and energy sources. Each of these influences was examined in detail.

3.1   Effect of Initial Glucose Concentration

Several batch culture experiments were carried out at initial glucose concentrations of

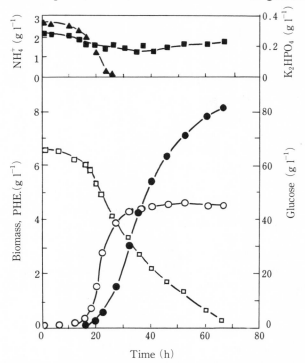

FIG. 2. Batch culture of *E. coli* NST 74 for L-phenylalanine production. The initial glucose concentration was 65 g l⁻¹. Symbols used for the various concentrations are: biomass (○), glucose (□), L-phenylalanine (●), NH₄⁺ (■), and K₂HPO₄ (▲).

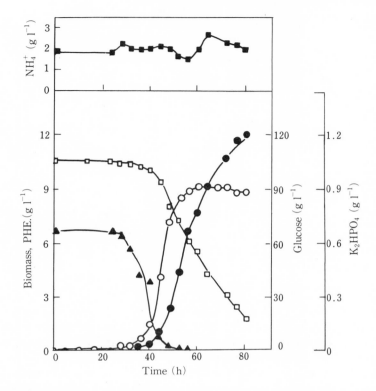

FIG. 3. Batch culture of *E. coli* NST 74 for L-phenylalanine production. The initial glucose concentration was 107 g l⁻¹. Symbols as in Fig. 2.

25 g l⁻¹, 65 g l⁻¹, and 107 g l⁻¹ in order to evaluate the effect of relatively high glucose levels on batch culture kinetics. Typical sets of data are shown in Figs. 2 and 3. Also shown are the phosphate (K_2HPO_4) and NH_4^+ concentrations in the medium, illustrating that the final cell concentration was controlled by phosphate limitation. Other authors, for example Choi and Tribe (1982), have suggested that maximal L-phenylalanine production occurs under phosphate limiting conditions.

A comparison of various kinetic parameters is given in Table 2, and from the results it is clear that increasing the glucose concentration brought about a significant decline in the specific rates of growth and L-phenylalanine production, as well as increasing the length of the lag-phase. It was also noted that the yield of L-phenylalanine ($Y_{p/s}$) was maximal at the lowest glucose concentrations, suggesting that increased glucose levels may repress some of the enzymes involved in the specific production of L-phenylalanine.

3.2 Effect of L-phenylalanine Concentration on Batch Culture Kinetics

A series of batch culture experiments was carried out to determine whether or not L-phenylalanine was likely to cause inhibition. In these experiments the biomass concentrations were maintained at 1.5–2.0 g l⁻¹ by using low PO_4^{3-} concentrations (KH_2PO_4 level of 0.1 g l⁻¹) and various concentrations of L-phenylalanine in the range 0–26 g l⁻¹ were added

TABLE 2. Comparison of kinetic parameters for various cell density batch cultures.

Parameter	Final cell density (g l^{-1})		
	1.5	4.5	9.0
Concentrations			
Phe (g l^{-1})	3.8	8.2	12.1
Glucose (g l^{-1})			
initial	25.0	65.0	107.0
residual	6.0	3.0	13.2
Fermentation Time (h)			
Culture	52.0	66.0	80.0
lag phase	0.5	15.0	36.0
Overall yields (g g^{-1})			
$Y_{p/s}$	0.2	0.13	0.13
$Y_{x/s}$	0.08	0.07	0.09
Specific rates (g g^{-1} h^{-1})			
μ	0.55	0.44	0.28
q_p	0.11	0.08	0.06
Overall productivity (g l^{-1} h^{-1})			
Q_v	0.073	0.124	0.151

to the media. In all cases to minimize the additional effect of L-phenylalanine formed during growth, the maximum concentration likely to be produced was kept below 2 g l^{-1} by controlling the final biomass concentration.

The results of kinetic analysis are shown in Table 3, and it is evident from this information that increasing the concentration of L-phenylalanine leads to a progressive decrease in the specific growth rate and product yield ($Y_{p/s}$). At an added L-phenylalanine concentration of 26 g l^{-1}, the specific rate of growth was reduced to 70% of its maximum value, and the product yield to 60%, indicating that product inhibition is important at L-phenylalanine levels likely to be achieved in fed-batch fermentation.

TABLE 3. Comparison of kinetic parameters for batch cultures on a medium containing various initial concentrations of phenylalanine.

Initial Phe	(g l^{-1})	0	8	13	20	26
Concentrations	(g l^{-1})					
Biomass Produced		1.6	1.8	2.0	1.5	1.8
Phe Consumed		2.7	2.2	1.8	1.0	1.2
glucose		13.5	13.1	12.5	7.7	10.3
Overall yields	(g g^{-1})					
$Y_{p/s}$		0.20	0.17	0.15	0.13	0.12
$Y_{x/s}$		0.12	0.14	0.16	0.20	0.17
Specific growth rate						
μ	(g g^{-1} h^{-1})	0.55	0.55	0.50	0.43	0.38
Overall productivity						
Q_v	(g l^{-1} h^{-1})	0.09	0.07	0.06	0.03	0.04

3.3 Effects of Various Macronutrient Limitations on the Hyperproduction of L-phenylalanine

Although it is well known that limitation of specific nutrients can influence cell morphology and macromolecular composition (Dean and Rogers, 1967; Tempest and Dicks, 1967; Koch, 1971; Tempest and Neijssel, 1978; Tempest and Wouters, 1981), there have been only a few studies on the effect of nutrient limitation on metabolite overproduction.

For the hyperproduction of L-phenylalanine by *E. coli*, Choi and Tribe (1982) reported that maximal product yields could be achieved under phosphate-limited conditions in continuous culture. The specific role of PO_4^{3-} in cell metabolism relating to L-phenylalanine production was unclear, although it is possible that a "metabolite overflow" may have occurred under excess carbon conditions (Hueting et al., 1979; Neijssel and Tempest, 1979).

In the present study the work of Choi and Tribe (1982) has been extended. Detailed kinetic studies were carried out in batch culture in which the carbon source (glucose), the

TABLE 4. Comparison of kinetic parameters for various nutrient-limited batch cultures.

Limiting nutrient	Glucose	NH_4^+	PO_4^{3-}
Max. concentrations (g l^{-1})			
Biomass	10.1	13.5	9.0
Phe	1.9	5.6	12.1
Yields on glucose (g g^{-1})			
$Y_{p/s}$	0.06	0.07	0.13
$Y_{x/s}$	0.36	0.15	0.09
Specific rates (g g^{-1} h^{-1})			
μ	0.56	0.35	0.28
q_p	0.06	0.06	0.06
Overall productivity			
Q_v (g l^{-1} h^{-1})	0.033	0.147	0.151

TABLE 5. Comparison of kinetic parameters for various nutrient-limited chemostats at $D = 0.1$ h.$^{-1}$

Limiting nutrient		K^+	Mg^{++}	SO_4^{2-}	PO_4^{3-}
Concentrations (g l^{-1})					
Biomass		7.8	8.0	8.15	8.4
Phe		2.6	0.6	2.17	4.8
Residual					
glucose		2.9	21.0	32.5	21.0
Yields	(g g^{-1})				
$Y_{p/x}$		0.33	0.08	0.27	0.57
[a] $Y_{p/s}$		0.06	0.01	0.05	0.11
[a] $Y_{x/s}$		0.19	0.18	0.19	0.19
Specific rates	(g g^{-1} h^{-1})				
q_s		0.54	0.55	0.52	0.52
q_p		0.03	0.01	0.03	0.06
Volumetric productivity					
Q_v	(g l^{-1} h^{-1})	0.26	0.06	0.22	0.48

[a] Product yield based on glucose consumed.

nitrogen source (NH_4^+), and the PO_4^{3-} concentration were each depleted in turn. Kinetic analyses of this data are summarized in Table 4 and it is clear from the kinetic parameters that the yields of L-phenylalanine (based on glucose) were appreciably higher under PO_4^{3-} limiting conditions. It was interesting to note also that L-phenylalanine production ceased once the carbon or nitrogen sources were depleted, which is not surprising since both carbon and nitrogen are required for its synthesis. However, L-phenylalanine production continued well into the stationary phase of growth following PO_4^{3-} depletion in the media (see also Figs. 2 and 3), indicating the possible use of an internal phosphate pool to facilitate further metabolism.

The enhanced L-phenylalanine yield when phosphate ions limit growth was confirmed in chemostat studies (Table 5). The continuous culture results have been presented much more fully in a recent publication (Park and Rogers, 1986).

4. Fed-Batch Culture

Fed-batch culture is an effective means of overcoming inhibition from high initial substrate concentrations, and many authors have reported the use of programmed nutrient feeding to increase yields and productivities of cells and metabolites (e.g., Pirt, 1975; Yu and Saddler, 1983; Yamane and Shimizu, 1984).

For L-phenylalanine production by *E. coli* NST 74 the following feeding profile was designed based on batch culture data and other published information:

(1) The glucose concentration should be maintained between 5–25 g l^{-1} to avoid any substrate limiting or inhibition effects. As shown in Table 2, increasing the glucose concentration from 25 g l^{-1} caused significant declines in the specific rates of growth and L-phenylalanine production.

(2) The ammonium concentration was designed to be between 1.5 and 3.0 g l^{-1}. Previous studies by Thompson et al. (1985) with *E. coli* reported evidence of growth inhibition above 3.6 g l^{-1} NH_4^+ levels, with complete inhibition occurring above 8.64 g l^{-1}. Kole et al. (1985) found that in excess of 0.1 g l^{-1} ammonium ions should be maintained to avoid substrate limitation.

(3) The concentration of phosphate ions was designed to be low during the L-phenylalanine producing phase of the culture with values of K_2HPO_4 typically below 0.2 g l^{-1}. From the batch and continuous culture data presented earlier (Tables 4 and 5), it was established that conditions of phosphate limitation favored an increased yield of L-phenylalanine.

(4) An excess dissolved oxygen concentration (40–50% air saturation) was maintained to stimulate the rapid aerobic growth of *E. coli* and to prevent loss of yield due to the formation of anaerobic by-products.

The results of two typical fed-batch cultures are shown in Figs. 4 and 5. In one case (Fig. 4), the medium initially contained 3.5 g l^{-1} K_2HPO_4. This was found to fall rapidly to zero after 16 h following cell growth to 25–30 g l^{-1} concentration. The level of phenylalanine continued to increase during the onset of the stationary phase of growth, despite the fact that the medium was depleted of phosphate ions. Under these conditions a phenylalanine concentration of 18.4 g l^{-1} was achieved in 40 h with a yield of 0.15 g g^{-1} based on glucose utilized.

In the other experiment (Fig. 5), a concentrated glucose solution and a mixture of mineral salts were added continuously to give a final cell concentration of approximately 50 g l^{-1}.

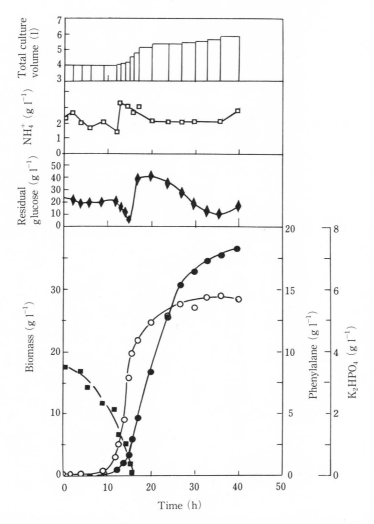

FIG. 4. Fed-batch culture of *E. coli* NST 74 for L-phenylalanine production. Glucose feeding (50% solution) was continuous, and the NH_4^+ concentration was maintained within the prescribed limits by pH control. See details in text. Symbols used for various concentrations are: biomass (○), glucose (◆), L-phenylalanine (●), NH_4 (□), and K_2HPO_4 (▲).

In both fed-batch cultures, the ammonium ion concentration in the culture medium was maintained close to the prescribed limits by the alternative addition of 9N NH_4OH or a mixture of 3.5 N NaOH and 1 N NH_4OH for pH control. During the exponential phase of growth, DOT was controlled to 40–50% air saturation either by sparging a mixture of pure oxygen (1 part) and air (1 part) at a maximum flow rate of 10 l min^{-1} (approximately 2–2.5 vvm), or by changing the agitation speed between 1,000 and 2,000 rpm. As shown in Fig. 5, phosphate-limited conditions prevailed after 25 h. The maximum concentrations of biomass and L-phenylalanine were 50 g l^{-1} and 22.4 g l^{-1}, respectively, and these concen-

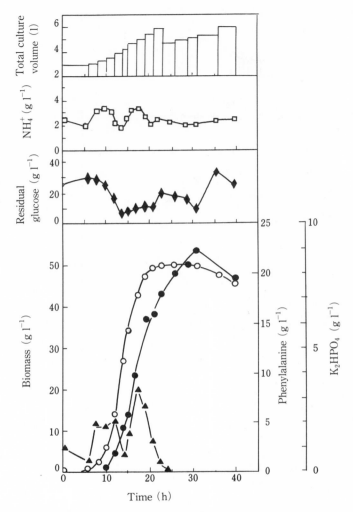

FIG. 5. Fed-batch culture of *E. coli* NST 74 for L-phenylalanine production. During the culture, continuous addition of a 50% glucose solution and a salts solution (15 ml h⁻¹) occurred. K_2HPO_4 solution was added at several points until phosphate limitation occurred at 24 h. Symbols as in Fig. 4.

trations decreased slightly at the end of the fermentation. The decrease in the concentrations was considered to be caused by dilution of the culture broth due to the continued addition of supplements. The L-phenylalanine yield was estimated at 0.11 g g⁻¹ and the overall productivity was 0.72 g l⁻¹ h⁻¹ for a 31 h fermentation. Acetic acid (ca. 2.4 g l⁻¹) was the only fatty acid detected in the fermentation broth.

5. Repeated Fed-Batch Culture

As a means of increasing overall productivity, repeated fed-batch fermentation was evaluated. Cells were retained in the culture vessel using a hollow fiber membrane cartridge and approximately 80% of the culture broth was removed as permeate via the membrane cartridge.

The results are shown in Fig. 6 and an analysis is provided in Table 6. Nutrients were also maintained within the limits prescribed for the optimum production of phenylalanine by *E. coli* in fed-batch culture, that is, the glucose concentration was kept between 5–25 g l^{-1} by adding a 50% glucose feed, the NH_4^+ concentration between 1.5–3.0 g l^{-1} by adding NH_4OH for pH control, and the phosphate concentration below 0.2 g l^{-1} for most of the fed-batch culture. Approximately 20 g l^{-1} of new biomass was produced for each cycle, a concentration corresponding to the amount of phosphate contained in the additional medium. The highest L-phenylalanine concentration obtained was 22.8 g l^{-1} in the second cycle at a maximum biomass concentration of 46.1 g l^{-1}.

Since the cell recycling system used in the present investigation was relatively large (approximately 1.0 l) compared to the culture volume (2.5 l), considerable quantities of concentrated cells were retained in the cell recycle system during ultrafiltration. This resulted in a drop in biomass concentration at the end of each cycle. This effect would be reduced significantly with a larger fermentor volume in proportion to the volume of the hollow fiber membrane module.

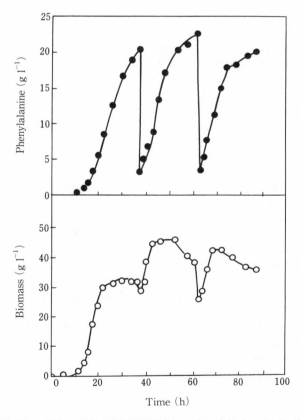

FIG. 6. Repeated fed-batch culture of *E. coli* NST 74 for L-phenylalanine production. Cell recycling was achieved using a hollow fiber membrane module.

TABLE 6. Repeated fed-batch culture with cell recycle using a hollow fiber membrane module.

Cycle No.	0	1	2
Biomass (g l^{-1})			
initial	0.07	28.8	26.1
final	32.0	46.1	42.5
Phenylalanine (g l^{-1})			
initial	0.0	3.4	3.6
final	20.5	22.8	20.1
Culture volume (l)			
initial	2.5	2.5	2.5
final	3.5	4.0	4.0
50 % Glucose (l)			
feed volume	0.91	1.40	1.32
Yield on glucose			
$Y_{p/s}$ (g g^{-1})	0.15	0.11	0.10
Overall Productivity			
Q_v (g l^{-1} h^{-1})	0.56	0.86	0.74

From the results shown in Table 6, it is evident that higher productivities at sustained high L-phenylalanine levels can be achieved by combining fed-batch culture with a cell recycle system. One problem that became evident, however, was the decline in yield of L-phenylalanine from 0.15 g g^{-1} in the first fed-batch culture to 0.10 g g^{-1} after a further two fed-batch cultures. Some loss of strain activity has occurred due to either the increased cell concentrations in the repeated batch cultures (with an increase in possible metabolic inhibitors) or to the effect of extended exposure of the cells to relatively high L-phenylalanine concentrations.

6. Discussion and Conclusions

Comparison of data on the production of L-phenylalanine by *E. coli* NST 74 with published information on other microorganisms (Table 7) illustrates the possible advantages of an *E. coli*-based fermentation.

From Table 7, it is evident that comparable concentrations of L-phenylalanine can be produced by *E. coli* at rates faster than for strains of *C. glutamicum* and *B. lactofermentum*. With *E. coli* NST 74 the yield values are comparable, if not better, at relatively low product concentrations. However, at higher concentrations a significant decline in yield occurs. This phenomenon has been reported by other investigators with high-density cultures of different *E. coli* strains (Shiloach and Bauer, 1975; Bauer and White, 1976; Landwall and Holme, 1977 a, b; Reiling et al., 1985). As the economic feasibility of a fermentation process such as L-phenylalanine production is very sensitive to raw material costs, and hence product yields, future attention needs to be given to identifying the reason(s) for the decline in yields with *E. coli* at high product concentrations.

In the present study, gas chromatographic analysis of the fermentation broth for volatile and non-volatile fatty acids revealed that acetic acid was the only fatty acid accumulated in measurable quantities. The highest concentration of acetic acid was 3.6 g l^{-1} for an experiment with a final cell density of 50 g l^{-1}. The accumulation of acetic acid at this

TABLE 7. Comparison of kinetic parameters from various studies on L-phenylalanine production.

Culture system	Organism	Final Phe conc. (g l^{-1})	Product[a] yield (g g^{-1})	Volumetric productivity (g l^{-1} h^{-1})	Reference
Batch	*E. coli*	12.1	0.13	0.15	This work
	E. coli	3.9	0.08	0.05	Tsuchida and Sano (1983).
	C. glutamicum	16.8	0.11	0.23	Nakayama et al. (1973).
	B. lactofermentum	23.1	0.18	0.32	Tsuchida (1975).
	B. lactofermentum	17.1	0.14	0.29	Akashi et al. (1979).
Continuous	*E. coli*	16.4	0.1	0.49	This work (Dilution rate: 0.03 h^{-1})
	E. coli	8.7	0.19	0.44	Choi and Tribe (1982). (Dilution rate: 0.05 h^{-1})
Fed-batch	*E. coli*	15.0	0.11	0.31	Hwang et al. (1985)
	E. coli	22.4	0.11	0.72	This work
	E. coli	22.8	0.11	0.86	This work[b]

[a] Product yield based on glucose consumed.
[b] Repeated fed-batch culture with cell recycle.

concentration would only cause a small reduction in yield, and is not likely to cause inhibition of cell growth. Landwall and Holme (1977 a, b) found that acetate alone was not growth inhibitory to *E. coli* at pH 7.0 and at a concentration of 20 g l^{-1}.

The possibility also exists that significant quantities of related amino acids could cause reduction in the yield at high L-phenylalanine concentrations. As shown in Fig. 1, tryptophan and tyrosine can be produced from the same metabolic pathway via chorismate as an intermediate. High performance liquid chromatography was carried out to quantify these other aromatic amino acids; however, the results were consistent with those found previously by Choi and Tribe (1982) who reported that the level of tyrosine was relatively minor (less than 0.5 g l^{-1}) and tryptophan production was barely at the level of detection.

Data from the present study also compares favorably with recent reports on L-phenylalanine production by an auxotrophic, regulatory mutant derived from the wild-type strain *E. coli* W 3110 (Park et al., 1984; Hwang et al., 1985). The study was extended also to a pilot-scale investigation using a 500 l fermentor and a strategy of intermittent feeding of D-glucose plus phosphate (Gil et al., 1985). Under these conditions, L-phenylalanine yields of 0.10–0.11 g g^{-1} glucose were achieved together with productivities of 0.35–0.40 g l^{-1} h^{-1} at L-phenylalanine titers of 14–15 g l^{-1}. These pilot-scale results, taken together with the improved yields and L-phenylalanine concentrations found in the present study, indicate that a direct fermentation process for L-phenylalanine may have some commercial potential. Further advantages of an *E. coli*-based process are the simpler media requirements (glucose-salts media) and the additional opportunities for genetic manipulation and strain improvement.

Acknowledgments

Financial support is acknowledged from Austgen-Biojet International Pty. Ltd. for part of this project. One of the authors (N.H.P.) gratefully acknowledges the support of his company (Cheil Sugar Company, Republic of Korea) during his postgraduate studies.

REFERENCES

1. Akashi, K., Shibai, H., Hirose, Y., "Effect of oxygen supply on L-phenylalanine, L-proline, L-glutamine and L-arginine fermentations." *J. Ferment. Technol.*: **57**, 321–327 (1979).
2. Asai, T., Aida, K., Oishi, K., "On the enzymatic preparation of L-phenylalanine." *J. Gen. Appl. Microbiol.*: **5**, 150–152 (1959).
3. Bauer, S., White, M.D., "Pilot scale exponential growth of *Escherichia coli* W to high cell concentration with temperature variation." *Biotechnol. Bioeng.*: **18**, 839–846 (1976).
4. Camakaris, H., Pittard, J., "Regulation of tyrosine and phenylalanine biosynthesis in *Escherichia coli* K-12: properties of the *tyrR* gene product." *J. Bacteriol.*: **115**, 1135–1144 (1973).
5. Choi, Y.J., Tribe, D.E., "Continuous production of phenylalanine using an *Escherichia coli* regulatory mutant." *Biotechnol. Letters*: **4**, 223–228 (1982).
6. Dayan, J., Sprinson, D.B., "Enzyme alterations in tyrosine and phenylalanine auxotrophs of *Salmonella typhimurium*." *J. Bacteriol.*: **108**, 1174–1180 (1971).
7. Dean, A.C.R., Rogers, P.L., "The cell size and macromolecular composition of *Aerobacter aerogenes* in various systems of continuous culture." *Biochim. Biophys. Acta*: **148**, 267–279 (1967).
8. Gil, G.H., Kim, S.R., Bae, J.C., Lee, J.H., "Pilot-scale production of L-phenylalanine from D-glucose." *Enzyme Microb. Technol.*: **7**, 370–732 (1985).
9. Goto, A., Ishihara, M., Sakurai, M., Enei, H., Takinami, K., "Method for producing L-phenylalanine by fermentation." Jap. Pat. 56–64793 (1981). (in Japanese)
10. Hagino, H., Nakayama, K., "L-phenylalanine production by analog-resistant mutants of *Corynebacterium glutamicum*." *Agr. Biol. Chem.*: **38**, 157–161 (1974).
11. Hueting, S., Lange, T., Tempest, D.W., "Energy requirement for maintenance of the transmembrane potassium gradient in *Klebsiella aerogenes* NCTC 418: a continuous culture study." *Arch. Microbiol.*: **123**, 183–188 (1979).
12. Hwang, S.O., Gil, G.H., Cho, Y.J., Kang, K.R., Lee, J.H., Bae, J.C., "The fermentation process for L-phenylalanine production using an auxotrophic regulatory mutant of *Escherichia coli*." *Appl. Microbiol. Biotechnol.*: **22**, 108–113 (1985).
13. Im, S.W.K., Davidson, H., Pittard, J., "Phenylalanine and tyrosine biosynthesis in *Escherichia coli* K-12: Mutants derepressed for 3-deoxy-D-arabinoheptulosonic acid 7-phosphate synthetase (phe), 3-deoxy-D-arabinoheptulosonic acid 7-phosphate synthetase (tyr), chorismate mutase T-prephenate dehydrogenase, and transaminase A." *J. Bacteriol.*: **108**, 400–409 (1971).
14. Im, S.W.K., Pittard, J., "Tyrosine and phenylalanine biosynthesis in *Escherichia coli* K-12: complementation between different *tyrR* alleles." *J. Bacteriol.*: **115**, 1145–1150 (1973).
15. Kitai, A., Kitamura, I., Miyachi, N., "The formation of L-amino acid (L-alanine and L-phenylalanine) through the conjugated reaction system." *Amino acids*: **5**, 61–65 (1962). (in Japanese)
16. Klausner, A., "Building for success in phenylalanine." *Bio/Technol.*: **3**, 301–307 (1985).
17. Koch, A.L., "The adaptive responses of *Escherichia coli* to a feast and famine existence." *Adv. Microbiol. Physiol.*: **6**, 147–217 (1971).
18. Kole, M.M., Thompson, B.G., Gerson, D.F., "Controlled-batch fermentations of *Escherichia coli* at low ammonium concentrations." *J. Ferment. Technol.*: **63**, 211–214 (1985).
19. Landwall, P., Holme, T., "Removal of inhibitors of bacterial growth by dialysis culture." *J. Gen. Microbiol.*: **103**, 345–352 (1977a).
20. Landwall, P., Holme, T., "Influence of glucose and dissolved oxygen concentrations on yields of *Escherichia coli* B in dialysis culture." *J. Gen. Microbiol.*: **103**, 353–358 (1977b).
21. Nakayama, K., Sagamihara, H., Hagino, H. "Process for preparing L-phenylalanine." US Pat. 3,759,790 (1973).
22. Neidhardt, F.C., Bloch, P.L., Smith, D.F., "Culture medium for Enterobacteria." *J. Bacteriol.*: **119**, 736–747 (1974).
23. Niejssel, O.M., Tempest, D.W., "The physiology of metabolite overproduction." In *Microbial Technology: Current State, Future Prospects* A.T. Bull, D.C. Elwood, C. Ratledge (eds.), p. 55–82. Camb. Univ. Press, 1979.
24. Park, S.H., Hong, K.T., You, S.J., Lee, J.H., Bae, J.C., "L-phenylalanine production by auxotrophic regulatory mutants of *Escherichia coli*." *Korean J. Chem. Eng.*: **1**, 65–69 (1984).
25. Park, N.H., Rogers, P.L., "L-phenylalanine production in continuous culture using a hyperproducing mutant of *Escherichia coli* K-12." *Chem. Eng. Commun.*: **45**, 185–196 (1986).

26. Pirt, S.J., *Principles of Microbe and Cell Cultivation* Blackwell Scientific Publications, Oxford, U.K., 1975.
27. Reiling, H.E., Laurila, H., Fiechter, A., "Mass culture of *Escherichia coli*: Medium development for low and high density cultivation of *Escherichia coli* B/r in minimal and complex media." *J. Biotechnol.*: **2**, 191–206 (1985).
28. Shiloach, J., Bauer, S., "High-yield growth of *E. coli* at different temperatures in a bench scale fermentor." *Biotechnol. Bioeng.*: **17**, 227–239 (1975).
29. Tempest, D.W., Dicks, J.W., "Interrelationships between potassium, magnesium, phosphorus and ribonucleic acid in the growth of *Aerobacter aerogenes* in a chemostat." In *Microbial Physiology and Continuous Culture* E.O. Powell, C.G.T. Evans, R.E. Strange, D.W. Tempest (eds.), p. 140–154 (1967).
30. Tempest, D.W., Niejssel, O.M., "Ecophysiological aspects of microbial growth in aerobic nutrient-limited environments." *Adv. Microbial Ecology*: **2**, 105–153 (1978).
31. Tempest, D.W., Wouters, J.T.M., "Properties and performance of microorganisms in chemostat culture." *Enzyme Microb. Technol.*: **3**, 283–290 (1981).
32. Thompson, B.G., Kole, M., Gerson, D.F., "Control of ammonium concentration in *Escherichia coli* fermentations." *Biotechnol. Bioeng.*: **27**, 818–824 (1985).
33. Tribe, D.E., Camakaris, H., Pittard, J., "Constitutive and repressible enzymes of the common pathway of aromatic biosynthesis in *Escherichia coli* K-12: Regulation of enzyme synthesis at different growth rates." *J. Bacteriol.*: **127**, 1085–1097 (1976).
34. Tsuchida, T., Matsui, H., Enei, H., Yoshinaga, F., "Method of producing L-phenylalanine by fermentation." US Pat. 3,909,353 (1975).
35. Tsuchida, T., Sano, K., "Producing L-phenylalanine by fermentation." UK Pat. GB 2,053,906A (1980).
36. Tsuchida, T., Sano, K., "Method for producing L-phenylalanine by fermentation." US Pat. 4,407,952 (1983).
37. Yamane, T., Shimizu, S., "Fed-batch techniques in microbial processes." *Adv. Biochem. Eng.*: **30**, 147–194 (1984).
38. Yu, E.K.C., Saddler, J.N., "Fed-batch approach to production of 2, 3-butanediol by *Klebsiella pneumoniae* grown on high substrate concentrations." *Appl. Environ. Microbiol.*: **46**, 630–635 (1983).

III-4
Production of *Bacillus thuringiensis* Insecticides

R. Ertola

Centro de Investigación y Desarrollo en Fermentaciones Industriales, Facultad de Ciencias Exactas UNLP, 47 y 115, 1900 La Plata, Argentina

1. Introduction

The microbiological control of insects is considered an important aspect of biological control and can be defined as the use of entomopathogenic microorganisms for insect control. In the last fifteen years an increased interest in the use of microbial insecticides has arisen. This is mainly due to the problems associated with extensive use of chemical insecticides, not only because these affect human beings and beneficial insects but also because chemical insecticides have been producing increased target insect resistance, leaving also in the environment harmful residues. It has been recently mentioned (Lisansky, 1984) that the total number of species of Arthropoda with reported cases of resistance through 1980 reached 269 to Dieldrin and BHC, and 229 to DDT.

According to the nature of the pathogenic organism, several kinds of bioinsecticides based on bacteria, viruses, fungi, and protozoa can be considered. The last three types are beyond the scope of this chapter, which is only concerned with a group of bacterial insecticides.

Most insect pathogenic bacteria belong to the families Pseudomonadaceae, Enterobacteriaceae, Lactobacillaceae, Micrococcaceae, and Bacillaceae, the latter being the only spore forming family. Among 100 *Bacillus* species mentioned as potential entomopathogens, only four have been studied as insecticides, *Bacillus thuringiensis*, *Bacillus popilliae*, *Bacillus lentimorbus*, and *Bacillus sphaericus*. *Bacillus thuringiensis* is the most important from the commercial point of view with an estimated market in western countries of about $30–$45 million which corresponds to nearly 90% of the present market for microbial insecticides (Lisansky, 1984).

Many papers, reviews, and several symposia and books have covered different aspects related to bacterial insecticides. Examples of reviews are those of Ignoffo and Anderson (1979) and Krieg and Miltenburger (1984). Symposia such as "Candidates for Microbial Control of Mosquitoes" (Myers and Yousten, 1981) can be cited, and finally books (Burges and Hussey, 1971, and Burges, 1981). However, very few are particularly concerned with the fermentation process.

The aim of this paper is to review the main aspects of production of *Bacillus thuringiensis* preparations particularly those concerned with medium design and fermentation process, including brief considerations about the strains used, toxin evaluations, and downstream processes. Some prospects are also discussed.

187

2. Strains

The group of *B. thuringiensis* strains, which are aerobic, motile, gram-positive, endospore forming rods, are closely related to *Bacillus cereus* and have the characteristic of producing a crystalline body in the cell during the sporulation step which is commonly called delta endotoxin.

The first strain known of the *Bacillus thuringiensis* group was *Bacillus sotto*, isolated in Japan in 1901 by Ishiwata. In 1911 Berliner, in Thüringen isolated a similar bacterium from death flour moth larvae and called the microorganism *Bacillus thuringiensis* (Rogoff, 1966).

In 1977, a new *Bacillus* that was proved to be toxic to mosquitoes and black flies was isolated in Israel and named *Bacillus thuringiensis* var. *israelensis* (Goldberg and Margalit, 1977). Since then, several other strains of *B. thuringiensis* have been isolated which are active against certain members of the orders Lepidoptera, Diptera, Hymenoptera, Coleoptera, and Orhoptera. For example, one strain recently reported, *B. thuringiensis* var. *San Diego* (Herrnstadt et al., 1986), has a toxic activity against coleopteran insects. Some reversionless asporogenous mutants of *B. thuringiensis* have also been used by several workers (Wakisaka et al., 1982 a; Lüthy and Studer, 1985).

de Barjac and Bonnefoi (1962) proposed a classification of *B. thuringiensis* strains based on the flagellar H-antigens. This classification now includes 23 serotypes with some subtypes. Thus, the strain commonly used for lepidoptera control, *B. thuringiensis* var. *kurstaki*, belongs to serotype 3a 3b and the variety *israelensis* to serotype 14. The variety *galleriae* corresponds to the serotype 5a 5b and so on.

The selection of the most adequate strain depends on the particular use of the final preparation, whether its toxicity is based on the endo- or exotoxin, the latter being only produced by some serotypes, independent of spore and crystal formation. Different varieties and strains have differing spectra of insecticidal activity. Although the activity is largely directed at insects in the order Lepidoptera, Diptera and Coleopteran insects are also affected by different strains. Most of Lepidoptera are susceptible to some *B. thuringiensis* strains, but there is a considerable variation of toxicity to the host species between serotypes and even within the same serotype. For this reason it is necessary to define the specificity spectrum of the strain and also that of the host insect in order to succeed in the practical use of a *B. thuringiensis* preparation (Carlberg, 1985).

3. Medium Design

Production of bioinsecticidal preparations require the design of an adequate medium for growth, sporulation, and endotoxin formation. In this respect it is important to know the nutritional requirement of the microorganism used. One of the more complete studies on minimal nutritional requirements for growth, sporulation, and crystal formation of *B. thuringiensis* was that of Nickerson and Bulla (1974). They tested 18 strains with a defined medium composed of glucose, $(NH_4)_2SO_4$, and minerals which included apart from K^+ salts the presence of Mg^{2+}, Ca^{2+}, Zn^{2+}, Cu^{2+}, and Fe^{2+}. They proved that *B. thuringiensis* does not grow in minimal glucose salts media, but the addition of citrate, aspartate, or glutamate allows the growth of all strains employed. They also showed that the *B. thuringiensis* strains tested, with the exception of the var. *serotoxicus*, do not require vitamins for growth or for sporulation. Although the results obtained are clear as far as growth is concerned, it should be remembered, as the authors mentioned, that it is much harder to define nutritional requirements for sporulation because the medium is no longer defined after vegetative growth is completed. Other papers related to medium requirements for growth and sporulation are those of Dubois (1968) and Dulmage (1970).

Glucose seems to be one of the best carbon sources, although others like starch, sucrose, and glycerol can be used. Smith (1982) tested glucose, sucrose, and glycerol for growing *B. thuringiensis* var. *israelensis* and proved that glucose gave the highest spore count but glycerol was preferred because a higher mosquito toxicity effect was achieved with that source. According to Arcas (1985) higher yields of colony forming unit (CFU, as obtained on agar plates after treatment of samples at 60°C for 20 min) and higher levels of delta en-dotoxin were obtained by using glucose as carbon source for a *B. thuringiensis* var. *kurstaki* strain.

An adequate organic nitrogen source is also essential for growth. Although several types of peptones, corn steep, or soymeal can be used, yeast extract seems to be the most con-venient. The presence of a particular amino acid seems to be essential as was proved with cystine (Rajalakshmi and Shethna, 1977). The use of yeast extract or casamino acids, both treated with H_2O_2, according to Lyman et al. (1946) which destroy cystine and methionine gives poor growth of *B. thuringiensis* var. *kurstaki* (Arcas et al., 1984).

In connection with mineral requirements several papers have emphasized the importance of Mn^{2+}, K^+, Ca^{2+}, and Zn^{2+} and in some cases Cu^{2+} and Fe^{2+} as well (Nickerson and Bulla, 1974; Vasantha and Freese, 1979). Although nothing has been published in con-nection with the composition of *B. thuringiensis*, it is interesting to point out the differences found in the composition of vegetative cells and spores of a *Bacillus cereus* strain that is closely related to *B. thuringiensis* (Table 1) (Ribbons, 1970).

It can be seen that spores contain as much as three to twenty times more of several ions. The presence of Mn^{2+} has been considered essential for endospore formation for several species of *Bacillus* (Charney et al., 1951). The lack of this ion in the medium affects the activity of phosphoglycerate phosphomutase because this enzyme requires Mn^{2+} as a co-factor (Vasantha and Freese, 1979). Potassium is also a fundamental ion needed for growth, sporulation, and endotoxin formation (Wakisaka et al., 1982 b). Limitation of growth by Ca^{2+} produces less refractive and less heat-resistant spores, as has been observed for *B. megaterium* (Grelet, 1952). The importance of Zn^{2+} supplementation may be related to the high content of spore of this element as in *B. cereus*.

When the supplementation of trace elements is considered it is important to remember that a component of a nondefined medium may contribute unknown amounts of some elements, as is the case of yeast extract. Analysis of the minor element composition of yeast extract like that reported by Grant and Pramer (1962) revealed that the content of some elements like Cu^{2+}, Fe^{2+}, and Zn^{2+} may be sufficient to support growth if enough concentration of yeast extract is used. Arcas et al. (1984) have shown that for growing *B. thuringiensis* var. *kurstaki* in a yeast extract (Oxoid 4 g l^{-1}) medium, it was not necessary to add either Fe^{2+} and Cu^{2+} or Zn^+ because the amount present in the yeast extract was sufficient for growth and spore formation. It is also necessary to point out that the possibil-

TABLE 1. Contents of some trace elements of vegetative cells and spores of *B. cereus* (g kg^{-1}).*

	Mn	Cu	Ca	Zn	Fe
Vegetative cells	0.48	0.034	0.28	0.24	0.21
Spores	1.44	0.74	0.56	2.57	0.28

* From Ribbons (1970).

ity exists for an ion not to be available in the medium owing to the presence of complexing agents. This is important when laboratory studies with defined or semi-defined media are applied to complex media used on industrial scale because these media contain a range of chelating, sequestering, or adsorbing materials such as amino acids, proteins, and other components that act to reduce the effective available ionic concentration (Jones et al., 1984).

The majority of nutritional requirement studies do not include the design and optimization of media for industrial processes. In this respect one of the following criteria can be adopted: 1) carrying out studies in batch culture by comparing the growth rate and the amount of biomass obtained as a result of the change in concentration or nature of medium components; 2) knowing the nutritional requirements, composition of microorganism,

TABLE 2. Composition of non-defined media and experimental conditions recommended for high yields of *B. thuringiensis*.

Microorganism	Composition of media (Components in g l⁻¹)	Experimental conditions	Reference
B. thuringiensis var. *thuringiensis*	Sucrose, 6.4; Corn starch, 68; Corn steep, 47; Casein, 19.5; Yeast, 6; PO_4 buffer 6.		Drake and Smythe (1963)
B. thuringiensis Berliner	Glucose, 30; Corn steep liquor, 5; Soy peptone, 2; Yeast extract, 4.5; ClK, 3; $(NH_4)_2SO_4$, 3; $MgSO_4$, 2; $CaCl_2 \cdot H_2O$, 36 mg; $FeSO_4 \cdot 7H_2O$, 13.5 mg; H_3PO_4, 7 ml; $CuSO_4 \cdot 5H_2O$, 7.5 mg; $ZnSO_4 \cdot 7H_2O$, 7.5 mg; $MnSO_4 \cdot 4H_2O$, 40 mg.	500 l fermentor Temp. 32°C Aeration 0.3 vvm Agitation 120–160 rpm.	Goldberg et al. (1980)
B. thuringiensis var. *thuringiensis* (exotoxin)	Molasses, 0–40; Soya, 20–60; KH_2PO_4, 5; K_2HPO_4, 5; $MgSO_4 \cdot 7H_2O$, 0.05; $MnSO_4 \cdot 4H_2O$, 0.03; $FeSO_4 \cdot 7H_2O$, 0.01; Na $(NH_4)PO_4 \cdot 4H_2O$, 1.5; $CaCl_2 \cdot 7H_2O$, 0.05.	7–8 l fermentor pH 6.5–8 Tem. 25–35°C Agitation 200–500 rpm Aeration 0.3–1.0 vvm.	Holmberg et al. (1980)
B. thuringiensis NCIB 9207	Cane molasses, 15; Corn-steep liquor, 40.	11 l fermentor pH 6.5 Temp. 30°C Agitation, variable speed. Aeration 0.56 vvm.	Moraes et al. (1980)
B. thuringiensis var. *israelensis*	Glucose, 15; Proflo, 10; Bacto-peptone, 2.0; Yeast extract, 2.0; $CaCO_3$, 1.0; $MgSO_4 \cdot 7H_2O$, 0.3; $FeSO_4 \cdot 7H_2O$, 0.02; $ZnSO_4 \cdot 7H_2O$, 0.02.	500-ml Erlen. flasks 100 ml of medium Temp. 29–30°C.	Smith (1982)
B. thuringiensis subsp. *kurstaki*	Glucose, 5; Potato starch, 10; Soybean meal, 15; Pharma media, 20; Powdery Pork meat, 10; Polypeptone, 2; $MgSO_4 \cdot 7H_2O$, 0.3; $CaCO_3$ 10; $ZnSO_4 \cdot 7H_2O$ 0.2; $FeCl_2$ 0.02.	30 l fermentor Temp. 28°C Agitation 400 rpm.	Wakisaka et al. (1982b)
B. thuringiensis	Glucose, 56; Yeast extract, 28; $(NH_4)_2SO_4$, 7; $MgSO_4 \cdot 7H_2O$, 7; KH_2PO_4, 7; K_2HPO_4, 7; $MnSO_4 \cdot 4H_2O$, 0.21; $CaCl_2 \cdot H_2O$, 0.28.	4 l fermentor pH 7.2 Temp. 30°C Agitation 550 rpm Aeration 0.5 vvm.	Arcas et al. (1985)

and the yield values (g biomass/g element) it is possible to design and prepare a medium for an estimated biomass yield (Pirt, 1975); 3) the use of factorial design experiments and statistical modeling techniques; 4) the continuous culture procedure of pulse addition (Mateles and Battat, 1974). Of all criteria mentioned it is considered that the best approach for optimization of media corresponds to the continuous culture procedure of pulse addition. This technique was applied by Goldberg et al. (1980) for optimization of media used for growing a *B. thuringiensis* strain (see Table 2).

In the case of bacterial insecticides, the cost of raw materials has a strong influence on the final cost of the preparation. For this reason low-cost components for media are to be preferred from the commercial point of view. In this respect beet or cane molasses, soybean meal, and corn steep liquor are generally employed. Malt sprouts, a byproduct of the malt industry, have been also recommended for replacing yeast extract (Arcas et al., 1984). Table 2 shows the composition of some media and experimental conditions which were recommended for obtaining high yields of spore-crystal preparations of *B. thuringiensis*.

4. Toxin Evaluation

The toxin activity associated with *B. thuringiensis* commonly called delta endotoxin takes place in the parasporal crystal formed in the mother cell during sporulation. The crystal is a glycoprotein which normally represents between 20–30% of the cell dry weight (Bulla et al., 1979). The so-called exotoxin is a nucleotide composed of an adenosine derivate linked through a glucose moiety to the 5′ position of phosphoallaric acid.

Owing to a supposed correlation between spore count and toxin level, *B. thuringiensis* preparations used to be analyzed by CFU, which can be considered equivalent to spore count. Although now CFU is still considered a useful parameter of a process, this determination is not reliable enough because toxicities can vary in relation to spore count in different strains and media. Smith (1982) has proved, for instance, that of three strains of *B. thuringiensis* var. *israelensis* tested, H-14-A, H-14-B, and H-14-C, only the last one showed spore counts proportional to insecticidal toxicity. For this reason other methods were developed. The first and most obvious way of testing insecticidal activity is against a chosen insect and this is the base of biological tests. Currently the toxicity of strains of *B. thuringiensis* active against Lepidoptera larvae is determined with the cabbage looper *Trichoplusia ni* (Splittstoesser and McEwen, 1961) and compared to an international standard designated E-61 which is prepared at the Institut Pasteur, Paris. Other bioassays proposed the use of *Ephestia kühniella* (Burgerjon and Yamvries, 1959), *Pieris brassicae* (Burgerjon, 1959), *Spodoptera littoralis Boisd* (Goldberg et al., 1980), and *Galleria mellonella* (Arcas et al., 1984). In the case of *B. thuringiensis* strains presenting activity against mosquitoes like the var. *israelensis* (serotype 14) there is an International Standard Protocol adopted by the WHO (de Barjac and Larget, 1980; McLaughlin et al., 1984).

Several methods have been proposed to replace insect bioassays. The rocket immunoelectrophoresis technique is a new approach which according to the authors is extremely accurate and constitutes a reliable alternative to insect bioassays (Andrews et al., 1980). Another approach, the *vitro* bioassay technique, involves the use of insect cell culture (Murphy et al., 1976; Johnson, 1981; Johnson and Davidson, 1984). Recently, a noncompetitive enzyme-linked immunosorbent assay (cited by Krieg and Miltenburger, 1984) has also been recommended for titration of crystal toxin during the fermentation process.

It is evident that the techniques mentioned proposed as alternatives to insect bioassays are very useful for the determination of toxin but cannot replace insect bioassays. The latter is a necessary step, both in the laboratory and in field conditions.

5. *Fermentation Process*

In any fermentation process, fundamental aspects like the effects of several variables and the dynamic behavior of growth, sporulation, and toxin formation have to be taken into account when production processes are considered. There are very few reports concerning the influence of process variables on growth or spore formation or on toxin formation.

With the exception of a work on the fed-batch process (Arcas et al., 1985) other references are concerned with batch cultures carried out in different sized fermentors. The experimental conditions used in some work are indicated in Table 2. According to the majority of papers it seems that the most important process variables studied were sugar concentration, pH, and aeration rate.

5.1   *Influence of Sugar Concentration*

The effect of glucose concentration in a semi-defined medium containing yeast extract and mineral salts was studied by Scherrer et al. (1973). They found that maximum yield of the endotoxin as obtained with glucose concentration not higher than 8 g l^{-1} but they did not study limitation of growth or toxin formation by other medium components. According to Holmberg et al. (1980) a dependence of exotoxin yield on initial sugar content was clear. They showed that toxin production reached a maximum with increasing initial sugar content up to 23 g l^{-1} (30 g l^{-1} of molasses). They also reported that higher initial concentrations seem to inhibit exotoxin production. Goldberg et al. (1980) mentioned also yields of 4×10^9 CFU ml^{-1} in the culture broth with a medium containing 30 g l^{-1} of glucose. They also showed that the increase of CFU had a corresponding increase in the toxin concentration, measured by the larval mortality.

Arcas et al. (1985) have shown that the concentration of glucose (in a glucose yeast extract salts medium) can be increased up to 56 g l^{-1} provided the rest of medium components are increased accordingly. They reported yields of CFU from 1.08×10^9 CFU ml^{-1} to 7.36×10^9 CFU ml^{-1}, when the sugar content was increased from 8 to 56 g l^{-1} (Table 3).

TABLE 3. Influence of high concentration of nutrients in media on parameters of growth and yield values of *B. thuringiensis* var. *kurstaki.**

Medium and osmolarity[a]	Lag period L (h)	Maximum biomass dry weight x_m(g l^{-1})	Specific growth rate μ (h^{-1})	Biomass yield[b] (Y_{X_m/O_2})	CFU (N° l^{-1})·10^{12}	CFU yield[c] ($Y_{CFU/S}$)
1(128)	1.7	4.02	0.77	1.50	1.08	0.135
3(358)	1.7	12.53	0.74	1.17	3.22	0.134
5(602)	3.5	20.07	0.72	1.01	5.31	0.133
7(808)	3.9	26.60	0.71	0.80	7.36	0.131
9(1.068)	5.7	32.46	0.64	0.74	6.92	0.096
11(1.294)	7.2	27.96	0.47	0.56	4.92	0.056

Medium 1 (Components in g l^{-1}): Glucose, 8; Yeast extract, 4; $(NH_4)_2SO_4$, 1; $MgSO_4 \cdot 7 H_2O$, 1; KH_2PO_4, 1; K_2HPO_4, 1; $MnSO_4 \cdot 4H_2O$, 0.03; $CaCl_2 \cdot H_2O$, 0.04. The other media contained 3, 5, 7, 9, and 11 times the concentration of all components of medium 1.
[a] In milliosmols. [b] g of biomass dry weight per g of O_2 consumed. [c] CFU per g of glucose consumed.
* From Arcas et al. (1985)

5.2   Effect of pH

The influence of buffering on mosquito toxin production by three strains of *B. thuringiensis* H-14 was studied by Smith (1982), who found no differences in the toxin level between runs presenting final pH values of 5.7 to 8.1. On the contrary, spore counts showed great variations within the range of pH values tested. Holmberg et al. (1980) also found no significant variations in the endotoxin concentration of their strain working with pH values between 6.5 and 8.

Although more research is needed concerning this variation it seems that no great differences in the toxin level are to be expected by variations in the pH between 5.8 and 8.

5.3   Influence of Aeration Rate

Several authors mentioned that high aeration rates are essential for spore and toxin formation (Scherrer et al., 1973; Moraes et al., 1980; Pendleton, 1969; Zamola et al., 1981; Arcas, 1984; Foda et al., 1985), but very few give data on this aspect.

Moraes et al. (1980) studied the influence of oxygen concentration on a *B. thuringiensis* culture. They found that the respiration rate increased when oxygen concentration increased from 2.5 to 10% of the saturation value and decreased thereafter. However no data were reported about spore count or toxin concentration achieved. Exotoxin production was found insensitive to changes in the aeration rate within the intervals studied, mentioned in Table 2 (Holmberg et al., 1980).

According to Arcas (1984) sporulation was related to the oxygen transfer coefficient k_La. The percentage of sporulation increases from 70 to 100 (with a ratio of spore to crystal protein equal to 1) when the k_La increased from 38 to 220 (h^{-1}).

Arcas et al. (1985) have also observed that using low concentration of glucose (8 g l^{-1}) the oxygen dissolved pressure remains higher than 50% of saturation value working with 4 l of medium in a laboratory fermentor aereated at 0.5 vvm and agitated at 550 rpm. When the concentration of glucose and the rest of medium components was increased 5 to 7 times, the oxygen dissolved pressure decreased to 0% of saturation, but this appears not to have influence on either sporulation or on toxin formation.

Foda et al. (1985) reported studies on *B. thuringiensis* var. *entomocidus* and mentioned the failure of the organism to survive or sporulate under low aeration levels mainly in the presence of high sugar (25 g l^{-1}) concentrations. However, it is necessary to point out that these authors defined low aeration levels in Erlenmeyer flasks located in a shaker operated at 100 rpm and working with a ratio of 3:2, air-medium, which means that extremely poor aeration rates were employed.

5.4   Parameters of Growth, Yield Values, and Process Dynamics

Very few references related to bacterial insecticide production are concerned with parameters of growth or with yield values of CFU or those related to toxin formation. Moraes et al. (1980) gave some data on the specific growth rate and its relationship with the respiration rate and dissolved oxygen concentration. Values reported varied from 0.35 to 0.62 h^{-1} for a batch process conducted in a 11 liter fermentor.

The values of the maximum specific growth rate given by Holmberg et al. (1980) for an exotoxin producing strain ranged from 0.70 to 1.90 h^{-1} for runs performed under different conditions. They also give values of a yield coefficient which varied from 6 to

40×10^7 cells (mg of sugar consumed).$^{-1}$ Lag periods observed by these authors varied from 1 to 3 h.

Arcas et al. (1985) gave data on parameters of growth and yield values achieved in their batch process in which they studied the effect of high levels of nutrients on growth, sporulation, and toxin level (Table 3).

As can be seen, the concentration of cell (x_m) increased with the concentration of nutrients up to nine times the values of medium 1. A further increase of the medium components seems to be harmful to the microorganism, because a decrease in the cell population occurs.

The specific growth rate was slightly affected up to medium 7. The average generation time calculated was 57 min which agrees with values mentioned by other workers (Goldberg et al., 1980).

The increase observed in the lag period from media 3 to 11 can be explained by changes in the physical environment due to the increase concentration of media components. The decrease obtained in the yield coefficient Y_{X_m/O_2} is correlated with the corresponding increase in the osmolarity of media, in agreement with Watson (1970).

A linear increase of the CFU values is observed up to medium 7. Higher concentration of nutrient affects these values negatively, in spite of the fact that the cell dry weight continues to increase as with medium 9; Holmberg et al. (1980) found the same effect. This may be related to the presence of an inhibitor of sporulation or to high concentrations of organic nutrients known to reduce the percentage of sporulating cells (Murrell, 1961) present in the medium. The value of the yield coefficient of CFU based on glucose consumed is remarkably constant up to medium 7 and is similar to those mentioned by Goldberg et al. (1980). The CFU values obtained were directly associated with endotoxin production. The bioassay carried out with *Galleria mellonella* indicated that the same insecticidal activity was attained with different media when the same number of CFU per gram of larvae diet was employed. The values obtained using an immunoelectrophoretic method demonstrated a 7-fold increase of the endotoxin level from medium 1 to 7, which were 1.05 mg ml^{-1} and 6.85 mg ml^{-1} of culture broth, respectively.

As far as the author is aware, the only papers concerned with the dynamic behavior of toxin production are those of Sommerville (1971) and Holmberg et al. (1980). They measured the time functions of bacterial, exotoxin, and degree of sporulation. Sommerville (1971) found that endotoxin formation started at the stationary phase of growth and was strongly linked to sporogenesis.

According to Holmberg et al. (1980), exotoxin formation started during the exponential growth phase and spore formation when the bacterial growth had ceased. Exotoxin production was complete at this point, which indicates that there is no correlation between exotoxin production and sporulation, and this could perhaps be used as an indication that the process is completed. They also studied the dependence of the parameters on the initial sugar concentration. The maximum specific growth rate appears to be independent of the initial sugar concentration but the yield coefficient and the Michaelis-Menten constant are dependent on it. Monod's model was used for describing the relationship between bacterial growth and substrate consumption. Concerning exotoxin production they applied a similar model proposed by Pirt (1975) for penicillin formation in which case the product formation rate is linearly related to bacterial growth. As the determinations of viable count and toxin are time consuming, the authors proposed another model for estimating microbial growth and toxin formation based on oxygen consumption. The estimation algorithm seems to give a better fit for growth than Monod's model.

It is evident that more data are needed, particularly those related to the kinetics of toxin production, in order to optimize the fermentation process. Although the models proposed by Holmberg et al. (1980) based on oxygen consumed or heat evolution are interesting, it is clear that models based on toxin determination must be developed.

5.5   Fed-batch Process

The operation procedure employed by Arcas et al. (1985) for their fed-batch process is based on the gradient-feed method reported by Srinivasan et al. (1977). Two connected reservoirs containing the first $10.5 \times$ the concentration of components of medium 1 (Table 3) and the second $31.5 \times$ the concentration of components of the same medium were used for feeding a constant flow rate of 142 ml h^{-1}. The experiments were carried out in a 6–1 LKB fermentor at 30°C. The initial volume was 2 1 and the final volume 4 1. The final concentration of nutrients coming into the fermentor was sufficient to give $11 \times$ the initial concentration in the fermentor. In the described experiments a total of 336 g of glucose and proportionate amounts of all the components were fed as a gradient. The feeding period lasted for 14 h starting at 6 h after inoculation and finishing at 20 h. After that, a period of 24 h without feeding was carried out in order to facilitate the formation of crystals and spores.

The fed-batch data obtained were as follows: lag period, L (corresponding to the batch step) 1.7; x_m, 56.43 g l^{-1}; μ, 0.77 (h^{-1}); Y_{X_m/O_2}, 1.29; CFU l$^{-1} \cdot 10^{12}$, 12; $Y_{CFU/S}$, 0.136. It is clear that the use of fed-batch process allows the production of very high yields of spore-crystal preparations of *B. thuringiensis*.

6. Downstream Processes

Downstream processes are concerned with separation of cells and formulation of commercial preparations. The separation step is currently carried out by centrifugation. Zamola et al. (1981) studied the influence of factors which affect the efficiency of separation and also the economical performance of continuous centrifugal separation equipment by employing different flow rates. The optimization of the separation step depends on the flow rate at which the broth passes through the separator. The paste-like material obtained is processed to obtain the commercial products which are generally in the form of emulsions or powders.

In order to obtain commercial liquid preparations, the process described by Cords et al. (1965) is one of the most commonly employed. The pH of the pasty cellular material is adjusted to 3.5–4.5 and mineral salts and oils are added for stabilization. To prevent contamination of liquid preparations, xylene or sorbitol are added (Smirnoff and Valeró, 1983).

Power preparations are produced by spray of the total fermentation broth or spray of the cells and spore-crystal material (Couch and Ross, 1980). This can be also dried by blending it with a water absorbing inert material such as bentonite.

It should be emphasized that the correct formulation step is of crucial importance. As mentioned by Couch and Ross (1980), the blame for failure of an insecticidal preparation is frequently assigned to the microbe used when very often the cause is an inadequate formulation. Lacey (1984) cited a recommendation of Burges who mentioned that formulation requirements should be also considered at all steps of the fermentation process.

As in any fermentation process, it is important to remember that integration of down- and up-stream processes has to be considered in order to attain global optimization of the process.

7. Prospects

It has been estimated that bioinsecticides could capture as much as 50% of the world insecticide market by the year 2000. Yet even a 5–10% share of the total market would represent a significant sector, since the production of insecticides, fungicides, and other pesticides amounted to a third of a million tons in 1981 in the EEC countries alone (Kristiansen and Lewis, 1986).

In order to meet market demands, higher production capacity and improved bioinsecticide quality are essential. In this respect the following aspects have to be considered: 1) changes in philosophy—although microbial insecticides cost less to develop, many companies expect them to bear their share of chemical development costs (Lisansky, 1984) and consequently a change in attitude is required if these insecticides are to be developed by the industry. 2) Improvement in the education of users, related to the mode of application and action. 3) Improvement in the efficiency of strains used by obtaining strains resistant to chemical and physical factors or the creation of asporogenous strains. (some successful results on both aspects have been already reported (Salama et al., 1984)). 4) Studies on genetics and mechanism of action. Genetic recombinant has already allowed gene coding for this toxin to be cloned within *B. subtilis* cells with the possible result of increasing toxin production per unit of fermentor volume. 5) The increase of production will also need more fundamental studies in process dynamics in order to increase productivity and to improve process optimization. 6) Improvement in formulation and efficiency of application in order to obtain better control of pests. 7) Overcoming pest resistance; some adverse finding of resistance which developed rapidly in the field have been recently mentioned in connection with the Indiana meal moth *Plodia interpunctella* (Kristiansen and Lewis, 1986), a pest of stored grain. It is expected that modern techniques of genetic manipulation will give rise to modified toxic organisms which will allow the bacteriologist to keep one step ahead of the pest.

If the market share of bioinsecticides increases as estimated, a considerable reduction in the use of chemical products is to be expected. It is reasonable, however, to think that although new preparations of bioinsecticides can be developed and used, the effectiveness of these biological agents might be realized in combination with synthetic chemicals as part of an integrated pest management strategy (Kristiansen and Lewis, 1986).

Acknowledgments
The author wishes to thank Dr. Yantorno and Dr. Silva for revision of the manuscript.

REFERENCES
1. Andrews, R.E. Jr., Iandolo, J.J., Campbell, B.S., Davidson, L.I., Bulla, L.A. Jr., "Rocket immunoelectrophoresis of the entomocidal parasporal crystal of *Bacillus thuringiensis* subsp. *kurstaki*." *Appl. Environ. Microbiol.*: **40**, 897–900 (1980).

2. Arcas, J., "Producción de bioinsecticidas." Thesis, Universidad National de la Plata, Argentina, 1985.
3. Arcas, J., Yantorno, O., Arrarás, E., Ertola, R., "A new medium for growth and delta-endotoxin production by *Bacillus thuringiensis* var. *kurstaki.*" *Biotechnol. Lett.*: **6**, 495–500 (1984).
4. Arcas, J., Yantorno, O., Ertola, R., "Production of delta-endotoxin of *Bacillus thuringiensis* by batch and fed-batch cultures." Paper presented at the 7th GIAM Conference, Helsinki, Finland, 1985.
5. Bulla, L.A.Jr., Davidson, L.I., Kramer, K.J., Jones, B.L., "Purification of the insecticidal toxin from the parasporal crystal of *Bacillus thuringiensis* subsp. *kurstaki.*" *Biochem. Biophys. Res. Commun.*: **91**, 1123–1130 (1979).
6. Burges, H.D., In *Microbial control of pests and plant diseases.* p. 1970–1980. Academic Press, London, 1981.
7. Burges, H.D., Hussey, N.W., In *Microbial control of insects and mites.* Academic Press, London, New York, 1971.
8. Burgerjon, A.,"Titrage et définition d'une unité biologique pour les préparations de *Bacillus thuringiensis* Berliner." *Entomophaga*: **4**, 201–206 (1959).
9. Burgerjon, A., Yamvrias, C., "Titrage biologique des préparations à base de *Bacillus thuringiensis* Berliner vis-à-vis de *Anagasta* (*Ephestia*) kuhniella Zell." *Comp. Rend. Acad. Sci. Paris*: **249**, 2871–2872 (1959).
10. Carlberg, G., "Biological control of pest insects with *Bacillus thuringiensis.*" Paper presented at the 7th GIAM Conference, Helsinki, Finland, 1985.
11. Charney, J., Fisher, W.P., Hegarty, C.P., "Manganese as an essential element for sporulation in the genus *Bacillus.*" *J. Bacteriol.*: **62**, 145–148 (1951).
12. Cords, MM. H., Fisher, R.A., Briggs, J.D., "Suspensions insecticides bactériennes concentrées stables." French Pat. N 1.393.646 (1965).
13. Couch, T.L., Ross, D.A., "Production and utilization of *Bacillus thuringiensis.*" *Biotechnol. Bioeng.*: **22**, 1297–1304 (1980).
14. de Barjac, H., Bonnefoi, A., "Essai de classification biochemique et serologique de 24 souches de *Bacillus* du type *thuringiensis.*" *Entomophaga*: **7**, 5–31 (1962).
15. de Barjac, H., Larget, I., "Proposals for the adoption of a standardized bioassay method for the evaluation of insecticidal formulations derived from a serotype H-14 of *Bacillus thuringiensis.*" Mimeograph document WHO/VBC 79. 744, Geneva, 1980.
16. Drake, B.B., Smythe, C.V., "Process for making pesticidal compositions." U.S. Pat. 3,087,865 (1963).
17. Dubois, N.R., "Laboratory batch production of *Bacillus thuringiensis* spores and crystals." *Appl. Microbiol.*: **16**, 1098–1099 (1968).
18. Dulmage, H.T., "Production of the spore-δ endotoxin complex by variants of *Bacillus thuringiensis* in two fermentation media." *J. Invertebr. Pathol.*: **16**, 385–389 (1970).
19. Foda, M.S., Salama, H.S., Selim, M., "Factors affecting growth physiology of *Bacillus thuringiensis.*" *Appl. Microbiol. Biotechnol.*: **22**, 50–52 (1985).
20. Goldberg, L.J., Margalit, J., "A bacterial spore demonstrating rapid larvicidal activity against *Anopheles sergentii, Uranotaenia unguiculata, Culex univitattus, Aedes aegypti* and *Culex pipiens.*" *Mosquito News*: **37**, 355–361 (1977).
21. Goldberg, I., Sneh, B., Battat, E., Klein, D., "Optimization of a medium for a high yield production of spore-crystal preparation of *Bacillus thuringiensis* effective against the Egyptian cotton leaf worm *Spodoptera littoralis* Boisd." *Biotechnol. Lett.*: **2**, 419–426 (1980).
22. Grant, C.L., Pramer, D., "Minor element composition of yeast extract." *J. Bacteriol.*: **84**, 869–870 (1962).
23. Grelet, N., "Nutrition azotée et sporulation de *Bacillus cereus* var. *mycoides.*" *Ann. Inst. Pasteur*: **88**, 60–75 (1952).
24. Herrnstadt, C., Soares, G.G., Wilcox, E.R., Edwards, D.L., "A new strain of *Bacillus thuringiensis* with activity againt coleopteran insects." *Bio/Technol.*: **4**, 305–308 (1986).
25. Holmberg, A., Sievänen, R., Carlberg, G., "Fermentation of *Bacillus thuringiensis* for exotoxin production: Process analysis study." *Biotechnol. Bioeng.*: **22**, 1707–1724 (1980).
26. Ignoffo, C., Anderson, R., "Bioinsecticides." In *Microbial Technology.* 2nd ed. Vol. 1, Peppler, H., Perlmann, D. (eds.), p. 1–28. Academic Press, London, 1979.
27. Johnson, D.E., "Toxicity of *Bacillus thuringiensis* entomocidal protein toward cultured insect tissue." *J. Invertebr. Pathol.*: **38**, 94–101 (1981).

28. Johnson, D.E., Davidson, L.I., "Specificity of cultured insect tissue cells for bioassay of entomocidal protein from *Bacillus thuringiensis*." *IN VITRO:* **20**, 66–70 (1984).
29. Jones, R.P., Greenfield, P.F., "A review of yeast ionic nutrition. Part I: Growth and fermentation requirements." *Process Biochem.*: **19**, 48–60, April 1984.
30. Krieg, A., Miltenburger, H., "Bioinsecticides: I. *Bacillus thuringiensis*." *Adv. Biotechnol. Processes*: **3**, p. 273–290. Alan R. Liss, Inc., New York, 1984.
31. Kristiansen, B., Lewis, C., "Bacterial insecticides: Recent developments." *TIBTECH*, 56–58, March 1986.
32. Lacey, L.A., "Production and formulation of *Bacillus sphaericus*." *Mosquito News*: **44**, 153–159 (1984).
33. Lisansky, S.G., "Biological alternatives to chemical insecticides." In *Proceedings of Biotech. '84 Europe* p. 455–466. Online Publications Ltd., Pinner, U.K., 1984.
34. Lüthy, P., Studer, D., "Control of blacklies and mosquitoes with *Bacillus thuringiensis* subsp. *israelensis*." Paper presented at the 7th GIAM Conference, Helsinki, Finland, 1985.
35. Lyman, C.M., Moseley, O., Wood, S., Hale, F., "Note on the use of hydrogen peroxide-treated peptone in media for the microbiological determination of amino acids." *Arch. Biochem. Biophys.*: **10**, 427–431 (1946).
36. Mateles, R.I., Battat, E., "Continuous culture used for media optimization." *Appl. Microbiol.*: **28**, 901–905 (1974).
37. McLaughlin, R.E., Dulmage, H.T., Alls, R., Couch, T.L., Dame, D.A., Hall, I.M., Rose, R.I., Versoi, P.L., "U.S. standard bioassay for the potency assessment of *Bacillus thuringiensis* serotype H-14 against mosquito larvae." *Bulletin of the ESA*: **30**, 26–29 (1984).
38. Moraes, I.O., Santana, M.H.A., Hokka, C.O., "The influence of oxygen concentration on microbial insecticide production." In *Advances in Biotechnology Proceed.* Sixth Internat. Ferm. Symp. London, Canada: Vol. **1**, 75–79 (1980).
39. Murphy, D.W., Sohi, S.S., Fast, P.G., "*Bacillus thuringiensis* enzyme-digested delta endotoxin: Effect on cultured insect cells." *Science*: **194**, 954–956 (1976).
40. Murrell, W., "Spore formation and germination as a microbial reaction to the environment." In *Microbial reaction to environment*. Eleventh Symposium of the Society for General Microbiol. p. 100, University Press, Cambridge, 1961.
41. Myers, P., Yousten, A., "Toxic activity of *Bacillus sphaericus* for mosquito larvae." *Develp. Indust. Microb.*: **22**, 41–52 (1981).
42. Nickerson, K.W., Bulla, Jr. L.A., "Physiology of sporeforming bacteria associated with insects: minimal nutritional requirements for growth, sporulation, and parasporal crystal formation of *Bacillus thuringiensis*." *Appl. Microbiol.*: **28**, 124–128 (1974).
43. Pendleton, I.R., "Insecticides of crystal-forming bacteria." *Process Biochem.*: 29–32, December 1969.
44. Pirt, S.J., "Design of a culture medium" and "Product formation in microbial cultures." p. 133–134 and 156–169. In *Principles of Microbe and Cell Cultivation*. Backwell Scientific Publications, Oxford, 1975.
45. Rajalakshmi, S., Shethna, Y., "The effect of amino acids on growth sporulation and crystal formation in *Bacillus thuringiensis* var. *thuringiensis*." *J. Indian Inst. Sci.*: **59**, 169–176 (1977).
46. Ribbons, D.W., "Quantitative relationships between growth media constituents and cellular yields and composition." In *Methods in Microbiology*. Vol. 3A. Norris, J.R., and Ribbons, D. (eds.), p. 299–304. Academic Press, London and New York, 1970.
47. Rogoff, M.H., "Production and formulations of *Bacillus sphaericus*." *Mosq. News*: **44**, 153–159 (1966).
48. Salama, H.S., Foda, M., Selim, M., "Isolation of *Bacillus thuringiensis* mutants resistant to physical and chemical factors." *Z. Ang. Ent.*: **97**, 139–145 (1984).
49. Scherrer, P., Lüthy, P., Trumpi, B., "Production of δ-endotoxin by *Bacillus thuringiensis* as a function of glucose concentrations." *Appl. Microbiol.*: **25**, 644–646 (1973).
50. Smirnoff, W.A., Valéro, J.R., "Preserving *Bacillus thuringiensis* concentrates and formulations without xylene." *J. Invert. Path.*: **42**, 415–417 (1983).
51. Smith, R.A., "Effect of strain and medium variation on mosquito toxin production by *Bacillus thuringiensis* var. *israelensis*." *Can. J. Microbiol.*: **28**, 1089–1092 (1982).
52. Somerville, H.J., "Formation of the parasporal inclusion of *Bacillus thuringiensis*." *Eur. J. Biochem.*: **18**, 226–237 (1971).
53. Splittstoesser, C.M., McEwen, F.L., "A bioassay technique for determining the insecticidal activity of preparations containing *Bacillus thuringiensis* berliner." *J. Insect Pathol.*: **3**, 391–398 (1961).

54. Srinivasan, V.R., Fleenor, M.B., Summers, R.J., "Gradient-feed method of growing high cell density cultures of *Cellulomonas* in a bench-scale fermentor." *Biotechnol. Bioeng.*: **19**, 153–155 (1977).
55. Vasantha, N., Freese, E., "The role of manganse in growth and sporulation of *Bacillus subtilis*." *J. Gen. Microbiol.*: **112**, 329–336 (1979).
56. Wakisaka, Y., Uno, J., Matsumoto, K., Ohdaira, O., Tanaka, K., "An insecticide and a process for its preparation." European Pat. 82101507.0 Shionogi & Co., Ltd. (1982a).
57. Wakisaka, Y., Masaki, E., Nishimoto, Y., "Formation of crystalline δ-endotoxin or poly-β-hydroxy-butyric acid granules by asporogenous mutants of *Bacillus thuringiensis*." *Appl. Environ. Microbiol.*: **43**, 1473–1480 (1982b).
58. Watson, T.G., "Effects of sodium chloride on steady-state growth and metabolism of *Saccharomyces cerevisiae*." *J. Gen. Microbiol.*: **64**, 91–99 (1970).
59. Zamola, B., Valles, P., Meli, G., Miccoli, P., Kajfež, F., "Use of the centrifugal separation technique in manufacturing a bioinsecticide based on *Bacillus thuringiensis*." *Biotechnol. Bioeng.*: **23**, 1079–1086 (1981).

IV. Measurement, Control and Design

IV-1
Algorithmic Monitors for Computer Control of Fermentations

Arthur E. Humphrey

Biotechnology Research Center, Lehigh University, Bethlehem, PA 18015, U.S.A.

Nearly twenty-five years ago, Dr. Nancy Millis and I joined Professor Aiba in offering at Tokyo University one of the very early biochemical engineering courses. Fermentation vessels in those days were only equipped with agitation, air flow, and antifoam control and pH and temperature sensors and control. We were experimenting with dissolved oxygen sensors and measuring cell, substrate, and product concentrations by off-line means. The three of us engaged in numerous discussions as to what was to be done if we were to have any kind of meaningful control of fermentation systems. It seemed to us that there were three possible approaches[1]: (1) devise sensors that could directly measure cell, substrate, and product concentrations on-line in a non-invasive manner; (2) develop metabolic models that could be expressed in measurable parameters; and (3) evolve algorithms based on combining various measurements in order to estimate the various control parameters by indirect means.

The first approach seemed most difficult to achieve. There were no non-invasive, sterilizable sensors even remotely available with the exception of optical density devices for cell concentration measurements.[11] Unfortunately, these were inoperable at the cell concentrations normally encountered in commercial fermentations due to quenching of the signal. It was practical to make off-gas analyses for O_2 and CO_2, but these were done, at that time, by long-path infrared analysis for CO_2 and paramagnetic wind measurements for O_2.[2] These measurements were most unsatisfactory in the early part of a fermentation as they were subject to large errors. The infrared analysis was used to monitor the progress of a fermentation but never to control the fermentation. If the two readings, CO_2 evolution and O_2 uptake rates, were combined in order to calculate the respiratory quotient, estimates made early in the fermentation were too unreliable for control. Off-gas analysis had to wait for the mass spectrometer to be developed in the mid-1970s.[3]

With the second approach, practical models for fermentations, in terms of measurable variables, did not exist. Even the most simple of models required the determination of cell and substrate or product concentrations. Furthermore, there was no knowledge base to relate the yield, maintenance, and inhibition constants with such fermentation parameters as temperature, pH, and medium composition.[10] This challenge, while still not satisfactorily resolved, is nearing solution.[5]

The third approach, that seemed most feasible at that time, was to combine various measurements, and, using some appropriate model or algorithm, to estimate some controllable parameter. The first thought we had at that time was to modify existing dissolved oxygen probes in order to render them sterilizable and then to use them to estimate oxygen uptake rate (OUR) and the volumetric mass transfer coefficient (k_La). Indeed, this was exactly what happened and over the next five years we and other biochemical engineers evolved the necessary techniques and relationships to provide on-line monitoring of OUR and k_La.[4,12]

1. Dynamic Measurement of OUR and k_La by Dissolved Oxygen Probes

Several dissolved oxygen probes, both polarographic and galvanic types, appeared in the early 1960s.[11] The problem in adapting them to fermentations was two-fold: first, using parts that would withstand sterilization and, second, incorporating rugged membranes that would give reasonable time responses to changes in oxygen concentrations. These problems proved readily susceptible to solution and soon sterilizable, dissolved oxygen probes with good response characteristics were routinely used in Professor Aiba's laboratory and others throughout the world.

It was quickly appreciated that the dissolved oxygen probe was more than just a simple sensor; indeed, if monitored under dynamic conditions both the OUR and k_La could be estimated (Fig. 1).[4]

At steady state (a good approximation for short times, i.e., over 1–5 min in most fermentations), the OUR can be estimated by

$$OUR = k_La(C^* - C_\infty)_{mean} \qquad (1)$$

where C^* is the dissolved oxygen concentration when the broth is saturated with or in equilibrium with the "average" air bubble in the fermentation broth.

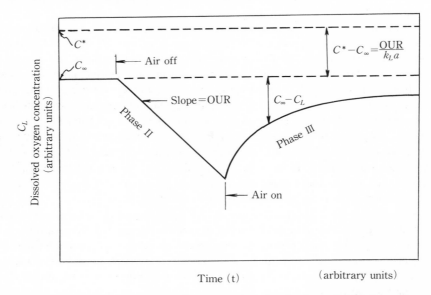

FIG. 1. Typical dissolved oxygen concentration versus time for the dynamic technique of mass transfer coefficient estimation.

C_∞ is the steady-state dissolved oxygen concentration at the beginning of the dynamic measurement (assumed to have the same value at the end of the dynamic measurement).

The general equation for the dissolved oxygen concentration changes with time is given by

$$\frac{dC_L}{dt} = k_L a(C^* - C_L)_{mean} - \text{OUR}. \tag{2}$$

For the steady-state period, $dC_L/dt = 0$, and Eq. (1) applies. This is approximately true because the OUR of a fermentation changes less than 5% in a 5-min period, the time of a dynamic measurement. Hence, OUR can be considered constant over this time. Also, since C^* is only about 7 mg l^{-1}, the accumulation term can be neglected during the time of a dynamic measurement.

For the non-aeration period (phase II in Fig. 1) $k_L a(C^* - C_L) = 0$ and the OUR is given by

$$\text{OUR}_{mean} = dC_L/dt \tag{3}$$

which is the slope of the C_L vs. time trace.

For the re-aeration period (phase III)

$$\frac{dC_L}{dt} = k_L a(C^* - C_L)_{mean} - \text{OUR}_{mean}. \tag{4}$$

Substituting for OUR_{mean} from Eq. (1), one obtains

$$\frac{dC_L}{dt} = k_L a(C^* - C_L)_{mean} - k_L a(C^* - C_\infty)_{mean} \tag{5}$$

or

$$\frac{dC_L}{dt} = k_L a(C_\infty - C_L)_{mean}, \tag{6}$$

rearranging

$$\frac{d \ln(C_\infty - C_L)}{dt} = -k_L a. \tag{7}$$

When one is working with high cell density fermentations $(C^* - C_\infty)$ is usually large and dC_L/dt in phase II changes very rapidly. In this case, data in phase II are not very accurate, so $k_L a$ is estimated from phase III data and OUR from Eq. (1). When one is working with low cell concentrations, $(C^* - C_\infty)$ is usually small and dC_L/dt in phase III changes rapidly, so one is left with estimating OUR from phase II data and cannot get a very good estimate of $k_L a$.

The OUR data can be checked by off-line measurements of cell dry weight with time by noting that

$$\text{OUR} = \frac{1}{Y_{X/O_2}} \frac{dX}{dt} + m_{O_2} X \tag{8}$$

where X is the dry weight concentration of cells, Y_{X/O_2} is the yield of cells per unit of oxygen consumed, and m_{O_2} is the maintenance requirement of oxygen by the cells. Generally, m_{O_2}

is small and can be neglected except at low growth rate conditions, i.e., less than 0.05 h^{-1}. Under these conditions,

$$\text{OUR} = \frac{1}{Y_{X/O_2}} \frac{dX}{dt} = \frac{1}{Y_{X/O_2}} \frac{X}{X} \frac{dX}{dt} = \frac{\mu X}{Y_{X/O_2}} \tag{9}$$

where μ is the specific growth rate. Because these measurements require only the use of a single dissolved oxygen probe, they are not susceptible to error propagation. However, the dynamic OUR measurement technique was never readily accepted as an on-line measurement for control. Most fermentor operators were afraid of turning off the air to a fermentation even for brief periods of 1–5 min. Demonstrations with laboratory fermentors that this was not detrimental to a fermentation failed to convince plant operators otherwise. So dynamic OUR and k_La measurements, although feasible, never became an acceptable measurement for control.

2. Off-Gas Analysis

With the advent of relatively cheap mass spectrometers for off-gas analysis approximately a decade ago, a new opportunity presented itself for computer control using parameters derived from indirect material balancing.[13] Because the mass spectrometer allows the monitoring of both CO_2 and O_2 simultaneously and relating these analyses to N_2 composition, error propagation can be minimized and no correction is necessary for gas leakage from the system, humidity, or pressure changes.[13] Therefore, the O_2 uptake rate (OUR) and CO_2 evolution rate (CER) can be determined using the following relationships:

$$\text{OUR} = F\, C_{N_2}\left[\left(\frac{C_{O_2}}{C_{N_2}}\right)_{in} - \left(\frac{C_{O_2}}{C_{N_2}}\right)_{out}\right] \tag{10}$$

and

$$\text{CER} = F\, C_{N_2}\left[\left(\frac{C_{CO_2}}{C_{N_2}}\right)_{out} - \left(\frac{C_{CO_2}}{C_{N_2}}\right)_{in}\right] \tag{11}$$

where F is the air flow rate determined at the same point where C_{N_2} is determined, and C_{O_2}, C_{CO_2}, and C_{N_2} are the concentrations of O_2, CO_2, and N_2, respectively, in the aeration gas.

To utilize these measurements for control, it is necessary to relate them to the cell concentration, X, and the cell growth rate, dX/dt, by the use of Eq. (8).

Rearranging Eq. (8),

$$\frac{dX}{dt} = Y_{X/O_2}(\text{OUR} - m_{O_2}X). \tag{12}$$

The cell concentration, X, as a function of time, t, can be calcualted from the integrated form of the above equation once the cell concentration is initialized, i.e., $X(0)$ is obtained. Usually $X(0)$ is estimated from the inoculum size. The resulting equation is:

$$\int_{x(0)}^{x(t)} dX = \int_0^t Y_{X/O_2}[\text{OUR} - m_{O_2}X(t)]\, dt. \tag{13}$$

Since OUR can be measured continuously (usually only 15–30-sec delays are involved in the measurements) and since on-line computational capability allows the solution of Eq. (13), then the substrate demand rate (SDR) can be estimated and used as a set point signal in feedback control of substrate feed in a fed-batch fermentation based on the following relationships:[7]

$$\text{SDR} = \frac{1}{Y_{X/S}} \frac{dX}{dt} + m_S X \qquad (14)$$

and

$$\Delta S = \int_{s(0)}^{s(t)} dS = \int_0^t Y_{X/S}(\text{SDR} - m_S X(t))\, dt \qquad (15)$$

where $Y_{X/S}$ is the yield of cells per unit of substrate consumed, and m_S is the maintenance requirement for substrate per unit of cells per unit time.

This latter equation permits the indirect estimation of substrate concentration in the fermentor. Unfortunately, the accuracy of the estimation of S depends upon the accuracy of the cell growth model, the constancy of the yield and maintenance coefficients, and the ability to estimate $X(0)$ and $S(0)$. However, up to the present time, this has been the best alternative.

3. Control by Direct Measurement of Substrate and Product Concentration

Recent advances in analysis using modern spectroscopy techniques, i.e., Fourier transform infrared spectroscopy, scanning laser fluorometry, etc., have opened the possibility of directly measuring both substrate and product concentrations and using this signal for direct feedback control.[6,8,9] Unfortunately, the accuracy of the information obtained from these signals to date is less than that obtainable using indirect estimates. However, the advantage of direct measurements is such that considerable effort is being expended to develop these techniques to a satisfactory level of accuracy.

In fluorometry, three approaches are being investigated. In the first, MIT researchers have noted that certain fluorophors added to fermentations can be used to monitor culture temperature, pH, dissolved oxygen, and phosphate concentrations. Indeed, for small (1 to 5 l) fermentors, the use of fluorophors in monitoring such culture parameters can be done at a cost of only a few dollars and appears to be a very attractive means to monitor such fermentations. Unfortunately, this technique is cost prohibitive in large-scale fermentors unless very cheap fluorophors become available.

In the Biotechnology Laboratories at Lehigh University, we have been investigating the other two approaches based on fluorescence measurements. In one case we monitor the reduced pyridine nucleotide fluorescence at 460 nm obtained upon exciting the whole broth at 366 nm. This signal is indicative of the cellular concentrations.[8] If properly corrected for effects of dissolved oxygen and energy substrate concentrations on cellular metabolism, attenuation of the signal by media components and cells, and the environmental influence of pH, temperature and dissolved oxygen, the fluorescent signal may provide a means of determining cell concentrations in most fermentations. This requires a knowledge base that includes information on the whole-culture background fluorescence and the relationship between metabolic activity and fluorescence response of a culture. While experimentally quantifying the relationships for each fermentation system takes time, nonetheless once this is done the fluorometric signal can be the most accurate, as well as an instantaneous and continuous, means for monitoring cell concentration on-line. Time differentiation of this signal will give cell growth rates.

In the second approach we feel that if one is to ultimately perform meaningful control of fermentation systems it will be necessary to monitor substrate and product concentrations as well as cell concentrations. As a consequence, we have been investigating laser scanning fluorescence spectra of aromatic amino acid containing proteins (peptone, invertase) and yeast in various media in an attempt to discern whether certain substrate and product concentrations can be continuously monitored.[8] We believe that with the use of

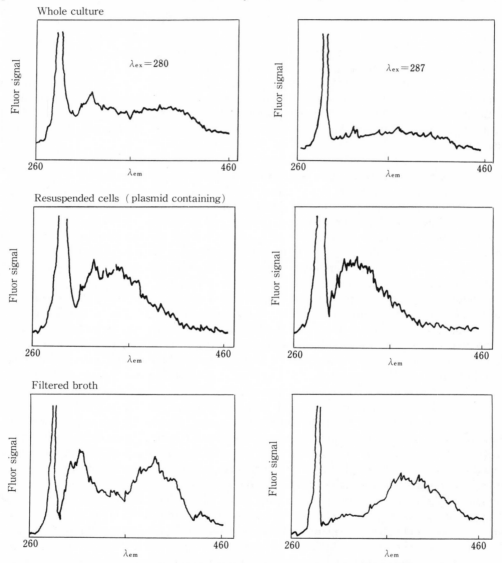

FIG. 2a. Fluorometric spectra for *S. cerevisiae* RTY 110 (pRB58) grown on GYP medium, where λex: excitation wavelength, nm, λem: emission wavelength, nm.

several light pipe systems and utilizing several unique excitation wavelengths and then monitoring perhaps a half-dozen unique fluorescent wavelengths, it will be possible to monitor amino acid (or protein) uptake and enzyme product formation as well as cell concentration and growth rate in certain yeast fermentations.

To demonstrate this concept we are using a yeast (*Saccharomyces cerevisiae*) auxotroph (RTY110, *leu⁻*, *uracil⁻*, *suc⁻*) that has a plasmid (pRB58) containing the uracil (*ura⁺*) and invertase (*suc⁺*) genes plus several other appropriate markers as a model system for studying the potential of scanning fluorometric monitoring. Since the clone expresses the *suc⁺* gene in the form of a glycosylated invertase, an active culture will excrete appreciable quantities of invertase. To date we have excited various plasmid-containing and plasmid-cured yeast cultures at 280 and 287 nm grown on both rich and defined media. There clearly is a difference between the plasmid-containing and plasmid-cured cultures (see Fig. 2a and 2b). Also, yeast and invertase separately contribute to the overall fluorescent spectra (Fig. 3).

We are now extending the excitation spectrum to higher (300 nm) and lower (260 nm) wavelengths in hopes that the excitation/emission spectra can be resolved using on-line computer techniques with sufficient accuracy to measure cell and invertase concentrations

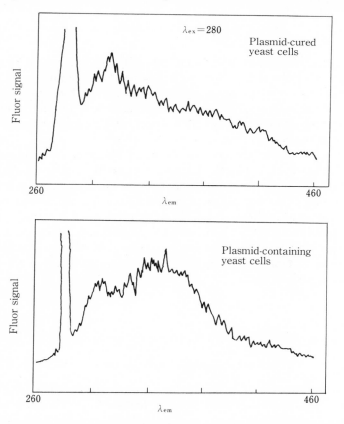

FIG. 2b. *S. cerevisiae* RTY110 grown on GYP medium - comparison of plasmid (pRB58)-containing and plasmid-cured yeast system.

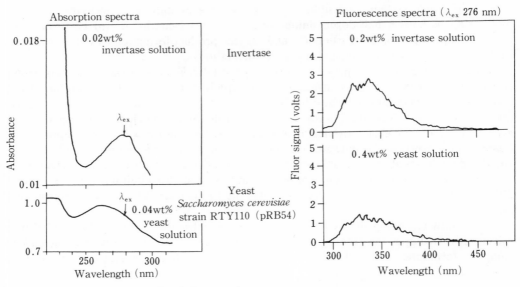

FIG. 3. Invertase/yeast fluorescence spectra.

for control purposes. We also are attempting to develop a tryptophan auxotroph of RTY110 in order to see if we can use fluorometric techniques for monitoring and controlling the tryptophan concentration of the broth when used as a growth limiting substrate. We are very optimistic on the basis of preliminary results and believe that in the near future these and other spectroscopy techniques will allow direct on-line measurement of cell, substrate, and product concentrations for control purposes.

The next step is to develop sensors that will allow one to monitor internal components of the cell. This will make it possible to use "structured" models in various bioreactor control algorithms. We have yet to apply the full potential of new spectroscopic techniques including such things as nuclear magnetic resonance and Raman spectroscopy. I feel confident that this will occur in the next decade. At that time we will see the many monitoring and control ideas initiated by Dr. Aiba twenty-five years ago finally come to fruition. On behalf of my many biochemical engineering colleagues, I thank Dr. Aiba for this and many other contributions to biochemical engineering and wish him well in his retirement. We will miss his advice and thoughts.

REFERENCES

1. Aiba, S., Humphrey, A.E., Millis, N.F., *Biochemical Engineering*, p. 267–278, Academic Press, New York (1965).
2. Aiba, S., Humphrey, A.E., Millis, N.F., *Biochemical Engineering*, 2nd Ed., p. 332–342, Academic Press, New York (1973).
3. Arminger, W.B., Humphrey, A.E., "Computer Applications in Fermentation Technology." In *Microbial Technology*, Vol. 2, 2nd Ed. (H.J. Peppler, D. Perlman) (eds.), p. 375–401, Academic Press, New York, 1979.
4. Bandyopadhyay, B., Humphrey, A.E., Taguchi, H., "Dynamic Measurement of the Volumetric Oxygen Transfer Coefficient in Fermentation Systems." *Biotechnol. Bioeng.*: **9**, 533–544 (1967).

5. Fieschko, J., Humphrey, A.E., "Statistical Analysis in the Estimation of Maintenance and True Growth Yield Coefficients." *Biotechnol. Bioeng.*: **26**, 394–396 (1984).
6. Harrison, D., Haimes, C., Humphrey, A.E., "Design Instrumentation and Control of Fermentation Systems." paper presented at 10th Intl. Congr. Microbiol., Mexico City (1970).
7. Humphrey, A.E., "Fermentation Process Modelling, An Overview." *Ann. N.Y. Acad. Sci.*: **326**, 17–33 (1979).
8. Humphrey, A.E., "The Use of Fluorometry in the Monitoring and Control of Cell Cultures." paper presented at 2nd Conference to Promote Japan/US Joint Projects and Cooperation in Biotechnology, Lake Biwa, Japan (1986).
9. Phillips, J.A., "Use of Infrared Spectroscopy for the Online Multicomponent Analysis and Control of Fermentations." Paper presented at 2nd Conference to Promote Japan/US Joint Projects and Cooperation in Biotechnology, Lake Biwa, Japan (1986).
10. Pirt, S.J., *Principles of Microbe and Cell Cultivation*, p. 63–97, Blackwell Scientific, Oxford (1975).
11. Solomons, G.L., *Materials and Methods in Fermentation*, p. 261–280, Academic Press, New York (1969).
12. Wang, D.I.C., Cooney, C.L., Demain, A.L., Dunnill, P., Humphrey, A.E., Lilly, M.D., *Fermentation and Enzyme Technology*, p. 220–236, John Wiley and Sons, New York (1979).
13. Zabriskie, D.W., Armiger, W.B., Humphrey, A.E., "Applications of Computers to the Indirect Measurement of Biomass Concentration and Growth Rate by Component Balancing." In *Workshop on Computer Applications in Fermentation Technology* (R.P. Jefferis) (ed.), GBF Mono. Ser. No. 3, p. 59–72, Verlag Chemie, Weinheim, 1976.

Transient D.O. Measurement Using a Computerized Membrane Electrode

Xiang-Ming Li and Henry Y. Wang

Department of Chemical Engineering, The University of Michigan, Ann Arbor, Michigan 48109, U.S.A.

1. Introduction

Our first contact with the field of biochemical engineering was through a book titled *Biochemical Engineering* written by Professor Shuichi Aiba and his colleagues.[1] Throughout the years, we have read very carefully any articles we came across by Professor Aiba and his co-workers. Even though one of us (H.Y. Wang) was in the U.S.A. and the other (X.M. Li) in China, our careers were deeply influenced by Professor Aiba. His contributions to the field of biochemical engineering and his ease in integrating basic and applied research, engineering, and biological sciences laid the foundations of this new discipline called biochemical engineering. One can encounter his work in photosynthesis, sterilization, kinetics, bioseparation, oxygen transfer, and recently, in the field of recombinant DNA technology of thermophilic *Bacillus*. In honor of his retirement, we describe here a research project on dissolved oxygen measurement in our laboratories, an area Professor Aiba touched upon in his earlier career.[1]

2. Dissolved Oxygen Measurement

Dissolved oxygen (D.O.) has been demonstrated to be a critical parameter for many aerobic fermentations and waste treatment processes. Many D.O. electrodes have been developed and are now commercially available. In general, most electrochemically based D.O. electrodes can be classified into two types: galvanic and polarographic. They are membrane electrodes composed of an anode and a cathode in contact with an electrolyte. An oxygen-permeable membrane is used to cover the electrode. As oxygen diffuses through the membrane and reaches the cathode surface, current or voltage output is generated when oxygen reduction occurs at the cathode surface. Like any other membrane electrodes, these D.O. probes are subject to fouling and are sensitive to the flow conditions of the bioreactor. Constant recalibration may be needed to obtain accurate D.O. measurement. Unfortunately, these may not meet many *in situ* monitoring and rigorous sterility requirements.

Galvanic probes are favored in fermentation and cell culture monitoring due to their ease in construction and also, to their feasibility of sterilization *in situ*. Basically, two modes of operations, steady state and transient, can be used to sample the current output of a galvanic D.O. electrode. The steady-state mode is based on measuring the steady-state current output from the electrode. This current is directly proportional to the oxygen flux reaching the cathode surface if no change occurs in the local environment.[2] Besides the bulk oxygen concentration, which dictates the oxygen flux and is the parameter to be measured, the oxygen transport resistance including resistance inside and outside of the electrode (electrolyte layer, membrane, fouling, and boundary layers) also influences the oxygen flux. Using thicker membrane will reduce the sensitivity of the measurement to the change of oxygen transport resistances via aggravation of the electrode both in sensitivity and in response time.

Transient-mode operation has been successfully applied to a polarographic electrode using pulse voltametric techniques.[3,4] In our laboratory, we have developed a computerized galvanic D.O. electrode for oxygen measurement. A relay controlled by a microcomputer was used to open and to close the electrical circuit of the oxygen electrode. The oxygen concentrations were determined using both computer simulation and a parameter estimation technique based on the measured current-time history. By measuring the transient current-time curves for rapid cycles of opening and closing the electrode circuit in an appropriate way, we are able to show this transient D.O. measurement is insensitive to fouling and to fluid flow conditions.

3. Theoretical Development

The tip of a membrane electrode consists of four layers as shown in Fig. 1: (1) electrolyte layer, (2) membrane, (3) fouled layer, and (4) boundary layer of fluid near the tip of the electrode. The oxygen transport processes in these layers can be described by the following equations and boundary conditions:

$$\frac{\partial a_i}{\partial t} = D_i \frac{\partial^2 a_i}{\partial x^2}, \quad \text{for } i = 1, 2, 3, 4. \tag{1}$$

Equation (1) is Fick's second law applied to these four different layers. The oxygen activity $a(= \gamma C)$ is considered as the transport driving force.

B.C.:

$$\frac{a_i}{S_i} = \frac{a_{i+1}}{S_{i+1}} \qquad \text{at } x_i, \quad i = 1, 2, 3 \tag{2}$$

$$D_i \frac{\partial a_i}{\partial x} = D_{i+1} \frac{\partial a_{i+1}}{\partial x}, \quad \text{at } x_i, \qquad i = 1, 2, 3 \tag{3}$$

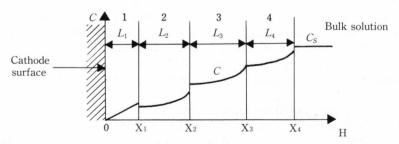

FIG. 1. Schematic diagram of the four-layer model.
1. Electrolyte layer, 2. Teflon membrane, 3. Fouling layer, 4. Mass transfer boundary layer.

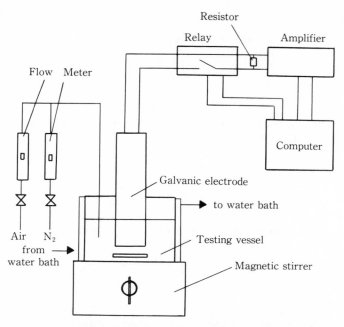

FIG. 2. Transient D.O. measurement system.

$$a_4 = \gamma_4 C_s \quad \text{at } x_4. \tag{4}$$

Equations (2) and (3) are equilibrium and mass conservation conditions at the boundaries between adjacent layers. In Eq. (4), the oxygen activity at the outer surface of the boundary layer is assumed equal to the product of the activity coefficient and the bulk concentration.

In the case of transient D.O. measurement, a relay controlled by a microcomputer is used to open and to close the electrode circuit (Fig. 2). When the electrode circuit is closed, the transient current-time curve can be monitored by the computer. After 1 second(s), the circuit is reopened and the reduction of oxygen at the cathode surface is stopped. The oxygen concentration is allowed to be replenished at the tip of the membrane electrode for a relatively short time period (about 1 min). Another oxygen measurement can be taken when the relay is closed again:

$$\text{during relay closed:} \quad a_1 = \gamma_1 C_1 = 0, \quad \text{at } x = 0 \tag{5}$$

$$\text{during relay open:} \quad \frac{\partial a_i}{\partial x} = 0, \quad \text{at } x = 0. \tag{6}$$

Equation (5) is the boundary condition at the cathode surface when the relay is closed. Since the reduction of oxygen at the cathode is much faster than oxygen transport to the surface, the oxygen concentration at the cathode surface can be safely assumed to be zero. Equation (6) is the impermeable condition at the cathode surface when the relay is opened. In this time period, the oxygen reduction stops, and oxygen concentration in the electrolyte layer and the membrane is replenished.

I.C.:

$$at\ t = 0, \qquad a_4 = \gamma_4 C_s; \tag{7}$$

$$a_3 = a_4/S_4 \times S_3; \tag{8}$$

$$a_2 = a_3/S_3 \times S_2; \tag{9}$$

$$a_1 = a_2/S_2 \times S_1. \tag{10}$$

These Eqs. (7–10) are the initial oxygen concentrations in different layers and are assumed to be in equilibrium with each other.

Current Output Equation

$$i = 4F\,A\gamma_1 D_1 \frac{\partial C_1}{\partial x}, \qquad at\ x = 0. \tag{11}$$

At steady state, the following analytical solution can be obtained:

$$i_{ss} = \frac{4F\,A\gamma_4 C_s}{\dfrac{L_4}{D_4} + \dfrac{L_3 S_4}{D_3 S_3} + \dfrac{L_2 S_4}{D_2 S_2} + \dfrac{L_1 S_4}{D_1 S_1}}. \tag{12}$$

The first and the second terms in the denominator on the right-hand side of Eq. (12) represent the oxygen transport resistances in the boundary and fouling layers, respectively. These terms change when the flow condition is changed (especially in viscoelastic fluid) or the membrane becomes fouled. The steady-state current output, i_{ss}, will remain nearly unchanged only if

$$\frac{L_4}{D_4} + \frac{L_3 S_4}{D_3 S_3} \ll \frac{L_2 S_4}{D_2 S_2} + \frac{L_1 S_4}{D_1 S_1}. \tag{13}$$

This means a very thick membrane should be used, which will compromise the sensitivity and the response time of the electrode. Our way to circumvent this problem is to operate the membrane electrode in a transient mode, meaning that the oxygen transport resistances outside the membrane will not affect the transient current output but only the length of the rest period required for oxygen to be replenished near the tip of the membrane electrode.

Based on the above mathematical model, a computer program has been developed in our laboratories to use transient-mode operation for dissolved oxygen measurement. During calibration, the electrode is placed in a solution with a known oxygen. The current-time curve is sampled in transient mode. The system parameters, such as the active surface of the cathode, thickness of the electrolyte layer, and the diffusivity of oxygen in the electrolyte, are estimated by using a parameter estimation technique[5] with the measured current-time curve. With an initial set of parameters, using a finite difference method to solve Eqs. (1–11) a theoretical current-time curve is generated. The simulated current-time curve is obtained by superimposing the residual current on the theoretical current-time curve. The parameter estimation routine is used to search for the best parameters that minimize the least square errors between the simulated results and the measured current-time curve. The best fitting parameters are kept in data files for further measurement application. The calibrations were performed in three temperature levels. During measurement, the unknown parameter is the oxygen concentration in the bulk solution. The system parameter files obtained in calibration are retrieved and used for estimating the oxygen concentration based on the measured current-time curve. If the temperature in the solution is not the same as the temperature in calibration, the system parameters will be interpolated using second-order Lagrange polynomials according to the temperature.

A specially constructed galvanic D.O. probe was used in conjunction with a COMPAQ portable II computer (Model 4, COMPAQ Computer Corp., Texas). A DASCON-1 interface board (MetraByte, Taunton, Maryland) with a terminal board (Model STA01) and a relay board (Model ERA01) were used for data acquisition.

4. Experimental Results

Preliminary experiments were carried out in a temperature-controlled vessel containing water saturated with a defined oxygen and nitrogen gas mixture. The gas mixture composition could be altered by mixing appropriate portions of compressed air and nitrogen. The water was constantly stirred with a magnetic stirrer and the D.O. membrane electrode was fixed at a constant location in the vessel. The water level in the vessel remained constant throughout the experiments. Parts of the measured current-time curves and the simulated current-time curves were compared and good agreement was obtained (Fig. 3).

The effect of replenishing time (when the relay is open) on the transient current output was also studied. Figure 4 shows three sequential current-time curves with a replenishing time of 30 s. The gradual drop of the current-time curves at sequential sampling cycles indicates that the replenishing time was not long enough to allow oxygen to reach equilibrium inside the tip of the membrane electrode before the next sampling was started. Figure 5 shows three sequential current outputs with a replenishing time of 60 s. The current-time curves remained identical in three sequential sampling cycles, indicating the replenishing time was more than adequate for transient D.O. measurement in these particular experi-

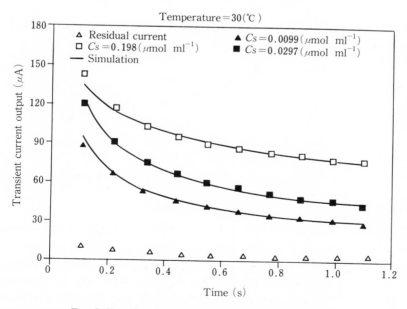

FIG. 3. Transient current output and simulation results.

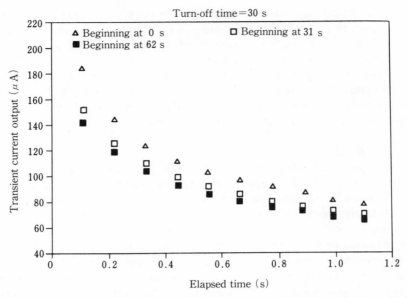

FIG. 4. Effect of replenishing time on transient current output (turn-off time = 30 s).

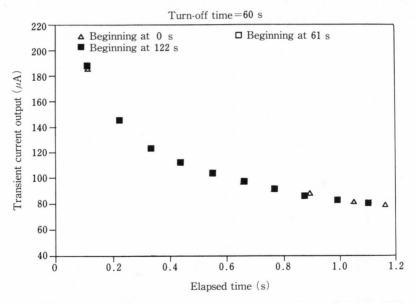

FIG. 5. Effect of replenishing time on transient current output (turn-off time = 60 s).

mental conditions. Additional replenishing time is required in the case of extremely fouled conditions. If the oxygen concentrations in the electrode layer and the membrane of the electrode reach equilibrium before each sampling cycle and the sampling time is short enough to prevent a change in concentration profile reaching the outer surface of the mem-

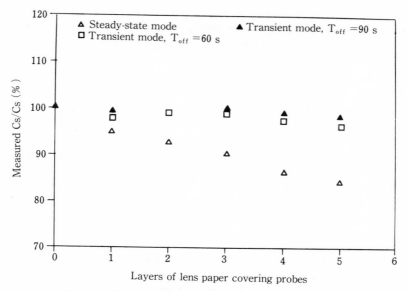

FIG. 6. Effect of simulated fouling on D.O. measurement.

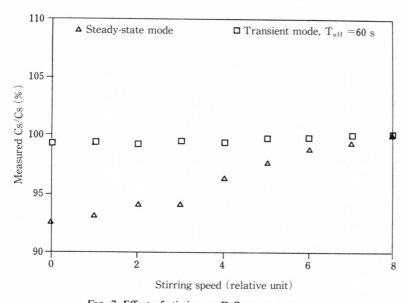

FIG. 7. Effect of stirring on D.O. measurement.

brane, fouling itself should not influence the transient current output but will influence the time required for oxygen to reach equilibrium in the membrane tip before next sampling.

When lens paper was used to cover a membrane D.O. electrode and to simulate fouling, the same electrode was operated under transient mode and steady-state mode to compare the fouling effect. Two replenishing times, 60 and 90 s, were used in the test and the results are shown in Fig. 6. When five layers of lens paper covered the electrode, the oxygen concentration predicted by the steady-state measurement was off by 15%. Those predicted by the transient mode measurement were off from the actual values by 2% (replenishing time = 90 s) and 3% (replenishing time = 60 s). This shows that the transient mode measurement is less prone to error caused by fouling when compared with the steady-state measurement technique. The replenishing time must be adjusted to be sufficient for the dissolved oxygen to reach equilibrium inside the membrane tip.

The effects of changing the fluid flow conditions on the steady-state and transient modes of measurements were studied using the same electrode to measure dissolved oxygen concentrations at different agitation speeds. The experimental results are shown in Fig. 7. Again, it was demonstrated that the transient mode measurement gave more reliable and reproducible results, and was less sensitive to stirring than steady-state measurement. This computerized D.O. measurement system is now undergoing testing in various fermentation processes and biological waste treatment plants.

5. Conclusion

We have demonstrated under laboratory conditions that transient dissolved oxygen measurement may be more appropriate than steady-state measurement in extremely fouled conditions and varied fluid flow systems. Our transient D.O. measurement system does require a microcomputer. However, as more and more computers are being used for fermentation and cell culture monitoring and control, the transient D.O. measurement function can be incorporated into existing computers with little modification of the hardware and software.

Nomenclature

A	active surface area of cathode, cm^2
a	activity of oxygen
C	oxygen concentration, g mol cm^{-3}
D	diffusivity of oxygen, cm^2 s^{-1}
F	Faraday constant, 9.65×10^4 Coulomb g mol^{-1}
i	current output, A
L	thickness, cm
S_i	solubility of oxygen in i-th layer, g mol cm^{-3} (cmHg)$^{-1}$
t	time, s
x	space coordinate, cm
γ	activity coefficient of oxygen

Subscript

i	for i-th layer
s	bulk solution
ss	steady state

Acknowledgments

We acknowledge partial financial support from the National Science Foundation. One of us (X.M. Li) also acknowledges the Michigan Biotechnology Institute (MBI) for a postdoctoral traineeship.

REFERENCES

1. Aiba, S., Humphrey, A.E., Millis, N.F., In *Biochemical Engineering*, 2nd Ed., p. 327–329, Academic Press, Inc., New York and London, 1973.
2. Lee, Y.H., Tsao, G.T., "Dissolved oxygen electrodes." *Adv. Biochem. Eng.*: **13**, 35–86 (1979).
3. Mancy, K.H., Okun, D.A., Reilley, C.N., "A galvanic cell oxygen analyzer." *J. Electroanal. Chem.*: **4**, 65–92 (1962).
4. Mancy, K.H., "In-situ measurement of dissolved oxygen by pulse and steady state voltamertic membrane electrode systems." In *Chemistry and Physics of Aqueous Gas Solutions*, p. 281–289, W.A. Adams (ed.), 1975.
5. Bech, J.V., Arnold, K.J., In *Parameter Estimation in Engineering and Science*, p. 117–129, John Wiley and Sons, New York, 1977.

IV-3
Calorimetry

Makoto Shoda

Research Laboratory of Resources Utilization, Tokyo Institute of Technology, Midori-ku, Yokohama, Japan

1. Introduction

Calorimetry has been used in the biological field to investigate the thermodynamics of various biochemical reactions. The heat evolution rate, which is proportional to the reaction rate in question, can be used for kinetic analysis and/or for monitoring a biological process. The total amount of heat, which is also proportional to the amount of organic materials consumed, can be used to estimate the energy efficiency of catabolism and/or which substrates are concerned. Acquisition of thermodynamic data is required before designing even simple, basic fermentation-process equipment such as heaters and/or coolers.

Despite the versatility of calorimetric investigation, the principal reason that has hindered its application to more diversified fields is that the analysis is too general. However, the calorimetric method is very useful whenever quick detection of an anomaly in a complex series of reactions, such as that in living cells or in soil materials, is needed. Reference books on biological calorimetry[1,2] have been published, and it is now possible to gain detailed insight into this technique. This paper will give a brief review of recent achievements.

2. Liquid Phase

2.1 Qualitative Analysis

The heat evolution rate from a culture broth, which is measured by calorimetry in "thermograms" or "power of heat-time curves" and is continuously recorded on a chart during cultivation, is dependent on the culture conditions. Hence, any chemical or physical change in the cultured broth is reflected *in situ* and on the thermogram. This sensitive response is useful when monitoring and/or control of a specific fermentation process is required. Figure 1 demonstrates clearly the differences in thermograms of carbon-, nitrogen- and oxygen-limiting cultivations, in which *Escherichia coli* was grown aerobically in a synthetic medium.[3] Discontinuities in the heat evolution rate such as shown in I-1 to III-3 throughout the figure correspond to points where nutritional exhaustion and/or commencement of by-product consumption by the cells might have occurred.

Lactic acid bacteria are employed in food preservation owing to their ability to lower the pH value, thus minimizing the growth of other contaminants in the food. A few species of lactic acid bacteria give rise to spoilage or foul fermentation even in a low pH juice or

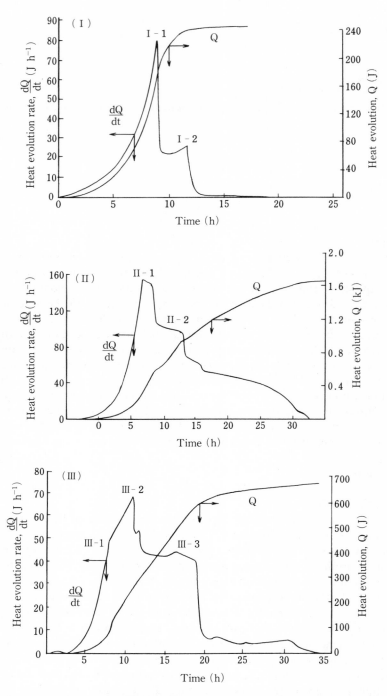

FIG. 1. Power-time curves in a glucose-limited (I), a nitrogen-limited (II), and an oxygen-limited (III) batch culture of *E. coli*. dQ/dt: heat evolution rate (J h^{-1}); Q: heat evolved (J).[3]

in alcoholic beverages. Monk[4] determined the rates of anaerobic as well as aerobic catabolism of glucose in terms of heat evolution rates by using a suspension of micro-aerophilic bacteria, *Streptococcus agalactiae*, under different pH conditions. Flow microcalorimetry was skillfully used in his experimentation. In an aerobic cultivation, a cell suspension containing glucose was aerated and pumped through a flow calorimeter. After confirming it was in a steady state, anaerobic conditions were established by cutting off the flow rate of the suspension, thus allowing the bacteria to exhaust completely the dissolved oxygen that was available at the time of flow stoppage through the calorimetric vessel. A constant heat evolution rate, if observed, would be due to anaerobic catabolism of glucose.

The effect of pH on heat evolution is shown in Fig. 2. DNP (2,4-dinitrophenol) used in the figure is an uncoupler of catabolic energy. Over a pH range from 5.7 to 6.7, the rate of glycolysis by this bacterium remained almost unchanged regardless of whether conditions were aerobic or anaerobic. The molar yield of lactic acid from glucose anaerobically was near 2.0 regardless of the presence or absence DNP. This implies essentially that homolactic fermentation was going on, whereas aerobically the same molar yield was 1.47 and 1.62 of lactic acid without and with the addition of DNP, respectively. The increased rate of heat evolution in an aerobic culture as noted in Fig. 2 might be due to conversion of about 20% of the glucose to unknown products other than lactate. Obviously, the presence of DNP increased the rate of glucose catabolism in both aerobic and anaerobic fermentations.

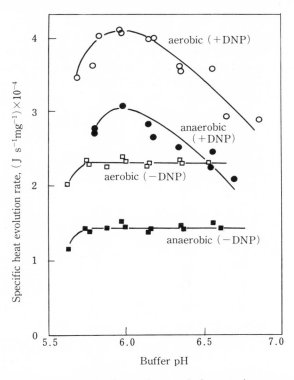

Fig. 2. Effect of pH of 0.2 M phosphate buffer on heat evolution rate in a suspension of *S. agalactiae* catabolizing aerobically with and without 250 μM DNP (○, □) and anaerobically with and without the same concentration of DNP (●, ■).[4]

Thermal behaviors associated with metabolism of the energy source in the aerobic grow-ing phase of microorganisms have been studied comprehensively elsewhere,[1,2] and those in the non-growing phase are also of interest. Nunomura and Fujita measured the heat evolution rate from the endogenous metabolism in yeast culture aerobically by transferring the washed cells into a buffer solution in the absence of any exogenous energy source.[5] The heat evolution rate was obtained by following the amount of energy source that had been stored in the cells. The amount varied depending on the pre-culture conditions and the strains used. An example of the heat evolution rate pattern is shown in Fig. 3. The average value of enthalpy change was estimated from the thermogram to be 719 J g dry cell^{-1}. This enthalpy change is equivalent to 46 mg of glucose, where the enthalpy change in complete combustion of glucose was taken as 15.6 J mg glucose^{-1}.

Upon the addition of glucose to the non-growing cells in the above-mentioned culture, the energy source was converted to ethanol quantitatively, i.e., $C_6H_{12}O_6 \rightarrow 2C_2H_5OH + 2CO_2$, and ethanol was consumed only after the glucose had been exhausted. Most of the heat evolved stemmed from both glycolysis and aerobic consumption of ethanol.[6]

The transport system of glucose in yeast cells was estimated in phosphate buffer by using non-metabolizable 3-O-methyl-D-glucose (3-O-MG).[7] When iodoacetic acid (IAA) was added as the glycolysis inhibitor, no heat evolved. This suggests that glucose transport is associated either with endogenous respiration or with glycolysis. The enthalpy change in the transport of 3-O-MG was -1660 KJ mol^{-1} and this value is almost equivalent to that of the active transport process of E. coli.[7]

A similar analysis was conducted in E. coli in phosphate buffer using a titrimetric calori-meter.[8] The heat evolution pattern is shown in Fig. 4 for different kinds of carbon sources. The first sharp peak for fructose was due to fructose uptake; the second and third peaks reflected glycolysis and respiration of cells, respectively. The uptake of glucose and fructose was carried out via phosphotransferase, while that of glucose-6-phosphate was via per-

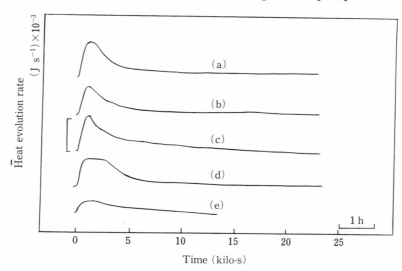

FIG. 3. Profiles on heat evolution rate vs. power-time among different yeast strains. (a) *Saccharomyces carlsbergensis* IAM 4727; (b) *S. cerevisiae* IAM 4171; (c) baker's yeast; (d) commercial dry yeast; (e) *Candida utilis* IAM 4961.[5]

mease. The different patterns in Fig. 4 indicate the differences in the transport systems.

Various extra-carbon sources were added to glucose-limited and steady state continuous cultures of *Klebsiella aerogenes*, and the response of power output (heat evolution rate) to each addition of carbon source was compared (Fig. 5).[9] Glucose and pyruvate belonged to Group 1. Glucose-1-phosphate, glucose-6-phosphate, sucrose, and all the C-4 compounds of the TCA cycle (malate, succinate, and fumarate) fell into Group 2. Formate, acetate, and glycerol were in Group 3, and citrate and α-ketoglutarate were in Group 4. No general conclusion has been drawn yet from the comparison. However, a more detailed analysis would merit further use of calorimetry.

Meanwhile, regarding transitory changes from one steady state to another in continuous cultures, those similar to the responses shown in Fig. 5 were observed, and hence the steady state could be easily confirmed by monitoring the levelling off of the heat evolution rate.[10]

2.2   Quantitative Analysis

The amount of heat evolved even from a definite microbial system is subject to change depending on the difference in medium composition and/or that of the measuring device. To minimize deviation, it is important to maintain the reaction conditions unaltered throughout and also to prevent any disturbances to the system. The simplest way of accomplishing these is to use a minimal amount of substrate (less than the amount of oxygen available) and in addition to minimize the amount of cells used. This precaution avoids oxygen deficiency and/or prevents extra heat generation, such as titration heat of acid produced from the cells.

Lovrien et al.[11] used a sugar concentration in a range less than 10^{-4} M and a diluted suspension of 5×10^7 cells ml^{-1} to facilitate the measurement of the heat evolution rate. Heat evolved was measured as a function of substrate and the evolution rate was extrapolated to the case where the substrate concentration was zero. The extrapolated value of

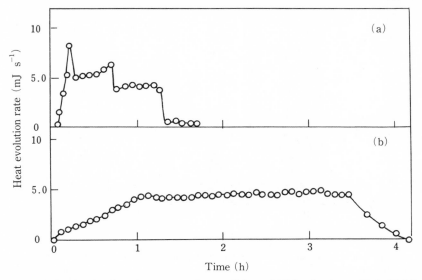

FIG. 4.   Heat evolution by *E. coli* under non-growing conditions. (a) 0.3 M fructose added; (b) 0.3 M glucose-6-phosphate added.[8]

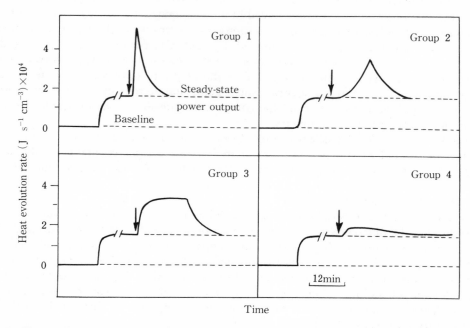

FIG. 5. Change of power output caused by the addition of extra carbon source (arrow) to a steady-state growing culture of *K. aerogenes* in a glucose-limited chemostat culture at D = 0.52h⁻¹.[9]

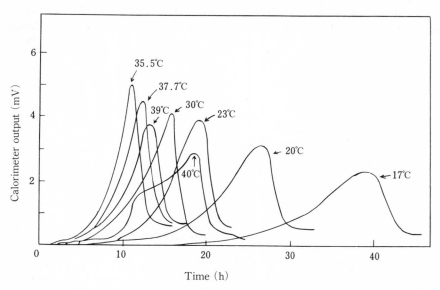

FIG. 6. Growth thermograms of baker's yeast at various temperatures. The ordinate is proportional to the heat evolution rate.[15]

TABLE 1. Maximum growth rate constant and substrate constant determined at various temperatures.[15]

Temp. (°C)	μ_m (h^{-1})	K_s (mM)
16.0	0.148	1.14
20.0	0.212	1.34
23.6	0.304	1.32
30.0	0.433	1.38
35.5	0.483	1.60
37.7	0.416	1.35
39.0	0.393	1.42

heat evolution, defined as the limiting-differential molar heat of metabolism, gives the heat of metabolism for which the least degree of correction of the data is required. According to this analysis, the molar enthalpy of glucose by *E. coli* in aerobic cultivation was -1273 kJ mol^{-1}.[11] This value is larger than those obtained separately by other workers, i.e., -818,[12] -1017,[13] and -1076[3] kJ mol^{-1}. In anaerobic cultivation, the value was -142 kJ mol^{-1}. This value of glucose enthalpy in anaerobic cultivation was consistent with that previously reported.[14]

Whatever the contradictory picture might be, calorimetry is a convenient means for monitoring the adaptation period when one species of sugar in a microbial culture is switched off to another, because the calorimetric response to the change is quite rapid.

Calorimetry was applied by Itoh and Takahashi to their kinetic studies on the growth of baker's yeast at various temperatures.[15] Figure 6 shows an example of the power-time response in growth at different temperatures. When the Monod equation was used to represent the power increasing phase in the thermograms, μ_m and K_s values as functions of temperature were assessed (Table 1). It is interesting to note that the maximum growth rate was dependent on temperature, while the saturation constant K_s was almost independent of it. The activation energy with regard to μ_m was 56.3 kJ mol^{-1}.

Calorimetric analysis can provide thermochemical constants of the microbial systems in question. The overall growth yield as correlated with true growth and maintenance that has been derived by Pirt is:

$$1/Y_c = m_e/\mu + 1/Y_c^{\max} \tag{1}$$

where m_e is the maintenance coefficient, Y_c the overall growth yield, $Y_c^{\max}$ the maximum value of Y_c or true growth yield, and μ the specific growth rate.

Within a limited range of μ (or D:dilution rate) in a given microbial growth system, it is generally accepted that $1/Y_c$ is linearly correlated with $1/\mu$ (or $1/D$). According to James et al.,[16] however, the linear relationship between $1/Y_c$ and $1/\mu$ was not apparent when they measured the heat evolution by *Klebsiella aerogenes* grown in a glucose-limited chemostat culture. ΔH_g (kJ g^{-1} (gen.)$^{-1}$), which is defined as enthalpy change in the formation of 1 g of cells for every generation time, could be correlated well with the Y_c values as shown in Fig. 7. Both plots of ΔH_g and Y_c against D^{-1} consisted of two lines, respectively, as is apparent from Fig. 8 in which two values of maintenance coefficient (m_e) are assessed. To account for this anomalous phenomenon, the following equation was derived by rearranging Eq. (1):

$$D/Y_c = D(1/Y_{c(\mathrm{th})}^{\max} + m_g) + m_e' \tag{2}$$

230   M. Shoda

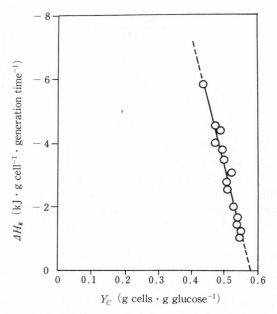

FIG. 7. ΔH_g (enthalpy change) as a function of the yield coefficient (Y_c) for cells of *K. aerogenes* grown in a glucose-limited chemostat culture at 37°C.[16]

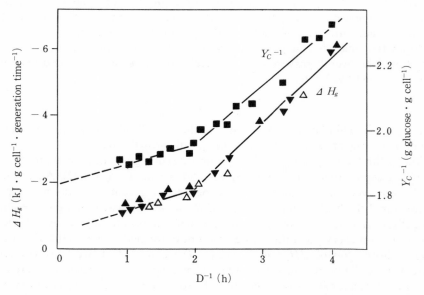

FIG. 8. Both plots of Y_c^{-1} and ΔH_g against D^{-1} for cells of *K. aerogenes*. For the plot of ΔH_g against D^{-1}, glucose concentrations were changed: ▲ 1.9; ▼ 4.0; △ 4.9 (mmol dm⁻³).[16]

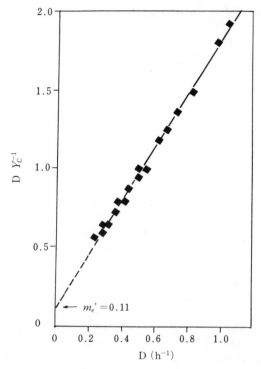

FIG. 9. Replot of DY_c^{-1} data in Fig. 8 against D.[16]

where m_g is the growth rate-dependent maintenance coefficient, g glucose (g cell)$^{-1}$; m'_e the growth rate-independent maintenance coefficient, g glucose (g cell)$^{-1}$h^{-1}; and $Y_{c(\text{th})}^{\max}$ an intercept on the abscissa in Fig. 7.

As shown in Fig. 9, data on D/Y_c that were replotted from Fig. 7 became linear against D and thereby the values of both m_g and m'_e could be obtained. It was found that the maintenance coefficient (or energy) was composed of two parts, one, $D \cdot m_g$, dependent on growth rate, and the other, m'_e, independent of it.

Values of enthalpy change obtained in batch and continuous cultures often do not agree. James et al.[17] also showed that ΔH_g was -2.6 kJ g^{-1} gen.$^{-1}$ in batch culture of *K. aerogenes* in a glucose-limited medium with generation time of 45 min, whereas the corresponding value in chemostat culture (D = 0.9 h^{-1}) was -1.05 kJ g^{-1} gen.$^{-1}$. The discrepancy suggests that an excess amount of carbon source present in the medium was wastefully consumed, thus liberating a large amount of heat. This inference was confirmed to be plausible in nitrogen-limited or magnesium-limited cultures; at a given dilution rate, ΔH_g values in nitrogen- or magnesium-limited cultures were higher than the corresponding value obtained from glucose-limited chemostat by a factor of about 3. Similar results have been obtained in the cultivation of *E. coli* (Table 2).[18] When azide, which acts on the site F$_1$-ATPase as an uncoupler between catabolism and anabolism, was added to glucose-limited culture, the specific heat evolution for *K. aerogenes* almost doubled and a 20–30%

TABLE 2. Heat evolution, ΔH_m (kJ mol glucose^{-1}), and fraction of catabolic energy used for anabolism, R(%), in a culture of E. coli.[18]

	Dilution rate, D (h^{-1})	ΔH_m (kJ mol^{-1})	R* (%)
	0.093	−949	62.2
Continuous	0.176	−970	60.9
culture	0.26	−903	60.8
	0.52	−460	53.4
	Glucose-limiting	−1074	61.7
Batch	Oxygen-limiting	−1220	48.7
culture	Nitrogen-limiting	−1689	36.8

* R = (heat of combustion of the cells) × 100/(heat of combustion of the cells + ΔH_m).

decrease in biomass was observed. When azide was added to nitrogen-limited culture, heat evolution increased only by 20%. This fact implies that there was high ATPase activity in the nitrogen-limited chemostat, supporting the high level of uncoupling.[9]

Most calorimetric devices for laboratory or industrial use are of the flow type. Technical difficulties such as oxygen deficiency, adhesion of cells to the inner surface of tubes, precipitation of cells, etc. are inherent in flow-type devices. Cooney et al. demonstrated the usefulness of dynamic calorimetry.[13] However, this system sometimes requires a sensitive temperature-measuring device and hence an accurate timer is required to evaluate the heat evolution rate accurately. In this context, Bayer et al. proposed the use of an impulse counter embedded in a microcomputer and platinum thermister.[19] They measured thermograms on Candida intermedia grown in a sucrose-limited chemostat culture. The temperature control of the reactor was suspended periodically, and temperatures at both ends of the noncontrol period were recorded. Although the heat evolution rate can be correlated with the rate of oxygen consumption,[20,21] they used instead the oxygen content in the exhaust gas monitored continuously with a gas analyzer and the temperature increase. An empirical equation for the growth of this yeast as functions of the oxygen content of exhaust gas and its temperature increase is:

$$P = 35.61 - 1.42 \, dT - 1.73 \, O_2 \tag{3}$$

where P is the biomass productivity (g l^{-1} h^{-1}); dT the net temperature increase (°C); and O_2 the oxygen content in the exhaust gas (%).

A quick response in this calorimeter is essential, especially when a change in substrate species occurs. When sucrose was pulsed into the reactor, the biomass productivity calculated from Eq. (3) was in good agreement with actual observations. If a computer were in the calorimetric system, calorimetry could be implemented on an industrial scale to establish heat balance promptly at any moment in a fermentation run, allowing the supply and/or withdrawal of the optimal amount of cooling water in aerobic as well as in anaerobic fermentations.

3. Solid Phase

3.1. Qualitative Analysis

As an example of the application of calorimetry to a complex microbial system, the activity of soil microbes was investigated.[22, 23] A specially designed multiplex calorimeter was used in the study of a soil system and the heat evolution rate due to microbial degradation of glucose was measured when pollutants such as mercury, cadmium, selenium, and iodoacetic acid were mixed with the soil. Power-time curves are shown in Figs. 10 and 11 for mercury and selenium, respectively. The reproducibility of power-time curves for soil was significantly high, once the soil was prepared for the measurement of heat. Although the mechanism of inhibition of these contaminants on microbes in soil has not been analyzed yet, calorimetry is a useful means to evaluate microbial degradation ability in soil for which other appropriate means have not been developed.

3.2. Quantitative Analysis

The calorimeter for the soil system mentioned above was used to characterize the effect

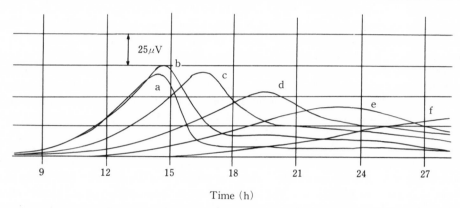

Time (h)

FIG. 10. Effect of ethyl mercuric phosphate on degradation of glucose in alluvial soil at 30°C.[22] Concentrations of Hg on the element base are: a) 0; b) 0.2; c) 0.4; d) 0.6; e) 0.8; and f) 1.0 ppm. The ordinate shows the output of the recorder chart. Calorimetric sensitivity under steady heat is 42.8 $\mu W \cdot \mu V^{-1}$.

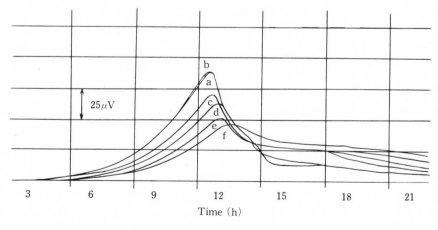

Time (h)

FIG. 11. Effect of selenic acid on degradation of glucose in soil at 30°C.[22] Concentrations of Se on the element base are: a) 0; b) 10; c) 30; d) 50; e) 70; and f) 90 ppm. See legend to Fig. 10.

of temperature in a range from 28° to 40°C on the microbial degradation rate of various sugars (glucose, sucrose, fructose, galactose, and mannose) in soil. From the data on specific growth rates at different temperatures, the apparent activation energies assessed ranged from 64 to 92 kJ mol^{-1}, depending on the sugar species.[24]

The biodegradability of straw was studied in a mixed bacterial culture.[25] The fermentation medium in the anaerobic degradation of straw at 30°C was introduced into the calorimetric vessel and the power-time curve was measured in accordance with the progress of anaerobic fermentation. The power-time curves for different cultivation periods (0–2 months) were almost the same, suggesting that the microbial composition, i.e., microbial florae in the anaerobic medium, remained unchanged. Ninety percent of recovered carbon in the cultivation products that were analyzed by gas chromatography were acetic acid, propionic acid, butanoic acid, and ethanol, while methane comprised only 0.3%. Because the addition of cellulase doubled the amount of heat evolved, the rate-limiting step for the degradation of straw was in the process of cellulose decomposition. The enthalpy change of cellulose monomer, $(C_6H_{10}O_5)$ was -120 kJ mol^{-1}. This value was close to that for glucose in a pure, anaerobic culture of E. coli.[14]

Microcalorimetry was found to be a useful tool for the study of degradation of heterogeneous and poorly defined substrates in a mixed culture. There is some possibility that this technique could be applied to screening from the natural environment specific microorganisms that could degrade cellulolytic substances most efficiently.

4. Conclusion

Slow but steady progress in calorimetric research was reviewed. The future development of calorimetry will focus on quick monitoring of microbial activities in mixed cultures such as in soil, in organic degradation in methane fermentation, in composting, in contaminated food, etc. Calorimetry for plant tissues or mammalian cells, still very rare at present, will also become important in the future. A new calorimetric system that contains immobilized enzymes inside the calorimeter and works as an enzyme thermister could be used further to detect continuously and on line many kinds of substrates or components in fermentation broth.

REFERENCES
1. Brown, H.D., *Biochemical Microcalorimetry*. Academic Press, New York, 1969.
2. Beezer, A.E., *Biological Microcalorimetry*. Academic Press, New York, 1980.
3. Ishikawa, Y., Nonoyama, Y., Shoda, M., "Microcalorimetric Study of Aerobic Growth of *Escherichia coli* in Batch Culture." *Biotechnol. Bioeng.*: **23**, 2825–2836 (1981).
4. Monk, P.R., "Calorimetric Measurement of the Effect of pH on the Anaerobic and Aerobic Catabolism of Glucose by *Streptococcus agalactiae*." *J. Gen. Appl. Microbiol.*: **30**, 329–336 (1984).
5. Nunomura, K., Fujita, T., "Calorimetric Study of the Endogenous Metabolism of Yeast." *J. Gen. Appl. Microbiol.*: **27**, 357–364 (1981).
6. Nunomura, K., Fujita, T., "Calorimetric Study of the Sugar Metabolism of Yeast under Nongrowing Conditions." *J. Gen. Appl. Microbiol.*: **28**, 479–490 (1982).
7. Nunomura, K., Fujita, T., "Heat Effects Associated with Glucose Transport in *Saccharomyces cerevisiae*." *J. Gen. Appl. Microbiol.*: **29**, 233–238 (1983).
8 Gonda, K., Wada, H., Murase, N., "Calorimetric Studies of Glycolysis of *E. coli* under Nongrowing Condition." *Thermochimica Acta*: **88**, 339–344 (1985).
9. James, A.M., Djavan, A., "Microcalorimetric Studies of *Klebsiella aerogenes* Grown in Chemostat Culture—3 Transition (Non-steady) State." *Microbios*: **34**, 17–30 (1982).

10. Ishikawa, Y., Shoda, M., "Calorimetric Analysis of *Escherichia coli* in Continuous Culture." *Biotechnol. Bioeng.*: **25**, 1817–1827 (1983).
11. Lovrien, R., Jorgenson, G., Ma, M.K., Sund, W.E., "Microcalorimetry of Microorganism Metabolism of Monosaccharides and Simple Aromatic Compounds." *Biotechnol. Bioeng.*: **22**, 1249–1269 (1980).
12. Eriksson, R., Holme, T., "The Use of Microcalorimetry as an Analytical Tool for Microbial Processes." *Biotechnol. Bioeng. Symp. Series No. 4*, 581–590 (1973).
13. Mou, D.G., Cooney, C.L., "Application of Dynamic Calorimetry for Monitoring Fermentation Processes." *Biotechnol. Bioeng.*: **18**, 1371–1392 (1976).
14. Belaich, A., Belaich, J.P., "Microcalorimetric Study of the Anaerobic Growth of *Escherichia coli*, Growth Thermograms in a Synthetic Medium." *J. Bacteriol.*: **125**, 14–18 (1976).
15. Itoh, S., Takahashi, K., "Calorimetric Studies of Microbial Growth: Kinetic Analysis of Growth Thermograms Observed for Bakery Yeast at Various Temperatures." *Agric. Biol. Chem.*: **48**, 271–275 (1984).
16. James, A.M., Djavan, A., "Microcalorimetric Studies of *Klebsiella aerogenes* Grown in Chemostat Culture—1 Glucose-limited Cultures." *Microbios*: **29**, 171–183 (1980).
17. James, A.M., Djavan, A., "Microcalorimetric Studies of *Klebsiella aerogenes* Grown in Chemostat Culture—2 C-limited and C-sufficient Cultures." *Microbios*: **30**, 163–170 (1981).
18. Shoda, M., Ishikawa, Y., In *Modelling and Control of Biotechnical Processes*, A. Halme, (ed.), p. 33. Pergamon Press, Oxford, 1983.
19. Bayer, K., Fuehrer, F., "Computer Coupled Calorimetry in Fermentation." *Process Biochem.* July/Aug., 42–45 (1982).
20. Wang, H., Wang, D.I.C., Cooney, C.L., "The Application of Dynamic Calorimetry for Monitoring Growth of *Saccharomyces cerevisiae*." *Eur. J. Appl. Microbiol. Biotechnol.*: **5**, 207–214 (1978).
21. Imanaka, T., Aiba, S., "A Convenient Method to Estimate the Rate of Heat Evolution in Fermentation." *J. Appl. Microbiol. Biotechnol.*: **26**, 559–567 (1976).
22. Kawabata, T., Yamano, H., Takahashi, K., "An Attempt to Characterize Calorimetrically the Inhibitory Effect of Foreign Substances on Microbial Degradation of Glucose in Soil." *Agric. Biol. Chem.*: **47**, 1281–1288 (1983).
23. Kimura, T., Takahashi, K., "Calorimetric Studies of Soil Microbes: Quantitative Relation between Heat Evolution during Microbial Degradation of Glucose and Changes in Microbial Activity in Soil." *J. Gen. Microbiol.*: **131**, 3083–3089 (1985).
24. Yamano, H., Takahashi, K., "Temperature Effect on the Activity of Soil Microbes Measured from Heat Evolution during the Degradation of Several Carbon Sources." *Agric. Biol. Chem.*: **47**, 1493–1499 (1983).
25. Fardeau, M.L., Plasse, F., Belaich, J.P., "Microcalorimetry: A Tool for the Study of the Biodegradability of Straw by Mixed Bacterial Cultures." *Eur. J. Appl. Microbiol. Biotechnol.*: **10**, 133–143 (1980).

Interaction of Electromagnetic Field with Enzymatic Reaction and pH Effects

D. Šnita and M. Marek

Department of Chemical Engineering, Prague Institute of Chemical Technology, 166 28 Prague 6, Czechoslovakia

1. Introduction

Transport of reaction components and its enzymatic kinetics jointly determine the course of most biological processes. In situations where the components have a nonzero electric charge (ions) or when the charge on molecules is unevenly distributed the electromagnetic field can also affect the course of reactions. These effects are often negligible in systems with usual intensities of the electric and magnetic fields. However, they can become evident in nonlinear reaction systems with high parametric sensitivity, as, for example, in excitable systems. Thus, it was experimentally demonstrated and explained by means of mathematical models that an electric field can qualitatively change the behavior of chemical waves (fronts, pulses) in excitable reaction media.[1-4]

Recently, there has been great interest in the development of various types of biosensors, in which the interaction of an electromagnetic field with transport processes and reaction often takes place. Thus detailed study of various aspects of the general problem of interaction of an electromagnetic field with reaction and transport processes is necessary.

We shall first briefly discuss one possible way to introduce Maxwell's equations into thermodynamic and balance relations and then illustrate several effects of interaction between an electrostatic field and a simple enzyme system. In these examples we shall assume the following:

1) The electrostatic field affects the transport of reaction components but the reaction kinetics remains unaffected; the interaction of reaction kinetics with an electrostatic field was discussed recently by Neuman.[5]

2) Transport of ionic components can be described by the Nernst-Planck equation (the system is isothermal, isobaric, very diluted, and no convection is present).

3) The condition of local electroneutrality holds. This assumption cannot be used in situations where a nonzero surface charge density on an interphase exists. Problems of external diffusion when a charge is bound to the reaction surface are discussed elsewhere.[6-13]

4) Diffusion coefficients (mobilities) of individual components are generally different. In the literature the study of this problem is often limited to the use of an effective diffusion coefficient for all components; however, this is correct only for ambipolar diffusion of two types of ions.[9,14-16] The choice of diffusion coefficients is also connected with the

237

discussion on the use of equilibrium and nonequilibrium distributions of the electrostatic potential.[11-13]

5) Dissociation equilibria are established in the reaction system.

6) The reaction rate is strongly dependent on pH (the dependence has a pronounced maximum), and H^+ ions are evolving in the reaction. Such a system was intentionally chosen because it has been both experimentally[17-19] and theoretically[20-25] demonstrated that activation-inhibition mechanisms connected with the concentration of H^+ ions cause various nonlinear effects (hysteresis, pattern formation, oscillations). An electrostatic field may be formed in systems with gradient of H^+ ions since H^+ ions have far greater mobility than other common ions. Hence, the study of pH effects may serve as a simple example of interaction between electromagnetic field and reaction.

7) An electric current due to the external electrostatic field can pass through the system.

2. Balance Equation

Let us consider a distributed reacting system described by fields of the intensive quantities:

$$c_1(\vec{x}, \tau), \ldots, \quad c_N(\vec{x}, \tau), \quad \vec{E}(\vec{x}, \tau), \quad \vec{H}(\vec{x}, \tau), \tag{2.1}$$

which represent the molar concentrations, the intensity of an electric field, and the intensity of a magnetic field, respectively. Here, N denotes the number of components, $\vec{x}$ the vector of spatial coordinates, and τ time. A generally valid local mass balance (2.2) and the Maxwell equation (2.3) can be set up;

$$\nabla \cdot \vec{j}_i = r_i - \partial c_i / \partial \tau \; ; i = 1, \ldots, N \tag{2.2}$$

$$\nabla \cdot \vec{D} = q, \tag{2.3}$$

$$\nabla \cdot \vec{B} = 0, \tag{2.4}$$

$$\nabla \times \vec{E} = -\partial \vec{B} / \partial \tau, \tag{2.5}$$

$$\nabla \times \vec{H} = \vec{i} + \partial \vec{D} / \partial \tau \tag{2.6}$$

where $\vec{j}_i$ is the density of molar flux of i-th component, r_i the source density of the i-th component (due to chemical reaction), $\vec{D}$ the vector of an electric induction, q the charge density, $\vec{B}$ the vector of a magnetic induction, and $\vec{i}$ the charge flux density (density of an electric current). The following definitions (2.7–2.13) can be introduced:

$$\vec{D} = \varepsilon_0 \vec{E} + \vec{P}, \tag{2.7}$$

$$\vec{B} = \mu_0 \vec{H} + \vec{M}, \tag{2.8}$$

$$q = \Sigma_i c_i q_i, \tag{2.9}$$

$$\vec{i} = \Sigma_i \vec{j}_i q_i, \tag{2.10}$$

$$\vec{P} = \Sigma_i c_i \vec{p}_i, \tag{2.11}$$

$$\vec{M} = \Sigma_i c_i \vec{m}_i, \tag{2.12}$$

$$\vec{j}_i = c_i \vec{v}_i. \tag{2.13}$$

Here ε_0 is the vacuum permittivity, μ_0 the vacuum permeability, q_i, $\vec{p}_i$, and $\vec{m}_i$ the molar charge, polarization, and magnetization of the i-th component, respectively, and v_i the velocity of the i-th component. Additional relations are given by the following constitutive equations:

$$\vec{p}_i = \vec{p}_i(c_1, \ldots, c_N, \vec{E}, \vec{H}) \quad \Big| \quad = \kappa_i \vec{E} \qquad (2.14)$$

$$\vec{m}_i = \vec{m}_i(c_1, \ldots, c_N, \vec{E}, \vec{H}) \quad \Big| \quad = \chi_i \vec{H} \qquad (2.15)$$

$$q_i = q_i(c_1, \ldots, c_N, \vec{E}, \vec{H}) \quad \Big| \quad = z_i F \qquad (2.16)$$

$$\vec{j}_i = \vec{j}_i(c_1, \ldots, c_N, \vec{E}, \vec{H}) \quad \Big| \quad = -D_i \nabla c_i \qquad (2.17)$$

$$r_i = r_i(c_1, \ldots, c_N, \vec{E}, \vec{H}) \quad \Big| \quad = r_i(c_1, \ldots, c_N). \qquad (2.18)$$

The form of the constitutive equations depends on the molecular properties of components. Typical examples of constitutive equations are given on the right-hand side of the broken line. Here, κ_i is the molar electric susceptibility, χ_i the molar magnetic susceptibility (κ_i, χ_i are scalar quantities in an isotropic medium, but generally they are tensors), D_i the diffusion coefficient, z_i the charge number, and F the Faraday's constant. The constitutive equations must satisfy certain thermodynamic constraints. Let us consider only a pseudostationary electromagnetic field (the dynamic terms in Eqs. (2.5–2.6) are neglected). Then the electric and magnetic components of the electromagnetic field mutually interact only via the terms q and $\vec{i}$. From Eqs. (2.4–2.5) follows the existence of fields $\varphi(\vec{x}, \tau)$ and $\vec{A}(\vec{x}, \tau)$ such that:

$$\nabla \varphi = -\vec{E}, \qquad (2.19)$$

$$\nabla \times \vec{A} = \vec{B}; \nabla \cdot \vec{A} = 0, \vec{v} \cdot (\nabla \vec{A}) = -\vec{v} \times \vec{B} \text{ for any } \vec{v} \qquad (2.20)$$

where φ is an electric potential and $\vec{A}$ a magnetic vector potential.

$$\varphi(\vec{x}, \tau) = \frac{1}{4 \pi \varepsilon_0} \int \frac{q}{r} \, dV, \qquad (2.21)$$

$$\vec{A}(\vec{x}, \tau) = \frac{\mu_0}{4\pi} \int \frac{\vec{i}}{r} \, dV \qquad (2.22)$$

where r is the absolute value of the vector of coordinates of a volume element dV with respect to $\vec{x}$, and the integration is performed over the entire space except for the point, x.

The change of Gibbs' free energy under isothermal and isobaric conditions is given as

$$dG = dW^C + dW^E + dW^B + dW^P + dW^M$$
$$= \Sigma_i \mu_i dc_i - \varphi \, dq - \vec{A} \cdot \vec{di} + \vec{E} \cdot \vec{dP} + \vec{H} \cdot \vec{dM} \qquad (2.23)$$

where dW is reversible work done on the system; it is composed of the chemical work (dW^C), electric work (dW^E), magnetic work (dW^B), polarization work (dW^P), and magnetization work (dW^M). The work terms are introduced on the basis of the electromagnetic field theory somewhat intuitively in agreement with Neuman.[5] We assume that the changes dq, di, $\vec{dP}$, and $\vec{dM}$ are caused by chemical reactions. When a positive change of the polarization $\vec{dP}$ (or the magnetization $\vec{dM}$) occurs, the field with a positive intensity of the electric field $\vec{E}$ (or magnetic field $\vec{H}$) must perform work and thus the energy of the mixture (i.e., of mutual chemical interactions) increases. On the other hand, whenever there is a positive change of charge density dq (or charge flux density $\vec{di}$) and the positive value of electric potential φ (or the magnetic vector potential $\vec{A}$), work against the field is performed and the energy content of the mixture decreases. More detailed treatments of the interactions of matter with the electromagnetic have been published elsewhere.[26,27] The main problem here is to form a convenient basis common to both the thermodynamics which will work with measurable quantities and the theory of the electromagnetic field which can also include nonmeasurable quantities.

All given forms of energy in Eq. (2.23) are intensive quantities related to unit volume; μ_i denotes the chemical potential of the i-th component. It follows from Eq. (2.23) that

$$\mu_i = \partial G/\partial c_i, \tag{2.24}$$

$$-\varphi = \partial G/\partial q, \tag{2.25}$$

$$-\vec{A} = \partial G/\partial \hat{i}, \tag{2.26}$$

$$\vec{E} = \partial G/\partial \vec{P}, \tag{2.27}$$

$$\vec{H} = \partial G/\partial \vec{M}. \tag{2.28}$$

Let us use a Legendre transformation and introduce a new thermodynamic potential function $\tilde{G}$ (a generalized Gibbs' free energy).

$$\tilde{G} = G + \varphi q + \vec{A} \cdot \hat{i} - \vec{E} \cdot \vec{P} - \vec{H} \cdot \vec{M} \tag{2.29}$$

$$d\tilde{G} = \Sigma_i \mu_i \, dc_i + q \, d\varphi + \hat{i} \cdot \vec{dA} - \vec{P} \cdot \vec{dE} - \vec{M} \cdot \vec{dH}. \tag{2.30}$$

Then

$$\mu_i = \partial \tilde{G}/\partial c_i, \tag{2.31}$$

$$q = \partial \tilde{G}/\partial \varphi, \tag{2.32}$$

$$\hat{i} = \partial \tilde{G}/\partial \vec{A}, \tag{2.33}$$

$$-\vec{P} = \partial \tilde{G}/\partial \vec{E}, \tag{2.34}$$

$$\vec{M}- = \partial \tilde{G}/\partial \vec{H}. \tag{2.35}$$

Differentiating Eqs. (2.31–2.35) with respect to c_i, we obtain:

$$\frac{\partial^2 \tilde{G}}{\partial \varphi \partial c_i} = \frac{\partial \mu_i}{\partial \varphi} = \frac{\partial q}{\partial c_i} = \frac{\partial}{\partial c_i} \Sigma_i c_i q_i \quad \bigg| \quad = q_i \tag{2.36}$$

$$\frac{\partial^2 \tilde{G}}{\partial \vec{A} \partial c_i} = \frac{\partial \mu_i}{\partial \vec{A}} = \frac{\partial \hat{i}}{\partial c_i} = \frac{\partial}{\partial c_i} \Sigma_i \vec{j}_i q_i \quad \bigg| \quad = q_i \vec{v}_i \tag{2.37}$$

$$\frac{\partial^2 \tilde{G}}{\partial \vec{E} \partial c_i} = \frac{\partial \mu_i}{\partial \vec{E}} = -\frac{\partial \vec{P}}{\partial c_i} = -\frac{\partial}{\partial c_i} \Sigma_i c_i \vec{p}_i \quad \bigg| \quad = -\vec{p}_i \tag{2.38}$$

$$\frac{\partial^2 \tilde{G}}{\partial \vec{H} \partial c_i} = \frac{\partial \mu_i}{\partial \vec{H}} = -\frac{\partial \vec{M}}{\partial c_i} = -\frac{\partial}{\partial c_i} \Sigma_i c_i \vec{m}_i \quad \bigg| \quad = -\vec{m}_i. \tag{2.39}$$

Quantities on the right-hand side of the broken line hold for an important special case, where q_i, $\vec{p}_i$, and $\vec{m}_i$ do not depend on the composition. Integrating Eqs. (2.36–2.39),

$$\mu_i(\varphi) - \mu_i(0) = \int_0^\varphi q_i d\varphi = \int_\infty^{\vec{x}} q_i \nabla \varphi \cdot d\vec{x} = - \int_\infty^{\vec{x}} q_i \vec{E} \cdot d\vec{x} \qquad (2.40)$$

$$\mu_i(\vec{A}) - \mu_i(\vec{0}) = \int_{\vec{0}}^{\vec{A}} q_i \vec{v}_i \cdot d\vec{A} = \int_{\vec{0}}^{\vec{x}} q_i \vec{v}_i \cdot \left(\frac{\partial \vec{A}}{\partial \vec{x}} d\vec{x} \right)$$

$$= \int_\infty^{\vec{x}} q_i (\vec{v}_i \nabla \vec{A}) \cdot d\vec{x} = - \int_\infty^{\vec{x}} q_i [\vec{v}_i \times (\nabla \times \vec{A})] d\vec{x}$$

$$= - \int_\infty^{\vec{x}} q_i (\vec{v}_i \times \vec{B}) \, d\vec{x} \qquad (2.41)$$

$$\mu_i(\vec{E}) - \mu_i(\vec{0}) = - \int_{\vec{0}}^{\vec{E}} \vec{p}_i \cdot d\vec{E} = - \int_\infty^{\vec{x}} \vec{p}_i \cdot \left(\frac{\partial \vec{E}}{\partial \vec{x}} d\vec{x} \right) = - \int_\infty^{\vec{x}} (\vec{p}_i \nabla \vec{E}) \cdot d\vec{x} \qquad (2.42)$$

$$\mu_i(\vec{H}) - \mu_i(\vec{0}) = - \int_{\vec{0}}^{\vec{H}} \vec{m}_i \cdot d\vec{H} = - \int_\infty^{\vec{x}} \vec{m}_i \cdot \left(\frac{\partial \vec{H}}{\partial \vec{x}} d\vec{x} \right) = - \int_\infty^{\vec{x}} (\vec{m}_i \nabla \vec{H}) \cdot d\vec{x}. \qquad (2.43)$$

Let us denote:

$$\mu_i(\varphi, \vec{A}, \vec{E}, \vec{H}) = \bar{\mu}_i = \mu_i(0, \vec{0}, \vec{0}, \vec{0},) + \int_0^\varphi q_i d\varphi$$

$$+ \int_{\vec{0}}^{\vec{A}} q_i \vec{v}_i \cdot d\vec{A} - \int_{\vec{0}}^{\vec{E}} \vec{p}_i \cdot d\vec{E} - \int_{\vec{0}}^{\vec{H}} \vec{m}_i \cdot d\vec{H}. \qquad (2.44)$$

Equation (2.44) expresses the relationship between the chemical potential of the i-th component unaffected (μ_i) and affected ($\bar{\mu}_i$) by the electromagnetic field.

A negative gradient of chemical potential of the i-th component (i.e., the generalized chemical potential) represents the thermodynamic driving force acting on a molar unit of the i-th component. It follows from Eqs. (2.40–2.44):

$$\vec{F}_i = - \nabla \bar{\mu}_i = - \Sigma_j (\partial \mu_j / \partial c_j) \nabla c_j + q_i \vec{E} + q_i \vec{v}_i \times \vec{B} + \vec{p}_i \nabla \vec{E} + \vec{m}_i \nabla \vec{H}, \qquad (2.45)$$

$$\vec{F}_i = \vec{F}_i^C + \vec{F}_i^E + \vec{F}_i^B + \vec{F}_i^P + \vec{F}_i^M. \qquad (2.46)$$

It can be seen that the electromagnetic forces acting on the charge ($\vec{F}_i^E$), moving charge ($\vec{F}_i^B$), electric dipole ($\vec{F}_i^P$), and magnetic moment ($\vec{F}_i^M$) are in agreement with the Coulomb and Biot-Savart laws.

Let us return to the constitutive equations (2.17–2.18). A general form of Eq. (2.17) obeys the principles of irreversible thermodynamics:

$$\vec{j}_i^* = \Sigma_j L_{ij} F_j; \quad \vec{j}_i^* = \vec{j}_i - \vec{j}_N; \, i = 1, \ldots, N - 1, \qquad (2.47)$$

where L_{ij} are phenomenological coefficients and $\vec{j}_i^*$ is a relative molar flux of the i-th component related to the reference flux (e.g., to the flux of the N-th component). From the thermodynamic point of view (N–1) driving forces are independent of the driving force for N. The Onsager reciprocity principle (symmetrical matrix L_{ij}) is generally accepted. However, when the action of external force fields is considered, forces acting on individual components can take generally any form and the reciprocity assumption does not hold.

We shall restrict ourselves to very dilute solutions. The component N is greatly in excess and its flux will thus approach zero in a system without convection. The interaction between the components will occur in such a way that a component may affect the field and the field then affects other components. Let us assume that the system behaves ideally without the field:

$$\frac{\partial \mu_i}{\partial c_i} = \frac{RT}{c_i} \; ; \quad \left(\frac{\partial \mu_i}{\partial c_j}\right)_{i \neq j} = 0, \quad L_{i,i} = \frac{D_i c_i}{RT} \; ; \; L_{i,\, j:j \neq 0} = 0. \qquad (2.48)$$

Then (cf. Eqs. 2.45, 2.47)

$$\vec{j}_i = D_i \!\left(- \nabla c_i + \frac{q_i c_i}{RT} \, \vec{E} + \frac{q_i \vec{j}_i}{RT} \times \vec{B} + \frac{\vec{p}_i c_i}{RT} \, \nabla \vec{E} + \frac{\vec{m}_i c_i}{RT} \, \nabla \vec{H}\right). \qquad (2.49)$$

If only the first two terms on the right-hand side (RHS) of Eq. (2.49) are considered, the well-known Nernst-Planck equation results.

The RHS of Eq. (2.49) contains the flux $\vec{j}_i$. It can be explicitly expressed as

$$\vec{j}_i = \left(\frac{\delta}{D_i} + \frac{q_i}{RT} \, \nabla \vec{A}\right)^{-1} \!\left(- \nabla c_i + \frac{q_i c_i}{RT} \, \vec{E} + \frac{\vec{p}_i c_i}{RT} \, \nabla \vec{E} + \frac{\vec{m}_i c_i}{RT} \, \nabla \vec{H}\right) \qquad (2.50)$$

where δ is a unit tensor. Equation (2.50) represents one of the important special forms of the constitutive equation (2.17). The main problem is how to express q_i, $\vec{p}_i$, and $\vec{m}_i$. As a first approximation we can use Eqs. (2.14–2.16).

The electromagnetic field may thus principally affect transport processes. It affects the chemical potential, can affect the chemical affinity, and thus also the rate of chemical reaction Eq. (2.18) and the equilibrium state.

Let us consider a reaction between the components B_i:

$$\Sigma_i \, \vartheta_i \, B_i = 0. \qquad (2.51)$$

Here ϑ_i are stoichiometric coefficients (for reagents, $\vartheta_i < 0$ and for products, $\vartheta_i > 0$). Let us define a total chemical affinity:

$$- A = \Sigma_i \, \mu_i \, \vartheta_i. \qquad (2.52)$$

A is equal to zero at equilibrium. Let us assume an ideal behavior of the system which is not affected by the electromagnetic field (Eq. (2.48)). Then

$$\bar{\mu}_i(\varphi, \vec{A}, \vec{E}, \vec{H}, c_i) = \bar{\mu}_i^*(\varphi, \vec{A}, \vec{E}, \vec{H}) + RT \ln c_i. \qquad (2.53)$$

Let us introduce the apparent equilibrium constant expressed in concentration.

$$K_c = \Pi_i \, c_i \, \vartheta_i. \qquad (2.54)$$

Using the integral form of the generalized Van't Hoff relation:

$$\ln K_c(Z) = \ln K_c(0) + \int_0^Z \frac{\Delta Z}{RT} \, dZ, \qquad (2.55)$$

where Z is one of the state variables, we obtain from Eqs. (2.36–2.39) the following equations:

$$Z = \varphi \qquad\qquad \Delta Z = - \Sigma_i \, \vartheta_i \, q_i, \qquad (2.56)$$

$$\vec{Z} = \vec{A} \qquad\qquad \Delta \vec{Z} = - \Sigma_i \, \vartheta_i \, q_i \vec{v}_i, \qquad (2.57)$$

$$\vec{Z} = \vec{E} \qquad\qquad \Delta \vec{Z} = \Sigma_i \, \vartheta_i \, \vec{p}_i, \qquad (2.58)$$

$$\vec{Z} = \vec{H} \qquad\qquad \Delta \vec{Z} = \Sigma_i \, \vartheta_i \, \vec{m}_i. \qquad (2.59)$$

These equations illustrate the possible effect of the electromagnetic field on chemical equilibrium. An analysis of the electromagnetic field possibly affecting the rate of reaction in the neighborhood of equilibrium can be made similarly, if we assume that a nonzero chemical affinity is a thermodynamic driving force of the reaction.

Let us accept additional simplifying assumptions in addition to Eq. (2.48):

$$\vec{H} = \vec{0};\ c_i \vec{p}_i \ll c_N \vec{p}_N;\ \vec{p}_N = \kappa_N \vec{E};$$
$$\vec{D} = (\varepsilon_0 + \kappa_N c_N)\vec{E} = \varepsilon \vec{E};\ q_i = z_i F,\ z_N = 0 \qquad (2.60)$$

where ε is a dielectric constant of the N-th component. We shall also assume that the system is in a stationary state and introduce characteristic quantities:

$$l_0 - \text{length},\ D_0 - \text{diffusivity},\ c_0 - \text{concentration},$$
$$r_0 - \text{reaction rate}. \qquad (2.61)$$

The characteristic quantities derived are:

$$
\begin{aligned}
j_0 &= D_0 c_0/l_0 && \text{– flux density,}\\
i_0 &= j_0 F && \text{– charge flux density } (\vec{i} = \Sigma_i \vec{j}_i q_i),\\
\varphi_0 &= RT/F && \text{– potential,}\\
\mathscr{H}_0 &= i_0 l_0/\varphi_0 && \text{– conductivity } \left(\mathscr{H} = \Sigma_i \frac{c_i D_i z_i^2}{RT}\right),\\
l_R &= (D_0 c_0/r_0)^{1/2} && \text{– reaction diffusion length,}\\
l_D &= (\varphi_0\,\varepsilon/c_0 F)^{1/2} && \text{– Debye length.} \qquad (2.62)
\end{aligned}
$$

Dimensionless quantities are:

$$
\begin{aligned}
\vec{X} &= \vec{x}/l_0, & I &= i/i_0\\
C_i &= c_i/c_0, & \Psi &= \varphi/\varphi_0\\
D_i &= D_i/D_0, & G &= \mathscr{H}/\mathscr{H}_0\\
R_i &= r_i/r_I, & \Phi &= l_0/l_R - \text{Thiele modulus,}\\
\vec{J}_i &= \vec{j}_i/j_0, & \Omega &= l_0/l_D - \text{electrostatic modulus.} \qquad (2.63)
\end{aligned}
$$

Substituting the above quantities into Eqs. (2.2, 2.3) we obtain after rearrangement

$$- D_i(\nabla^2 C_i + z_i \nabla C_i \cdot \nabla \Psi + z_i C_i \nabla^2 \Psi) = \Phi^2 R_i(C_1, \ldots, C_{N-1}), \qquad (2.64)$$

$$\nabla^2 \Psi = \Omega^2 \Sigma_i z_i C_i. \qquad (2.65)$$

From the definition in Eq. (2.10) and from Eqs. (2.49, 2.60), we obtain

$$\nabla \Psi = -(\vec{I} + \Sigma_i z_i D_i \nabla C_i)\frac{1}{G} \qquad (2.66)$$

where

$$G = \Sigma_i z_i^2 D_i C_i. \qquad (2.67)$$

Combining Eqs. (2.64–2.66) we have:

$$-D_i\left(\nabla^2 C_i - \frac{z_i}{G}\nabla C_i \cdot \vec{I} - \frac{z_i}{G}\nabla C_i \cdot \Sigma_j z_j D_j \nabla C_j\right)$$

$$= z_i C_i \Omega^2 \Sigma_j z_j C_j + \Phi^2 R_i(C_1, \ldots, C_{N-1}). \qquad (2.68)$$

Equation (2.68) does not contain φ and represents $N-1$ partial differential equations for $N-1$ unknowns, $C_1, \ldots, C_{N-1}$. We shall assume that the field $\vec{I}(\vec{X})$ is known; considering the conservation of charge in chemical reactions:

$$\Sigma_i q_i r_i = 0 \qquad (2.69)$$

hence

$$\nabla \cdot \vec{I} = 0. \qquad (2.70)$$

In a one-dimensional system the current density $\vec{I}$ is evidently constant and may be taken as a parameter of the system.

In many important cases, the term $(l_0/l_D)^2$ has an extremely high value ($l_D \approx 10^{-9}$m). Then the system of Eq. (2.68) exhibits "stiff" properties.

In a simplified approach, a local electroneutrality is often used:

$$\Sigma_i z_i C_i = 0. \qquad (2.71)$$

When Eq. (2.71) is introduced into (2.65), we obtain a Laplace equation for Ψ. This implies in the one-dimensional system that Ψ is linearly dependent on the space coordinate. However, the term $\Sigma z_i C_i$ may be multiplied by a large term $(l_0/l_D)^2$ in Eq. (2.65), which invalidates the approximation.

We apply the electroneutrality assumption in a different form. The term $\nabla^2 \Psi$ is rearranged from Eq. (2.66), using Eq. (2.70) as follows:

$$\nabla^2 \Psi = \frac{1}{G^2}(\vec{I} + \Sigma z_i D_i \nabla C_i) \cdot \nabla G - \frac{1}{G}\Sigma_i z_i D_i \nabla^2 C_i. \qquad (2.72)$$

After substitution of both Eqs. (2.66) and (2.72) into Eq. (2.64), we obtain the final set of model equations:

$$D_i \Sigma_j a_{ij} \nabla^2 C_i = \vec{f}_i \cdot \vec{g} + \Phi^2 R_i(C_1, \ldots, C_{N-1}); i = 1, \ldots, N-1, \qquad (2.73)$$

where

$$a_{ij} = \delta_{ij} - \frac{1}{G} z_i z_j C_i D_j \qquad (2.74)$$

$$\vec{f}_i = -z_i D_i (\nabla C_i - \frac{1}{G} C_i \Sigma_j z_j D_j \nabla C_j) \qquad (2.75)$$

$$\vec{g}_i = \frac{1}{G}(\vec{I} + \Sigma_j z_j D_j \nabla C_j) = -\nabla \Psi \qquad (2.76)$$

and δ_{ij} is Kronecker's δ.

Only $N-2$ equations from the set of Eq. (2.73) are independent. However, we can reduce the number of variables from $N-1$ to $N-2$ using the condition of Eq. (2.71). A number of variables in Eq. (2.73) can be further reduced when algebraic relations corresponding to the dissociation equilibria taking place in the system are considered as will be illustrated later on.

3. The Chemical System

Enzymes (as most proteins) have polyampholytic characters. Molecules of the enzyme (or of intermediates containing the enzyme) can accept or split off protons depending on hydrogen ion concentration. Different stages of ionization of the enzyme can then have different kinetic properties.

The general performance of H^+ ions affecting kinetics of an enzyme reaction is expressed in Scheme I in Fig. 1. Parameters γ_1, γ_2, α, and β characterize the difference in kinetic properties of the enzyme and enzyme-substrate complexes in different ionization states. Considering a quasi-steady state for the rate of reaction $S \xrightarrow{E,H^+} P$, we obtain:

$$r = \frac{k\, c^0_E\, c_S}{c_H/\alpha\, K_A + 1 + \beta\, K_B/c_H}$$

$$\times \left(\frac{\gamma_1}{c_S + K'_S\, \alpha\, K_A/c_H} + \frac{1}{c_S + K'_S} + \frac{\gamma_2}{c_S + K'_S\, c_H/\beta\, K_B} \right) \qquad (3.1)$$

where

$$K'_S = (c_H/K_A + 1 + K_B/c_H)/(c_H/\alpha\, K_A + 1 + \beta\, K_B/c_H)$$

$$c^0_E = c_E + c_{EH} + c_{EH_2} + c_{ES} + c_{EHS} + c_{EH_2S}.$$

Let us give several special forms of Eq. (3.1) for $\gamma_1 = \gamma_2 = 0$.

a) $\alpha = \beta = 1$ (Scheme II in Fig. 1)

$$r = \frac{kc^0_E}{K_S + c_S}\, F(c_H), \text{ where } F(c_H) = \frac{1}{(c_H/K_A + 1 + K_B/c_H)}. \qquad (3.2)$$

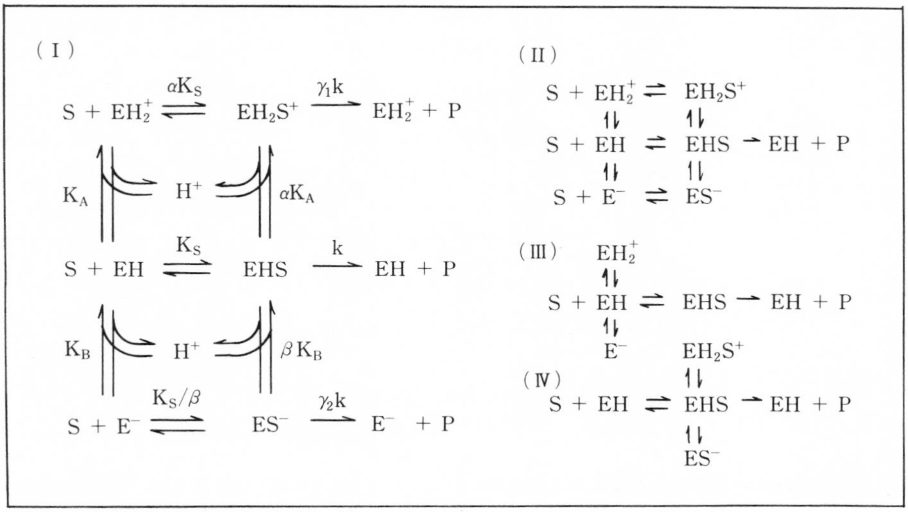

FIG. 1. Schematic diagram of mechanisms of action of hydrogen ions on a simple enzymatic reaction.

b) $\alpha \to \infty, \beta \to 0$ (Scheme III in Fig. 1)

$$r = \frac{k\, c_E^0\, c_S}{c_S + K_S/F(c_H)}\;: \tag{3.3}$$

c) $K_A \to \infty, K_B \to 0, \alpha K_A \to K_A', \beta K_B \to K_B'$ (Scheme IV in Fig. 1)

$$r = \frac{k\, c_E^0\, c_S}{K_S + c_S/F'(c_H)}, \text{ where } F'(c_H) = \frac{1}{(c_H/K_A' + 1 + K_B'/c_H)}. \tag{3.4}$$

d) $c_S \ll K_S$ then a) $\to$ b)

$$r = k' c_S\, F(c_H), \text{ where } k' = k\, c_E^0/K_S. \tag{3.5}$$

e) $K_S \ll c_S$ then a) $\to$ b)

$$r = k'' F(c_H), \text{ where } k'' = k\, c_E^0. \tag{3.6}$$

The last type is used in all subsequent examples. Hydrogen ions activate the reaction at low concentrations while inhibiting it at high concentrations. Hydrolysis of esters and oxidation of glucose to gluconic acid are just two examples from an important group of enzyme reactions involving consumption or generation of hydrogen ions. We shall discuss in another example the behavior of the system, where a weak or strong acid is generated in the presence of other ions.

Let us assume that the reaction system consists of nine components:

$$H^+, OH^-, AH, A^-, BH, B^-, K^+, S, \text{ and } H_2O \text{ in large excess} \tag{3.7}$$

The component K^+ represents either a cation ($C_K > 0$) or an anion ($C_K < 0$). The formalism involving a negative concentration reflects the fact that the component K^+ does not take part in chemical reactions and the concentration of this component is always multiplied by its charge in model equations.

The following reactions take place in the system.
Dissociation reactions:

$$H_2O \to H^+ + OH^-;\; r_1 = k_1 c_{H_2O} = k_1', \tag{3.8}$$

$$AH \to H^+ + A^-;\; r_2 = k_2 c_{AH}, \tag{3.9}$$

$$BH \to H^+ + B^-;\; r_3 = k_3 c_{BH}. \tag{3.10}$$

Association reactions:

$$H^+ + OH^- \to H_2O;\; r_4 = k_4 c_H c_{OH}, \tag{3.11}$$

$$H^+ + A^- \to AH;\; r_5 = k_5 c_H c_A, \tag{3.12}$$

$$H^+ + B^- \to BH;\; r_6 = k_6 c_H c_B. \tag{3.13}$$

Enzyme reactions:

$$S \to AH;\; r_7 = r = k F(c_H). \tag{3.14}$$

4. The Continuous Stirred Tank Reactor

We shall assume that both in the reactor and in the inlet stream the dissociation equilibria ($r \ll r_1, \ldots, r_6$) are established:

$$k_1' \quad - k_4 c_H c_{OH} = 0, \tag{4.1}$$

$$k_2 c_{AH} - k_5 c_H c_A = 0, \tag{4.2}$$

$$k_3 c_{BH} - k_6 c_H c_B = 0. \tag{4.3}$$

An analogous expression can be written for c_{H_0}, c_{OH_0}, c_{A_0}, c_{AH_0}, c_{B_0}, c_{BH_0} in the inlet stream and the condition of electroneutrality is satisfied:

$$c_H + c_K - c_{OH} - c_A - c_B = 0 \tag{4.4}$$

(similarly, for c_{H_0}, c_{K_0}, c_{OH_0}, c_{A_0}, c_{B_0}).

A stationary state in CSTR is described by the balance relations

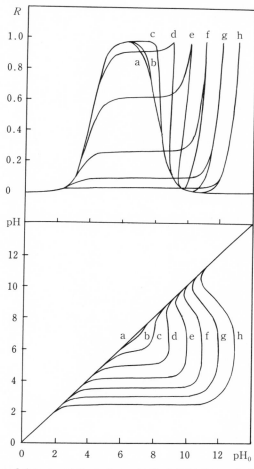

FIG. 2. CSTR-dependence of the apparent reaction rate R and the internal pH, respectively, on both the inlet pH (pH_0) and the Damköhler number (Da).
$K_{AH} \to \infty$, $c_{AHA_0} = 0$, $c_{BHB_0} = 0$;
a) $Da \leqslant 0.1$;   b) $Da = 1$;   c) $Da = 10$;
d) $Da = 100$;   e) $Da = 1000$;   f) $Da = 10^4$;
g) $Da = 10^5$;   h) $Da = 10^6$.

$$c_i - c_{i_0} = \tau \Sigma_j \, \vartheta_{ij} \, r_j; \quad i = H^+, OH^-, AH, A^-, BH, B^-, K^+, \tag{4.5}$$

where τ is residence time and ϑ_{ij} are stoichiometric coefficients. The dimensionless form of the combination of balance equations (4.5) for reactions (3.8–3.14) considering Eq. (4.1–4.4) can be expressed by:

$$(C_H - C_{H_0}) - \left(\frac{1}{C_H} - \frac{1}{C_{H_0}}\right) + C_{BHB_0} \frac{\underline{K}_{BH}(C_H - C_{H_0})}{(\underline{K}_{BH} + C_H)(\underline{K}_{BH} + C_{H_0})}$$

$$+ C_{AHA_0} \frac{\underline{K}_{AH}(C_H - C_{H_0})}{(\underline{K}_{AH} + C_H)(\underline{K}_{AH} + C_{H_0})} = \frac{\underline{K}_{AH} Da}{\underline{K}_{AH} + C_H} F(C_H) \tag{4.6}$$

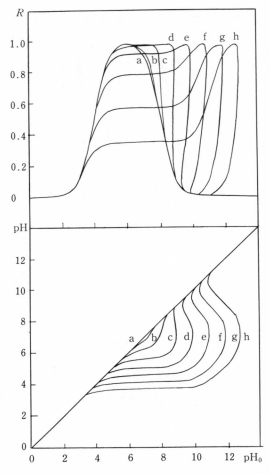

FIG. 3. CSTR-dependence of the apparent reaction rate R and the internal pH, respectively, on both the inlet pH (pH_0) and the Damköhler number (Da).

$pK_{AH} = 6$, $c_{AHA_0} = 0$, $c_{BHB_0} = 0$;

a) $Da \leqslant 0.1$;   b) $Da = 1$;   c) $Da = 10$;

d) $Da = 100$;   e) $Da = 1000$;   f) $Da = 10^4$;

g) $Da = 10^5$;   h) $Da = 10^6$.

where

$$C_i = c_i/c_0; \quad c_0 = 10^{-7} \text{ kmol m}^{-3}; \quad \underline{K}_{AH} = k_2/k_5 c_0$$

$$\underline{K}_{BH} = k_3/k_6 c_0; \quad \underline{K}_A = K_A/c_0; \quad \underline{K}_B = K_B/c_0; \quad Da = \tau\, k/c_0;$$

$$C_{AHA_0} = C_{A_0} + C_{AH_0}; \quad C_{BHB_0} = C_{B_0} + C_{BH_0};$$

$$F(C_H) = 1/(C_H/\underline{K}_A + 1 + \underline{K}_B/C_H). \tag{4.7}$$

Equation (4.6) contains C_{H_0}, C_{AHA_0}, C_{BHB_0}, $\underline{K}_A$, $\underline{K}_B$, $\underline{K}_{AH}$, $\underline{K}_{BH}$, and Da as parameters. Examples of solutions to Eq. (4.6) as functions of C_{H_0} are obtained numerically (see Figs. 2–6).

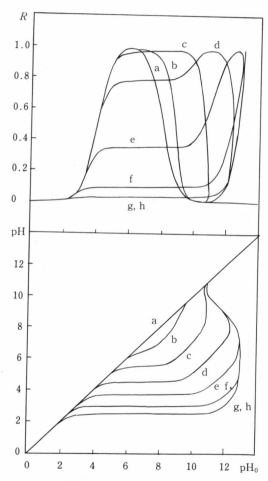

FIG. 4. CSTR-dependence of the apparent reaction rate R and the internal pH, respectively, on both the inlet pH (pH$_0$) and the dissociation constant K_{AH}.

$c_{AHA_0} = 0$, $c_{BHB_0} = 0$, $Da = 10^6$;

a) $pK_{AH} = 14$;   b) $pK_{AH} = 12$;   c) $pK_{AH} = 10$;
d) $pK_{AH} = 8$;   e) $pK_{AH} = 6$;   f) $pK_{AH} = 4$;
g) $pK_{AH} = 2$;   h) $pK_{AH} \leqslant 0$.

K_A and K_B values used in computations were fixed, respectively at

$$K_A = 10^{-8} \text{ kmol m}^{-3} \text{ and } K_B = 10^{-4} \text{ kmol m}^{-3}$$

pH values and the reaction rate R inside the continuous stirred tank reactor (CSTR) as functions of both the inlet pH (pH_0) at a high value of the product dissociation constant (strong acid, $K_{AH} \to \infty$) and the Damköhler number are shown in Fig. 2. It is evident that both activation and inhibition of the reaction occur as a consequence of the release of H^+ ions in the reaction. Multiple stationary states can exist as a result of the activation for higher values of pH_0 and Da.

An analogous situation at the lower value of the product dissociation constant (weak

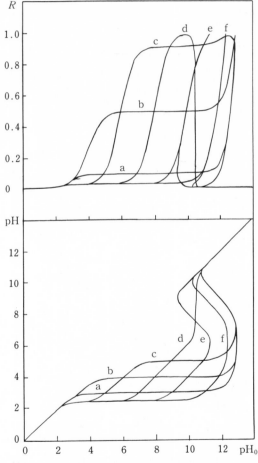

FIG. 5. CSTR-dependence of the apparent reaction rate R and the internal pH, respectively, on both the inlet pH (pH_0) and the dissociation constant K_{BH}.
$c_{AHA} = 0$, $pc_{BHB_0} = 1$, $K_{AH} \to \infty$, $Da = 10^6$;
a) $pK_{BH} = 2$;   b) $pK_{BH} = 4$;   c) $pK_{BH} = 6$;
d) $pK_{BH} = 8$;   e) $pK_{BH} = 10$;   f) $pK_{BH} = 12$.

acid, $pH_{AH} = 6$) is illustrated in Fig. 3. Curves are modified but the overall picture is pre-served. The dependencies of the reaction rate and pH in the reactor on pH_0 for various values of the product dissociation constant K_{AH} for various values of the product dis-sociation constant K_{AH} are presented in Fig. 4. It can be seen from Fig. 4 that both activa-tion and inhibition disappear when K_{AH} decreases. The presence in the reactor of the second acid BH (BH takes part only in the dissociation equilibria but not in the main reaction) is documented in Fig. 5. Figure 5 manifests dependencies of pH and R values in the reactor on pH_0 for various values of the dissociation constant for BH. It is evident that the presence of BH can qualitatively change the course of the presented dependencies (buffering effect). This fact is also illustrated in Fig. 6, which can be compared with Fig. 2.

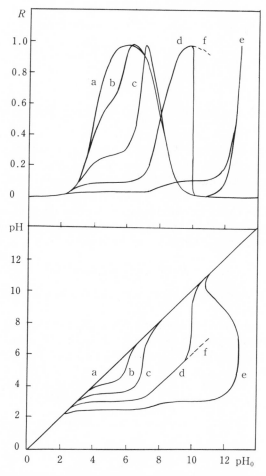

FIG. 6. CSTR-dependence of the apparent reaction rate R and the internal pH, respectively, on both the inlet pH (pH_0) and the Damköhler number (Da).
$K_{AH} \to \infty$, $K_{BH} = 8$, $pc_{BHB_0} = 2$, $c_{AHA_0} = 0$;
a) $Da = 1000$;   b) $Da = 10^4$;   c) $Da = 10^5$;
d) $Da = 10^6$;   e) $Da = 10^5$—results of Engasser and Horvath.[16]

Similar analysis in an analogous system was performed by Engasser and Horvath.[16] However, they did not consider properly the effects of OH^- ions and thus, their results for higher values of pH (Curve f) are not correct.

5. *External Diffusion*

If the flux in/out of a homogeneous reacting system occurs via diffusion through a fixed layer (e.g., a membrane) and can be described by Fick's law with equal and constant diffusion coefficients for all components, the same description as that given previously for CSTR is valid. The criterion Da is then modified to

$$Da = \frac{l_0 k}{D\, c_0} \tag{5.1}$$

(surface reaction on one side of the layer of the thickness l_0)

or

$$Da = \frac{l_0 V k}{DA\, c_0} \tag{5.2}$$

(fixed layer of the surface A and the thickness l_0 surrounding the homogeneous reaction volume V)

where D is the diffusion coefficient, k is the rate constant for the surface reaction (Eq. (5.1)) or volume reaction (Eq. (5.2)). Equations (5.1) and (5.2) become trivial when there are no reactions in the layer, and hold also when dissociation equilibria in the layer are considered.

When ionic mobilities of various components differ, the description by means of Fick's law is insufficient because it violates the condition of local electroneutrality. Then, we can use the Nernst-Planck equation (2.49) in a one-dimensional form:

$$J_i = -\underline{D}_i\, (dC_i/dX + z_i C_i d\Psi/dX). \tag{5.3}$$

Equation (5.3) is in a dimensionless form; the definition of dimensionless variables was given earlier by Eqs. (2.61–2.63).

Let us assume that equilibrium reactions (3.8–3.13) and the transport of components occur within the diffusion layer. The enzyme reaction (3.14) takes place on one of the outer surfaces of the layer or in the surrounding medium. The stationary state satisfies local component balances:

$$\nabla \cdot \vec{J}_i = R_i;\; i = H^+,\, OH^-,\, AH,\, A^-,\, BH,\, B^-,\, K^+. \tag{5.4}$$

There exist just three linearly independent combinations of Eqs. (5.4) with the RHSs equal to zero:

$$\nabla \cdot (\vec{J}_A + \vec{J}_{AH}) = 0, \tag{5.5}$$

$$\nabla \cdot (\vec{J}_B + \vec{J}_{BH}) = 0, \tag{5.6}$$

$$\nabla \cdot \vec{J}_K \quad\quad = 0. \tag{5.7}$$

Hence in a spatially one-dimensional system,

$$J_A + J_{AH} = J_{AHA} = \text{const,} \tag{5.8}$$

$$J_B + J_{BH} = J_{BHB} = \text{const,} \tag{5.9}$$

$$J_K = \text{const.} \tag{5.10}$$

Boundary conditions determine magnitudes of the above constant. The fluxes J_{BHB} and J_K can be equal to zero (e.g., one of the boundaries is an impermeable wall) or they are nonzero resulting from different concentrations of the components B^-, BH, or K^+ on permeable boundaries or from migration in the electric field. The nonzero flux J_{AHA} can result, in addition, from reaction (3.14) which takes place either as a surface reaction on the boundary or as a homogeneous reaction in the bulk of the surrounding medium.

$$J_{AHA} = Da\ R. \tag{5.11}$$

If Eq. (5.3) is introduced into Eqs. (5.8–5.10), three first-order differential equations for eight variables, C_H, C_{OH}, C_{AH}, C_A, C_{BH}, C_B, C_K, and Ψ, are obtained. The derivative of the potential is eliminated from Eq. (5.3) according to Eqs. (2.66–2.67). The relations for dissociation equilibria and the local neutrality assumption (4.4) can be written in the form:

$$C_H C_{OH} = 1 \tag{5.12}$$

$$C_H C_A = C_{AH} \underline{K}_{AH} \tag{5.13}$$

$$C_H C_B = C_{BH} \underline{K}_{BH} \tag{5.14}$$

$$C_H + C_K - C_A - C_B - C_{OH} = 0. \tag{5.15}$$

Equations (5.8–5.10) and (5.12–5.15) then form a system of three differential and four algebraic equations for seven variables, C_H, C_{OH}, C_{AH}, C_A, C_{BH}, C_B, and C_K.

The number of variables in Eqs. (5.8–5.10) can be further decreased by means of nonlinear transformations using Eqs. (5.12–5.15). We can take the following set of functions of new variables, Y_1, Y_2, and Y_3:

$$C_H = 10^{-Y_1},\ C_A = C_{AHA} \underline{K}_{AH}/(\underline{K}_{AH} + C_H),$$

$$C_{OH} = 10^{Y_1},\ C_{AH} = C_{AHA} C_H/(\underline{K}_{AH} + C_H),$$

$$C_{AHA} = 10^{-Y_2},\ C_B = C_{BHB} \underline{K}_{BH}/(\underline{K}_{BH} + C_H),$$

$$C_{BHB} = 10^{-Y_3},\ C_{BH} = C_{BHB} C_H/(\underline{K}_{BH} + C_H),$$

$$C_K = C_A + C_B + C_{OH} - C_H;\ \text{where } Y_1, Y_2, Y_3 \text{ represent}$$

$$\text{pH}, pC_{AHA}, pC_{BHB}. \tag{5.16}$$

The variables $C_H, \ldots, C_K$ satisfy Eqs. (5.12–5.15) for any set of Y_1, Y_2, and Y_3. Thus, terms including $C_H, \ldots, C_K$ or their derivatives in Eqs. (5.8–5.10) can be expressed as functions of Y_1, Y_2, and Y_3, using the implicit function theorem. Then Eqs. (5.5–5.7) may be written in the form:

$$J_{AHA}(Y_1, Y_2, Y_3, dY_1/dX, dY_2/dX, dY_3/dX, I) = J_{AHA} \tag{5.17}$$

$$J_{BHB}(Y_1, Y_2, Y_3, dY_1/dX, dY_2/dX, dY_3/dX, I) = J_{BHB} \tag{5.18}$$

$$J_K(Y_1, Y_2, Y_3, dY_1/dX, dY_2/dX, dY_3/dX, I) = J_K. \tag{5.19}$$

Parameters in Eq. (5.17) are J_{AHA}, J_{BHB}, J_K, I, $\underline{K}_A$, $\underline{K}_B$, $\underline{K}_{AH}$, $\underline{K}_{BH}$, Da and the values of Y_1, Y_2, Y_3 on one of the boundaries (Y_{1S}, Y_{2S}, Y_{3S}); when they are known, then the initial value problem is well defined. If some of the fluxes J_{AHA}, J_{BHB}, J_K, are unknown, and information available for the second boundary is used in an iterative way.

5.1. External Diffusion without Imposed External Electrostatic Field

Two types of system shown in Fig. 7 are described formally by the same set of equations differing only in the definition of the Damköhler criterion (see Eqs. (5.1–5.2)). Let us assume that in all regions of these systems (i.e., in the regions S, D, W, or V) are established dissociation equilibria (5.12–5.14) and the condition of local electroneutrality (5.15) is fulfilled. Further, let the flux densities J_{BHB} and J_K and electric current density (no external electrostatic field is considered) be equal to zero. Boundary conditions for Eqs. (5.17–5.19) can be set up in the form:

$$Y_1(1) = Y_{1S} \tag{5.20}$$

$$Y_2(1) = Y_{2S} \tag{5.21}$$

$$Y_3(1) = Y_{3S} \tag{5.22}$$

$$\Psi(1) = 0 \text{ (or any other constant value)} \tag{5.23}$$

$$J_{AHA} = DaR(C_H(Y_1(0))). \tag{5.24}$$

Equations (5.17–5.24) were solved numerically using the continuation algorithm combined with the shooting method[28] such that values of $Y_1(1)$ and $Y_1(0)$ (i.e., values of pH_S and pH) satisfy Eq. (5.25):

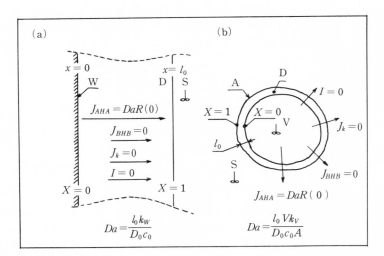

FIG. 7. Schematic representation of external diffusion without external electrostatic field (zero electric current density).
  a) surface reaction-classical external diffusion problem.
  b) volume reaction-primitive model of a cell:
    S–homogeneous environment, D–diffusion layer of thickness l_0, W–impermeable wall with an enzymatic reaction on the surface, V–homogeneous volume enclosed by a layer D (surface A).

$$Y_1^R(0) - Y_1(0) = 0. \tag{5.25}$$

The value of $Y_1^R(0)$ results from the integration of Eqs. (5.17–5.19) from $X = 0$ to $X = 1$; initial conditions are given by Eqs. (5.20–5.23) and J_{AHA} is defined in Eq. (5.24). When mobilities of individual ions differ, a potential difference ("diffusion potential") arises

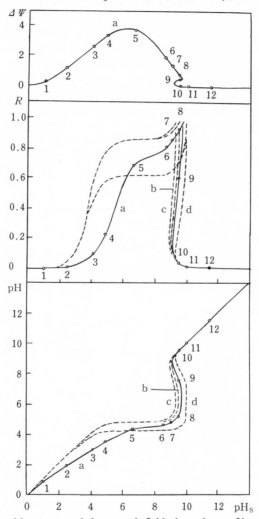

FIG. 8. External diffusion without external electrostatic field: dependence of internal pH, apparent reaction rate R, and difference of electrostatic potential between boundaries of the layer D ("diffusion potential") $\Delta\Psi$, respectively, on the external pH (pH$_S$).

a) $K_{AH} \to \infty$, $pc_{AHA} = 7$, $c_{BHB} = 0$, $Da = 1000$, $\underline{D}_H = 5$, $\underline{D}_{OH} = 3$, $\underline{D}_A = \underline{D}_{AH} = \underline{D}_B = \underline{D}_{BH}$
    $= \underline{D}_K = 1$.

b) the same as a) except $pc_{AHA} = 0$.

c) the same as a) except $\underline{D}_H = \underline{D}_{OH} = 1$.

d) the same as a) except $\underline{D}_H = \underline{D}_{OH} = \underline{D}_A = \underline{D}_{AH} = \underline{D}_B = \underline{D}_{BH} = \underline{D}_K = 5$.

Spatial profiles of pH and Ψ corresponding to the numbered points on curve a) are depicted in Fig. 9.

between the boundaries of layer D. This potential decreases with the increase in ionic strength (conductivity) in the system. We shall base our considerations on the fact that mobilities of ions $H^+ (OH^-)$ in an aqueous solution are approximately 5 to 3 times higher than those of all other common ions, and focus on the effects of the diffusion potential on the apparent reaction rate as illustrated in Fig. 8. Curves c) and d) correspond to cases where diffusion coefficients of all components are the same; c) $D_i = 1$ (the least mobile ions) and d) $D_i = 5$ (the most mobile ions). These dependencies are the same as those for the CSTR. Curves a) and b) correspond to the case of differing diffusion coefficients ($D_H = 5$, $D_{OH} = 3$, $D_{other\ ions} = 1$). In case b) there is in the environment (and thus also in the layer D) a high concentration of the fully dissociated product (A^-) and hence, also high ionic strength and conductivity. Therefore the resultant diffusion potential is very low. Curve b) is located approximately between curves c) and d). However, in case a), where the ionic strength in the environment is low, an appreciable diffusion potential arises and curve a) differs in the region of low values of pH from curves b), c), and d). Approximately, the inner boundary of the layer D (X = 0) assumes a negative potential, and hence it at-

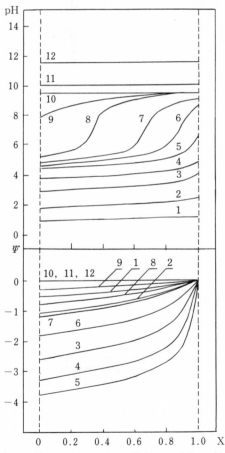

FIG. 9. External diffusion without external electrostatic field; spatial profiles of pH and Ψ in the layer D corresponding to numbered points on curve a) in Fig. 8.

tracts more H^+ ions, causing an increased inhibition of the reaction by H^+ ions in a certain pH range.

It is not surprising that due to the activation abilities of H^+ ions in the high pH region we can observe three steady states in the cases of external diffusion (a distributed system), similar to the above cases of the reaction in the CSTR (a lumped parameter system). Spatial profiles of pH and Ψ (electrostatic potential) corresponding to the numbered points on curve a) in Fig. 8 are presented in Fig. 9.

5.2 External Diffusion with Imposed External Electrostatic Field

Two types of system schematically shown in Fig. 10 are similarly described by the preceding equations which differ only in the definition of the Damköhler number. In both cases, the flux densities J_{BHB}, J_K and current density can be generally nonzero. In the special case, where both the composition and the potential on the left- and right-environments of the system are equal, $J_{BHB} = J_K = I = 0$ and then, the systems in Fig. 10 are reduced to the two systems in Fig. 7; the profiles of concentration and potential are symmetrical with respect to the plane $X = 0$. When the fluxes J_{BHB}, J_K, and I are nonzero due to the imposed external electrostatic field (potentials of the left- and right-environments of the system differ), the concentration and potential profiles cease to be symmetrical.

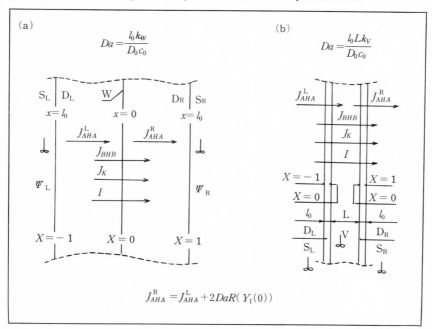

FIG. 10. Schematic representation of the external diffusion with external electrostatic field imposed (non-zero value of the electric current).
a) surface reaction.
b) volume reaction.
S_L, S_R–left and right homogeneous neighborhood of the system; D_L, D_R–left and right diffusion layer (both thicknesses are equal to l_0), W–permeable boundary between the layers D_L and D_R, where the surface reaction takes place; V–homogeneous volume between the layers D_L and D_R (distance L), where the volume reaction takes place.

Boundary conditions for Eqs. (5.17–5.19) can be stated in the following form:

$$Y_1(-1) = Y_1(1) = Y_{1S} \tag{5.26}$$

$$Y_2(-1) = Y_2(1) = Y_{2S} \tag{5.27}$$

$$Y_3(-1) = Y_3(1) = Y_{3S} \tag{5.28}$$

$$J_{AHA}^R = J_{AHA}^L + 2\, DaR(C_n(Y_1(0))). \tag{5.29}$$

Solutions to Eqs. (5.17–5.19 and 5.26–5.29) were obtained numerically using a continuation algorithm combined with the shooting method[28] such that values of J_{AHA}^L, J_{BHB}, J_K, and I satisfy Eqs. (5.30).

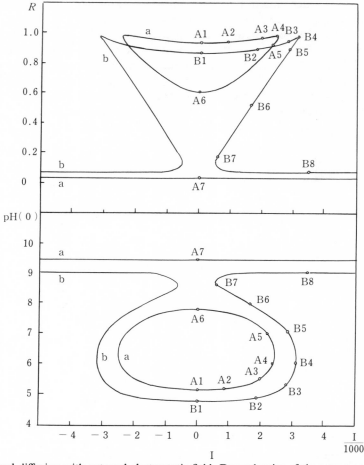

FIG. 11. External diffusion with external electrostatic field. Dependencies of the apparent reaction rate R and the internal pH on the electric current density.

a) $K_{AH} \to \infty$, $pc_{AHAs} = 7$, $c_{BHBs} = 0$, $Da = 1000$, $pH_S = 9.45$.

b) the same as a) except $pH_S = 9.1$.

Points B1, A1, A6 and A7 correspond to the points 7, 8, 9 and 10 on curve a in Fig. 8, respectively. The spatial profiles of pH corresponding to the denoted points on curves a and b ane depicted in Figs. 12 and 13, respectively.

$$Y_i^L(0) - Y_i^R(0) = 0; \quad i = 1, 2, 3. \tag{5.30}$$

The values of $Y_i^L(0)$ were obtained by the integration of Eqs. (5.17–5.19) from $X = -1$ to $X = 0$ and the values $Y_i^R(0)$, were determined by the integrated from $X = 1$ to $X = 0$. The effect of the electric field on the apparent reaction rate is illustrated in Fig. 11 for two qualitatively different cases: a) three stationary states at $I = 0$ (without an external field); and b) a single stationary state at $I = 0$. We have intentionally chosen parameter values at which strong activation by the hydrogen ions formed occurs and at which important pH gradients arise in the system. The effects of the electric field are then pronounced. Dependencies of the reaction rate and pH on the electric current density in case a) consist of two separate parts, where one forms an isola. Three stationary states exist for low values of the electric current density, but only one state exists for high current densities. Only a single stationary state exists at $I = 0$ in case b); three stationary states exist in this case in two intervals of current density symmetrically located with respect to the axis $I = 0$ and a single stationary state again exists for high values of I.

If the diffusion coefficients of all components are the same, the nonzero current flux does not affect the course of reaction (concentration profiles). If the diffusion coefficient of a certain component differs from those of other components, the concentration gradient of this component decreases with the increase in current density and eventually ceases to

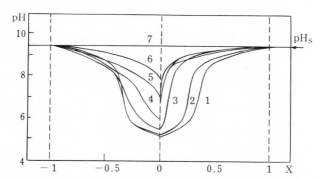

FIG. 12. External diffusion with external electrostatic field; spatial profiles of pH (1, ..., 7) correspond to the points A1, ..., A7 in Fig. 11.

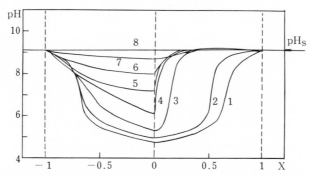

FIG. 13. External diffusion with an imposed electrostatic field: spatial profiles of pH (1, ..., 8) correspond to the points B1, ..., B8 in Fig. 11.

exist. The above statements follow from Eq. (2.68) when the local electroneutrality holds. Spatial pH profiles corresponding to denoted points in Fig. 11 are shown in Figs. 12 and 13. We can observe that the minimum value of pH is located always at the plane at $X = 0$.

6. Internal Diffusion

Systems with internal diffusion a): without an external electrostatic field (zero electric current density) and b): with an external electrostatic field imposed (nonzero value of electric current density) are schematically shown in Fig. 14. In the special case of b), where the composition and potential in the left and right environments of the system are the same, $J_{BHB} = J_K = I = 0$; then, we have in fact two systems a), located contiguous to each other such that spatial profiles of concentration and potential are symmetrical with respect to the plane $X = 0$. We shall assume that dissociation equilibria (Eqs. (5.12–5.14)) are established and the local electroneutrality condition is satisfied in all regions of the systems S, S_L, S_R, RD, and W.

Model equations for the description of the system with an internal diffusion were given in Section 2 (Eqs. (2.73–2.76, 2.71)). The number of variables in these equations are decreased similar to in Section 5, where new variables (Y_1, Y_2, Y_3) were introduced, (Eq. (5.16)) and the implicit function theorem as well.

Equations in the general form are:

$$\nabla \cdot (\vec{J}_A + \vec{J}_{AH}) = \nabla \cdot (\vec{J}_{AHA}) = \Phi^2 R(C_H), \tag{6.1}$$

$$\nabla \cdot (\vec{J}_B + \vec{J}_{BH}) = \nabla \cdot (\vec{J}_{BHB}) = 0, \tag{6.2}$$

$$\nabla \cdot \vec{J}_K = 0. \tag{6.3}$$

Equations (6.1–6.3) are transformed into:

$$\sum_{k=1}^{3} A_{i,k}(Y_1, Y_2, Y_3) \frac{d^2 Y_K}{dX^2} = B_i\left(Y_1, Y_2, Y_3, \frac{dY_1}{dX}, \frac{dY_2}{dX}, \frac{dY_3}{dX}, I\right) \quad i = 1, 2, 3. \tag{6.4}$$

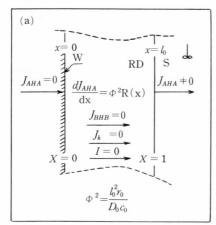
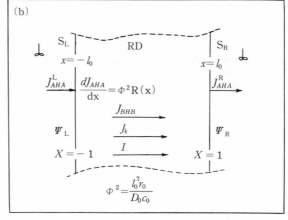

FIG. 14. Schematic representation of internal diffusion.
   a) without external electrostatic field (zero density of the electric current).
   b) with external electric field (nonzero density of the electric current): S(S_L, S_R)–homogeneous environment of the system (left, right), RD–reaction-diffusion layer (thickness a) l_0, b) $2l_0$, respectively) W-impermeable wall or symmetrical plane.

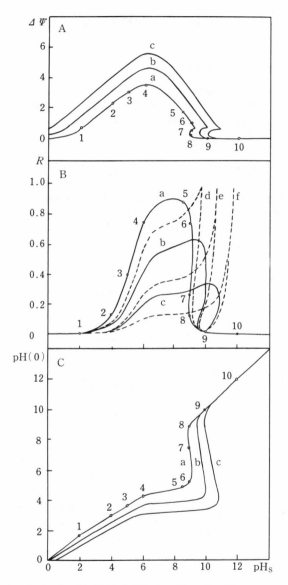

FIG. 15. Internal diffusion (a, b, c) and external diffusion (d, e, f) without external electrostatic field: dependence of diffusion potential $\Delta\Psi$, apparent reaction rate R and internal pH (pH(0)) on the external pH (pH$_S$), respectively.
a, d) $K_{AH} \rightarrow \infty$, $pc_{AHAs} = 7$, $c_{BHBs} = 0$, $Da = 1000$;
b, e) the same as a) except $Da = 10^4$;
c, f) the same as a) except $Da = 10^5$.
Spatial profiles of pH and Ψ corresponding to the numbered points on curve a) are depicted in Fig. 16.

This formulation represents a system of three second-order differential equations for three variables Y_1, Y_2, and Y_3. Boundary conditions to Eq. (6.4) are:

$$Y_1(-1) = Y_1(1) = Y_{1S}, \qquad (6.5)$$

$$Y_2(-1) = Y_2(1) = Y_{2S}, \qquad (6.6)$$

$$Y_3(-1) = Y_3(1) = Y_{3S}. \qquad (6.7)$$

Equations (6.4–6.7) were solved numerically by means of the continuation algorithm combined with the shooting method as done previously; i.e., derivatives $dY_1/dX(-1)$, $dY_2/dX(-1)$, $dY_3/dX(-1)$, and parameters (pH_S and/or I) must satisfy Eq. (6.8) below.

$$Y_i^L(1) = Y_{iS}, \quad i = 1, 2, 3. \qquad (6.8)$$

$Y_i^L(1)$ was assessed by integrating Eq. (6.4) from $X = -1$ to $X = 1$.

6.1 Internal Diffusion without External Electrostatic Field (cf. Fig. 14a)

The diffusion potential ($\Delta\Psi$), apparent reaction rate ($R = J_{AHA}(1)/Da$), and internal pH ($pH(0)$) as functions of the external pH (pH_S) are shown in Fig. 15, taking internal (a, b, c) as well as external (d, e, f) diffusion as parameters. The Thiele modulus Φ, which

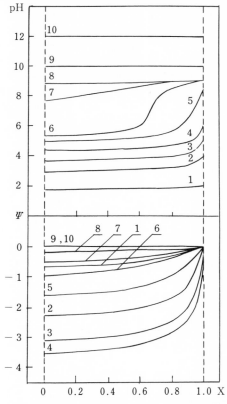

FIG. 16. Internal diffusion without external electrostatic field: spatial profiles of pH and Ψ corresponding to the numbered points on curve a) in Fig. 15.

is commonly used in the description of the internal diffusion, is correlated to the Damköhler criterion in the above formulations as follows:

$$\Phi = \sqrt{Da}. \tag{6.9}$$

We can infer from Figs. 15 and 8 that in this range of parameters the behavior of an internal diffusion system possesses most of the qualitative features observed in the system with external diffusion. Spatial profiles of pH and electrostatic potential corresponding to the denoted points on curve a) in Fig. 15 are given in Fig. 16.

6.2 Internal Diffusion with External Electrostatic Field Imposed (cf. Fig. 14b)

Effects of the electric current flux on the apparent reaction rate ($R = (J_{AHA}(1) - J_{AHA}(-1))/2Da$) are studied in Fig. 17. We can observe by comparison between Figs. 11 and

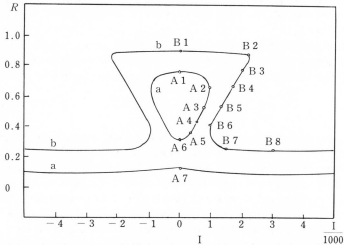

FIG. 17.   Internal diffusion with external electrostatic field: dependence of the apparent reaction rate R on the electric current density.
a) $K_{AH} \to \infty$, $pc_{AHAs} = 7$, $c_{BHBs} = 0$, $Da = 1000$, $pH_S = 9$.
b) the same as a) except $pH_S = 8.5$.
Points B1, A1, A6 and A7 correspond to those of 5, 6, 7 and 8 in Fig. 15. Spatial profiles of pH corresponding to the denoted points on curves a) and b) are depicted in Figs. 18 and 19.

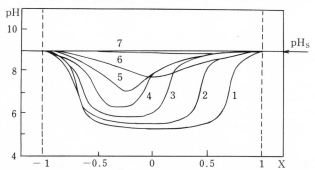

FIG. 18.   Internal diffusion in an imposed external electrostatic field: spatial profiles of pH (1, . . . , 7) correspond to points A1, . . ., A7 in Fig. 17.

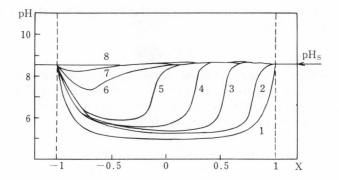

FIG. 19. Internal diffusion in an imposed external electrostatic field: spatial profiles of pH $(1, \ldots, 8)$ correspond to points $B1, \ldots, B8$ in Fig. 17.

17 that these effects are similar to those discussed above in the system-with external diffusion. Spatial profiles of pH corresponding to the denoted points in Fig. 17 are given in Figs. 18 and 19. The minimum value of pH in the layer can be in this case located off the plane at $X = 0$.

7. Conclusion

We have demonstrated in several examples that an electrostatic field may affect qualitatively the behavior of systems with non-linear enzyme reactions. For example, if two of the multiple stationary states depicted in Figs. 11 and 17 are stable, a temporarily applied external electric field can cause a switching between different stationary states and thus an electrically controlled chemical memory (a bistable flip-flop circuit) can be realized. However, such situations are just beginning to be considered for their experimental and more detailed theoretical verifications. Experimental studies on complex biological material are difficult to interpret and thus various model systems are used. Attempts to answer the following two basic questions are needed:

1) What electromagnetic field is formed due to the interaction of chemical reactions and transport? How it can be utilized for diagnostic and analytical purposes?

2) What types of behavior of nonlinear enzymatic systems in external fields are possible and how they can be used to better understand the action of enzymes?

REFERENCES

1. Schmidt, S., Ortoleva, P., "A new chemical wave equation for ionic systems." *J. Chem. Phys.*: **67**, 3771–3776 (1977).
2. Ševčíková, H., Marek, M., "Chemical waves in electric field." *Physica*: **9D**, 140–156 (1983).
3. Sevčíková, H., Marek, M., "Chemical front waves in an electric field." *Physica*: **13D**, 379–386 (1984).
4. Ševčíková, H., Marek, M., "Chemical waves in electric field-modelling." *Physica*: **21D**, 61–77 (1986).
5. Neuman, E., "Elementary analysis of chemical electric effect in biological macromolecules." In *Modern Bioelectrochemistry*, p. 97–175. F. Gutmann, H. Kreyzer (eds.), Plenum Publish. Corp., New York, 1986.
6. Hornby, W.E., Lilly, M.D., Crook, E.M., "Some changes in the reactivity of enzymes resulting from their chemical attachment to water-insoluble derivatives of cellulose." *Biochem. J.*: **107**, 669–674 (1968).

7. Shuler, M.L., Aris, R., Tsuchiya, H.M., "Diffusive and electrostatic effects with insolubilized enzymes." *J. theor. Biol.*: **35**, 67–76 (1972).

8. Hamilton, B.K., Stockmeyer, L.J., Colton, C.K., "Comments on diffusive and electrostatic effects with immobilized enzymes." *J. theor. Biol.*: **41**, 547–560 (1973).

9. Kobayashi, T., Laidler, K.J., "Theory of the kinetics of reactions catalyzed by enzymes attached to membranes." *Biotechnol. Bioeng.*: **16**, 77–97 (1974).

10. Engasser, J.-M., Horvath, C., "Electrostatic effects on the kinetics of bound enzymes." *Biochem. J.*: **145**, 431–435 (1975).

11. DeSimone, J.A., "Perturbations in the structure of the double layer at an enzymic surface." *J. theor. Biol.*: **68**, 225–240 (1977).

12. Kalthod, D.G., Ruckenstein, E., "Immobilized enzymes: Electrokinetic effects on reaction rates under external diffusion." *Biotechnol. Bioeng.*: **24**, 2189–2213 (1982).

13. Ruckenstein, E., Kalthod, D.G., "Immobilized enzymes: Electrokinetic effects on reaction rates in a porous medium." *Biotechnol. Bioeng.*: **24**, 2357–2382 (1982).

14. Ruckenstein, E., Rajora, P., "Optimization of the activity in porous media of proton-generating immobilized enzymatic reactions by weak acid facilitation." *Biotechnol. Bioeng.*: **27**, 807–817 (1985).

15. Liou, J.K., Rousseau, I., "Mathematical model for internal pH control in immobilized enzyme particles." *Biotechnol. Bioeng.*: **28**, 1582–1589 (1986).

16. Engasser, J.-M., Horvath, C., "Buffer-facilitated proton transport pH profile of bound enzymes." *Biochem. Biophys. Acta*: **358**, 178–192 (1974).

17. Caplan, S.R., Naparstek, A., Zabusky, N.J., "Chemical oscillations in a membrane," *Nature*: **245**, 364–366 (1973).

18. Naparstek, A., Thomas, D., Caplan, S.R., "An experimental enzyme-membrane oscillator." *Biochem. Biophys. Acta*: **323**, 643–646 (1973).

19. Naparstek, A., Romette, J.L., Kernevez, J.P., Thomas, D., "Memory in enzyme membranes." *Nature*: **249**, 490–491 (1974).

20. Kernevez, J.P., "Enzyme mathematics." In *Studies in Mathematics and its applications* Vol. **10**. North-Holland Publish. Co., Amsterdam, 1980.

21. Bunow, B., Colton, C.K., "Multiple steady states in cellular arrays with hydrogen ion-activation kinetics." In *Analysis and Control of Immobilized Enzyme Systems*. p. 41–60. D. Thomas, J.P. Kernevez (eds.), North-Holland Publish. Co., Amsterdam, 1976.

22. Bunow, B., "Enzyme kinetics in cells." *Bulletin of Mathematical Biology*: **36**, 157–169 (1974).

23. Kernevez, J.P., Joly, G., Duban, M.C., Bunow, B., Thomas, D., "Hysteresis, oscillations, and pattern formation in realistic immobilized enzyme systems." *J. Math. Biology*: **7**, 41–56 (1979).

24. Bailey, J.E., Chow, M.T.C., "Immobilized enzyme catalysis with reaction-generated pH change." *Biotechnol. Bioeng.*: **16**, 1345–1357 (1974).

25. Halwachs, W., Wandrey, C., Schügerl, K., "Immobilized α-chemotrypsin: Pore diffusion control owing to pH gradients in the catalyst particles." *Biotechnol. Bioeng*: **20**, 541–554 (1978).

26. Landau, L.D., Lifshic, E.M., *Theoretical Physics. Electrodynamics of Continua*. Moscow, 1982. (In Russian)

27. Babskij, B.G., Zhukov, M.J., Judovich, B.J., *Mathematical Theory of Electrophoresis*. Kijev, 1983. (In Russian)

28. Kubiček, M., Marek, M., *Computational Methods in Bifurcation Theory and Dissipative Structures*. Springer-Verlag, New York, 1983.

IV-5
High Cell Density Cultivation in Fed-Batch Systems

Takeshi Kobayashi

Department of Chemical Engineering, Faculty of Engineering, Nagoya University, Nagoya 464, Japan

A process controller with a microcomputer was constructed to keep nutrients and dissolved oxygen concentrations at optimal levels. With the help of this control system, a high concentration of biomass could be acquired in fed-batch culture of *Protaminobacter ruber* (85 g l^{-1}), *Escherichia coli* (125 g l^{-1}), and *Candida brassicae* (240 g l^{-1}). This control system was also applied to the production of metabolites; a high concentration of sorbose (628 g l^{-1}) was obtained from sorbitol in fed-batch culture of *Gluconobacter suboxydans*. A high concentration of lipids (83 g l^{-1}) was realized in *Lipomyces starkeyi*. This control system was further effective in cultivation of *E. coli* harboring a recombinant plasmid for overproduction of a gene product. The control system coupled with cross-flow filtration could separate continuously inhibitory metabolites from the broth and a high concentration of biomass was attained in fed-batch cultures of *Streptococcus cremoris* (82 g l^{-1}), *Lactobacillus casei* (49 g l^{-1}), *Bifidobacterium longum* (54 g l^{-1}), and *Propionibacterium shermanii* (227 g l^{-1}).

1. Introduction

In batch culture of microorganisms, the specific nutrients required become insufficient for microbial growth at some time during the culture and the maximum biomass concentration is usually less than 5 g l^{-1}. In chemostat culture, the nutrient species is programmed to be limiting for growth, and hence, the biomass concentration is rather low. In fed-batch culture, however, nutrients of any species are supplemented to the culture at any rate desired. If nutrients and dissolved oxygen (DO) concentrations in the broth can be kept at their respective optimal values, a high concentration of biomass can be obtained. Operation variables to be controlled are: concentrations of carbon, nitrogen and phosphorus sources including various metal ions, broth pH, and concentrations of DO as well as inhibitory metabolites. If a microorganism harboring recombinant plasmids requires a specific amino acid, the amino acid concentration is another variable to control.

The experiments reported here were designed to determine how to maintain optimal nutrient conditions in order to attain high biomass concentration. In the design, sensors appropriate for carbon sources were indispensable, and a process controller interfaced with a microcomputer was employed to guarantee the most favorable nutrient conditions. When metabolic products accumulate and become inhibitory, a control system coupled with cross-flow filtration of broth would be worth examining.

2. Materials and Methods

2.1 Microorganisms

The organisms used were *Escherichia coli* B, *Candida brassicae* E-17, *Gluconobacter suboxydans* ATCC 621, *Lipomyces starkeyi* IAM 4753, *Streptococcus cremoris* SBT 1306, *Lactobacillus casei* IFO 3245, *Bifidobacterium longum* YIT 4021, *Propionibacterium shermanii* PZ-3, *E. coli* C600, and *E. coli* HB101. Compositions of the culture media and details of the cultivation conditions have been described previously.[1-16]

2.2 Sensors for Carbon Sources

Silicone tubing was used to measure volatile carbon sources.[1] The tubing (1 mm in diameter, 0.25 mm thick; Fuji Kobunshi Co., Tokyo) was fitted with two hypodermic needles. Air was introduced from one end of the tubing and the exit gas from the other end, containing a volatile carbon source, was passed through a stainless steel tube to a flame ionization detector. Porous teflon tubing developed by Dairaku et al.[17] was also used.

Whenever a non-volatile carbon source was used, it was fed to the culture intermittently with the aid of a DO-stat[2] or pH-stat[12] as described previously.[2,12] Recently, we have developed an "on-line glucose analyzer" consisting of sampling, measurement, and operation units. Glucose concentration in the broth can be kept constant using this analyzer, the details of which will be presented elsewhere.[18]

2.3   DO-stat

The process controller[5,8,9] (Sanko Electronics Co., Nagoya) used in this study was equipped with the various interfaces needed for fed-batch culture. A schematic diagram

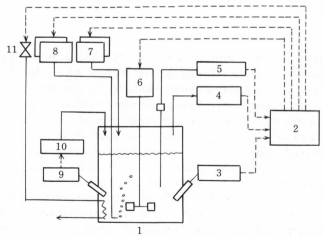

Fig. 1.   Schematic diagram of process control system in fed-batch culture.
1, jar fermentor; 2, process controller; 3, DO meter; 4, CO_2 and O_2 analyzers; 5, thermistor; 6, motor; 7, stepping motors; 8, mass-flow controllers; 9, pH meter; 10, pH controller; 11, on-off valve for cooling water.

of the process-control system in fed-batch culture is shown in Fig. 1. The jar fermentor had a six-bladed disc-turbine impeller and three baffles (working volume: 1~2 *l*). Temperature of the culture broth was measured by a thermistor (Type TXA-32, Takara Electronics Co., Tokyo) and kept constant by the "on-off" control of a cooling system. The sum of the air and oxygen flow rates was kept constant throughout, while the DO level was controlled by changing either the agitation speed or the proportion of air to oxygen gas with mass flow controllers (Type 5810, Ueshima Seisakusho Co., Tokyo).

2.4 Fermentor with Cross-Flow Filtration

A multihole-type Membralox (Toshiba Ceramics Co., Tokyo) was used as the cross-flow filter. The ceramic filter had an average pore size of 0.2 μm and filtration area of 0.18 m². The circulation rate of the cultured broth was 10 *l* min^{-1}. Fresh medium in the same volume as the filtrate was supplied via a feeding pump coupled to a liquid-level controller (Fig. 2). Both time and the intervals for backflush of the filter were regulated by a timer.

2.5 Analytical Method

The DO concentration was measured using an oxygen analyzer (Beckman Fieldlab, New York). Concentrations of oxygen and carbon dioxide in exhaust gas were detected using another oxygen analyzer (LC-700EK, Toray Co., Tokyo) and an infrared analyzer (VIA-300, Horiba Seisakusho Co., Kyoto), respectively. Cell growth was monitored by the optical density at 570 nm after dilution with 0.9% NaCl solution using a photometer (Spectronic 20, Shimadzu Co., Kyoto). The cell weight for the calibration was determined by drying the cells after centrifugation, rinsing them with water, and centrifuging them again. Concentrations of methanol and ethanol were measured by gas chromatography as described previously.[1] The amount of lactic acid was measured by the enzymatic method using F-kits for food analysis (Boehringer-Mannheim-Fujisawa Co., Tokyo). Concentrations of amino acids were determined using an automatic amino acid analyzer (HLC-6AH, Hitachi Co., Tokyo). Viscosity of the culture broth was measured at 36°C with a rotational

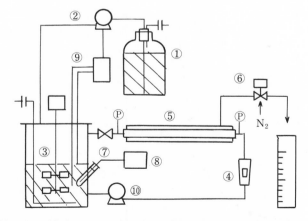

FIG. 2. Schematic diagram of fermentor equipped with cross-flow filter.
① reservoir for fresh medium; ② feed pump for fresh medium; ③ fermentor; ④ flow meter; ⑤ filter; ⑥ solenoid valve; ⑦ ball valve; ⑧ pH controller; ⑨ level controller; ⑩ pump for recycling culture broth; ⑫, pressure gauge.

viscometer (Visconic ED type, Tokyo Keiki Co., Tokyo). For measurement of osmotic pressure of the centrifuged supernatant, the supernatant (1 ml) was mixed with water (5 ml), and the osmotic pressure of this mixture was measured with an osmometer (Models 3L/3W, Advanced Instruments, New York). Vapor pressure of the supernatant was measured as described by Nesmeyanov,[19] and water activity was expressed as the ratio of the vapor pressure of the supernatant to that of water at 30°C. The volumetric oxygen-transfer coefficient was measured by the method reported by Cooper et al.[20]

3. *Results and Discussion*

Figure 3 shows a typical result[6] for cultivation of *P. ruber*, a methanol-utilizing bacterium. The DO level was maintained between 2 and 3 ppm, and methanol was fed by the tubing method or the DO-stat. Metal ions as well as nitrogen-phosphate compounds were supplemented during the cultivation. Final biomass reached 85 g l^{-1} and vitamin B$_{12}$ (2.7 mg l^{-1}) was produced intracellularly. In the usual batch cultivation (initial methanol concentration: 2.0 % v/v), the final biomass and vitamin B$_{12}$ concentrations were about 4 g l^{-1} and 0.13 mg l^{-1}, respectively. Both biomass and vitamin B$_{12}$ concentrations were enhanced about 21-fold in the fed-batch culture.

The data for *E. coli* B[2] are shown in Fig. 4. A very high biomass (125 g l^{-1}) was obtained after 11 h. For the initial period of 8 h, the specific growth rate was constant (0.73 h^{-1}), and then it decreased gradually. The DO level was controlled first at 2 to 3 ppm for 9 h,

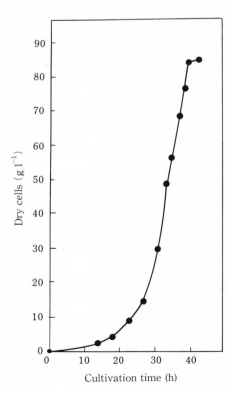

FIG. 3. Growth of *P. ruber* in fed-batch culture with DO-stat.

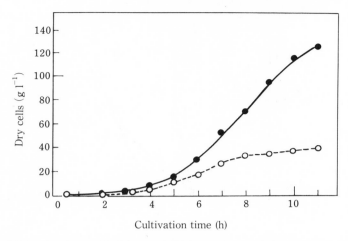

FIG. 4. *E. coli* in fed-batch culture with DO-stat. Solid line: mixture of air and pure oxygen. Broken line: air only.

and then it fell to nearly zero, although agitation was at the maximum speed and pure oxygen gas was supplied. The accumulation of acetic acid (11 g l^{-1}) interrupted the cell growth.[7] The broken line in Fig. 4 shows the result when only air was supplied, and the final biomass was 38 g l^{-1} at 11 h.

Examples of *C. brassicae*[9] growth are shown in Fig. 5. Here, the concentrations of ethanol, ammonia, and other nutritious components were maintained at the optimal range for cell growth. When the DO concentration was set at 0.2 ppm, the logarithmic growth phase continued until 8 h (specific growth rate: 0.41 h^{-1}). From 8 to 14 h, the growth curve was approximately expressed by another logarithmic phase with a specific growth rate of 0.22 h^{-1}, and then the specific growth rate deteriorated gradually. At 28 h, the optical density at 570 nm reached 356, from which the cell concentration was estimated at 224 g l^{-1}. This is the highest cell concentration ever reported in fed-batch culture as far as we know. The CO_2 concentration in the exhaust gas was less than 20 % throughout. Viscosity of the culture broth at 28 h was 23 mPa·s, somewhat higher than that of the initial medium (1 mPa·s), but lower than that expected from our previous results.[2]

When DO was set at 0.1 ppm, the growth was slower than at 0.2 ppm. Logarithmic growth with a specific growth rate of 0.35 h^{-1} was observed until 8 h. From 8 to 20 h, the growth curve was shown by another logarithmic phase with a specific growth rate of 0.10 h^{-1}. At 38.5 h, the optical density was 382, and the cell concentration was estimated at 240 g l^{-1}. The critical concentration of dissolved oxygen for the organism was most likely between 0.1 and 0.2 ppm.

When DO was set above 0.5 ppm, growth curves were similar to that at 0.2 ppm. However, growth was sharply retarded when optical density exceeded 200 (about 125 g l^{-1}). The growth retardation was not caused by CO_2 in the exhaust gas, because growth inhibition by CO_2 was not observed unless a concentration as high as 61 % was attained.[10] Thus the sharp growth retardation might have been caused by the accumulation of an unfavorable metabolite(s).

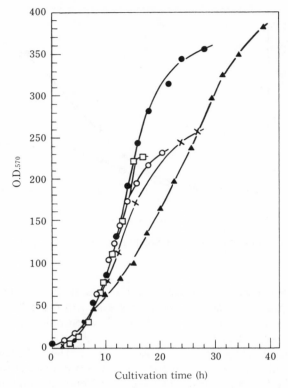

O.D.$_{570}$

Cultivation time (h)

FIG. 5. Fed-batch culture of *C. brassicae*. DO was set at various levels in ppm: (▲) 0.1, (●) 0.2, (×) 0.5, (○) 1, (□) 2 to 3.

Figure 6 shows a culture of *G. suboxydans*[3] fed with sorbitol that was oxidized to sorbose. This oxidation step is essential for the production of vitamin C. In this fed-batch culture, a high concentration of sorbose (628 g l^{-1}) was obtained at 14 h. In the initial 8-h stage, the specific growth rate was constant (0.30 h^{-1}). The growth rate decreased after 8 h and the biomass reached the maximal level (3.33 g l^{-1}) at 12 h and then decreased. The DO level was controlled between 2 and 3 ppm with the DO-stat, but after 12 h it fell to almost zero, although intensive agitation was maintained (900 rpm) and pure oxygen gas was supplied. The concentration of sorbose increased logarithmically for the first 14 h, and the conversion ratio of sorbitol to sorbose was maintained at about 1.0 during the run. Osmotic pressure reached a high level (538 mOsm) and water activity decreased to 0.748 at 14 h. As most microorganisms are unable to grow when water activity is less than 0.8, the decrease in water activity due to the accumulation of sorbose might have caused the growth cessation.

According to Uzuka et al.,[21] lipid formation by *L. starkeyi* was greatly affected by the composition of growth media, i.e., in ammonium-, iron-, or zinc-deficient medium; intracellular oil globules became exceptionally large compared to those in non-deficient medium. However, cell growth was very poor in these deficient media. Therefore, a two-stage culture was needed: at the first stage, the yeast was cultivated in a non-deficient medium and then oil formation was stimulated in the deficient medium. It was difficult to

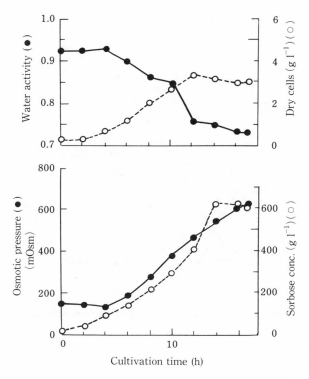

FIG. 6. Fed-batch culture of *G. suboxydans*.

attain zinc or iron deficiency at a particular period of growth, whereas it was feasible to prescribe ammonium deficiency by substituting a 10 NNaOH aqueous solution for a 33 v/v% ammonia water as the pH control agent, thus obtaining an ammonium-deficient fed-batch culture (Fig. 7).[4] In fact, ammonium ion in the broth became deficient at 48 h. Lipid production was greatly enhanced, while the growth quickly deteriorated and the total cell number remained almost constant. High concentrations of biomass (153 g l^{-1}) and lipid (83 g l^{-1}) were attained at 140 h. Fatty acids in the lipids were mainly palmitic and oleic acids.

To achieve a high-density cultivation of microorganisms that excrete metabolites inhibitory to growth, several fermentation processes with continuous removal of inhibitory metabolites, such as dialysis culture,[22] liquid-liquid extractive fermentation,[23] and vacuum or flash fermentation,[24] have been studied. Productivities of cell mass and metabolic products were improved by these processes, which appear to be quite promising in terms of industrial potential. However, there are several disadvantages in these processes. Removal of inhibitory metabolites by filtration is an alternative and simple method. Therefore, we studied high biomass cultivation of lactic acid bacteria and propionic acid bacteria using a fermentor equipped with a cross-flow ceramic filter.

In batch cultivation of *S. cremoris* with pH controlled at 6.5, the maximum biomass concentration was 2.8 g l^{-1} at 10 h (data not shown). Accumulation of lactate caused the

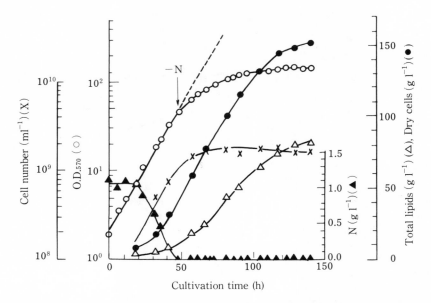

FIG. 7. Fed-batch culture of *L. starkeyi* (nitrogen in culture medium became deficient at O.D. = 50; see arrow.)

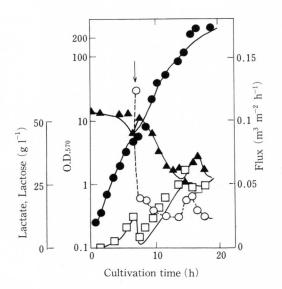

FIG. 8. Culture of *S. cremoris* with cross-flow filtration. The arrow points out the start of filtration. Symbols: (○), filtrate flux; (●), O.D.$_{.570}$; (▲), lactose; (□), lactate.

growth to cease. Figure 8 illustrates a fed-batch culture with cross-flow filtration.[14] The filtration was started at 6 h when the growth rate began to decrease as lactate accumulated, making it possible to remove the lactate and at the same time, to recycle completely the bacterial cells to the fermentor. The lactate concentration in the fermentor was kept at a low level and consequently logarithmic growth continued for 16 h. Finally, the cell concentration reached 82 g l^{-1} (3.5×10^{11} cells ml^{-1}). The cell concentration was 29 times as high as that without the cross-flow filtration. In another cultivation with cross-flow filtration of *L. casei*,[14] the maximum cell concentration was 49 g l^{-1}, whereas it was 2.2 g l^{-1} in batch cultivation (data not shown).

Bifidobacterium species are known to have several immunological functions, i.e., growth suspension of putrid bacteria, protection against infection, etc.[25] Tablets and foods containing *Bifidobacterium* cells are used clinically. The significance of a high biomass cultivation of the species (Fig. 9) is self-evident.[13] In this run, cross-flow filtration was started at 7 h when the growth rate began to decrease due to the accumulation of acetate and lactate. The organism did not grow for about 1 h after starting the filtration. Thereafter, it grew logarithmically for 9 h. However, a rapid decrease in the filtration rate owing to the high cell concentration was observed, and the subsequent accumulation of metabolites inhibited further growth. The final cell concentration was 54 g l^{-1}, seven times that achieved without filtration.

Accumulation of propionate and acetate inhibited the growth of *P. shermanii*. Removal of the metabolites allowed an extension of the logarithmic growth, and the maximum biomass was 227 g l^{-1} with a very high vitamin B_{12} content of 265 mg l^{-1} (data not shown).

Recent progress in gene engineering will permit the overproduction of industrially im-

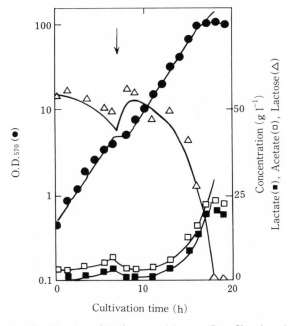

FIG. 9. Another example of cultivation of *B. longum* with cross-flow filtration. Arrow: start of filtration.

portant enzymes and hormones as well as miscellaneous drugs on an industrial scale. As for microorganisms that carry recombinant plasmids, small-scale cultivation at high cell density is preferable to large-scale culture at low biomass. We therefore studied the feasibility of our control system in the cultivation of *E. coli* C600, which harbors a recombinant plasmid.[12] Expression of the leucine gene from *Thermus thermophilus* in a recombinant plasmid was analyzed from the β-isopropylmalate dehydrogenase activity at 75°C. When *E. coli* was cultivated in a medium without leucine, biomass reached 15 g l⁻¹, and the specific activity became 0.082 U (mg protein)⁻¹. When leucine was fed throughout, the biomass reached 63 g l⁻¹. However, the specific activity decreased to 0.016 U (mg protein)⁻¹. When *E. coli* was cultivated in a fresh medium with leucine (1 g l⁻¹) but with no further feeding of leucine, the best result was obtained (Fig. 10). The specific activity remained virtually constant (about 0.13 U (mg protein)⁻¹) and biomass reached 32 g l⁻¹. The application of a pH-stat was also found useful for feeding amino acids as well as glucose.[12]

In the overproduction of gene products by gene engineering, the enhancement of plasmid copy-number, stability of plasmid, and, in addition, high efficiency of transcription are important. Several plasmid vector systems such as expression vectors and high copy-number plasmids have been developed by other workers. However, the overproduction of gene products sometimes gives rise to both deterioration of cell growth and instability of plasmids.[26] The use of inducible vectors that can render cell growth independent of gene product formation would be useful from the viewpoint of avoiding the above-mentioned defects. Runaway-replication plasmids belong to this category of vectors. Amplification of plasmid DNA is inducible by raising the culture temperature.[27] The copy number of these plasmids is restricted to 10 to 20 copies cell⁻¹ below 30°C, but the plasmid replicates without

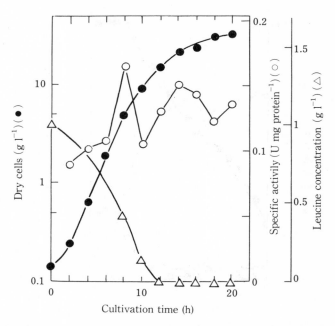

Fɪɢ. 10. Fed-batch culture of *E. coli* carrying a recombinant plasmid. Fresh medium contained leucine (1 g l⁻¹) but no further leucine was fed.

copy-number control, reaching as much as 2,000 copies cell⁻¹ at non-permissive temperatures above 35°C. The increased plasmid copy-number causes overproduction of gene products due to the gene dosage effect. However, the cultivation of host cells containing such large numbers of plasmid has been restricted only to the flask test. Thus, characteristics of a runaway-replication plasmid pCP3[28] were deemed worthy of study in a fed-batch culture or *E. coli* C600 (pcI857). Expression of the ampicillin resistance gene encoded on the plasmid was analyzed by β-lactamase activity.

E. coli harboring runaway-replication plasmids is usually cultured at 30°C, and runaway replication can be induced at 42°C in a small-scale experiment.[27,28] We attempted fed-batch culture[11] of *E. coli* (pCP3), following the same temperature shift previously used in batch culture. However, β-lactamase-specific activity decreased after the temperature shift (data not shown). This phenomenon indicates that neither the runaway replication nor the overproduction of gene product was successfully induced by temperature shift at a relatively high concentration of cells. We reexamined the growth and shift temperature in shaken flasks to transfer the information accrued therefrom to fed-batch culture. Optimal temperatures were 25°C for growth and 37°C for temperature shift. Fed-batch culture[11] with a temperature shift from 25° to 37°C in the mid-log phase is shown in Fig. 11. The specific activity of β-lactamase was kept minimal before the temperature shift. The temperature was elevated at 4.5 h and specific activity of the enzyme increased exponentially during the following period of about 4.5 h, finally reaching the level of 500 U (mg protein)⁻¹. Cell growth gradually deteriorated and stopped at 16 h. When the temperature was shifted up in the late-log phase, lower specific activity was observed, although a higher concentration of cells was recorded (data not shown). For scale-up of these cultivations, reexamination and optimization of culture conditions remain to be completed.

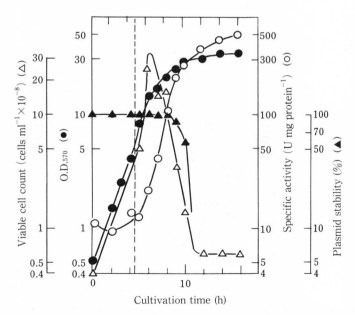

FIG. 11. Fed-batch culture of *E. coli* carrying runaway plasmid pCP3. Broken vertical line: temperature shifted from 25° to 37°C.

High-expression vectors with strong promoters are useful whenever bacterial growth can be segregated from overproduction of gene product. We cultured *E. coli* C600 cells harboring a recombinant plasmid pMCT98 with the β-galactosidase gene (*lacZ*) fused to the *trp* promoter to study the basic performance of this specific promoter.[15] In shaken flasks, the *trp* promoter functioned well in the presence of foreign genes when 3-β-indole acrylic acid (IAA) was added.[29] However, only slight functioning was observed in the fed-batch culture (data not shown). Control of the tryptophan level in the culture medium was more effective. Biosynthesis of β-galactosidase was stimulated by tryptophan deficiency and repressed by tryptophan. In the fed-batch culture, where tryptophan is absent, the *trp* promoter would be "switched-on" from the start. Upon feeding with nutrients, the maximum biomass reached 31 g l⁻¹, and the enzyme activity was 250 U ml⁻¹ (data not shown).

The switching[15] of the *trp* promoter from "off" to "on" by altering the tryptophan concentration in the medium is shown in Fig. 12. For the initial 3 h, fed-batch culture with feeding tryptophan and glucose was carried out. At 3 h, the feeding was stopped, while feeding of glycerol instead of glucose was begun to induce tryptophanase in *E. coli* to degrade the tryptophan rapidly. Thus, the tryptophan concentration in the medium dropped to almost zero within 3 h after the glycerol feeding, and β-galactosidase production was induced. In this run, the bacterium exhibited favorable growth and the final biomass was 48 g l⁻¹. The final level of β-galactosidase was 550 U ml⁻¹ or 35 U (mg protein)⁻¹. Growth and production phases, if segregated as seen in these two-stage examples, are a commendable means to deal with the culture of host cells that carry specific recombinant plasmids.

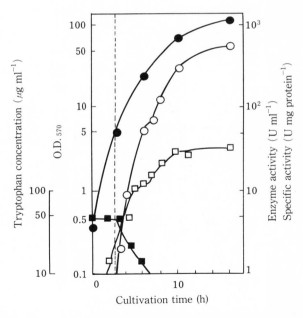

FIG. 12. Switching from "off" to "on" of *trp* promoter in fed-batch culture of *E. coli* carrying pMCT98 (*trp* promoter fused to *lacZ* gene). Broken line: induction time. Symbols: (●), O.D.₅₇₀; (○), β-galactosidase activity; (□), β-galactosidase specific activity; (■), tryptophan concentration.

REFERENCES

1. Yano, T., Kobayashi, T., Shimizu, S., "Silicone tubing sensor for detection of methanol." *J. Ferment. Technol.*: **56**, 421–427 (1978).
2. Mori, H., Yano, T., Kobayashi, T., Shimizu, S., "High density cultivation of biomass in fed-batch system with DO-stat." *J. Chem. Eng. Japan*: **12**, 313–319 (1979).
3. Mori, H., Kobayashi, T., Shimizu, S., "High density production of sorbose from sorbitol by fed-batch culture with DO-stat." *J. Chem. Eng. Japan*: **14**, 65–70 (1981).
4. Yamauchi, H., Mori, H., Kobayashi, T., Shimizu, S., "Mass production of lipids by *Lipomyces starkeyi* in microcomputer-aided fed-batch culture." *J. Ferment. Technol.*: **61**, 275–280 (1983).
5. Yano, T., Kobayashi, T., Shimizu, S., "Control system of dissolved oxygen concentration employing a microcomputer." *J. Ferment. Technol.*: **59**, 295–301 (1981).
6. Yano, T., Kobayashi, T., Shimizu, S., "Fed-batch culture of methanol-utilizing bacterium with DO-stat." *J. Ferment. Technol.*: **56**, 416–420 (1978).
7. Yano, T., Mori, H., Kobayashi, T., Shimizu, S., "Reusability of broth supernatant as medium." *J. Ferment. Technol.*: **58**, 259–266 (1980).
8. Yano, T., Mori, H., Kobayashi, T., Shimizu, S., "Control systems for dissolved oxygen concentration in aerobic cultivation." *J. Ferment. Technol.*: **57**, 91–98 (1979).
9. Yano, T., Kobayashi, T., Shimizu, S., "High concentration cultivation of *Candida brassicae* in a fed-batch system." *J. Ferment. Technol.*: **63**, 415–418 (1985).
10. Mori, H., Kobayashi, T., Shimizu, S., "Effect of carbon dioxide on growth of microorganisms in fed-batch cultures." *J. Ferment. Technol.*: **61**, 211–213 (1983).
11. Mizutani, S., Iijima, S., Kobayashi, T., "Fed-batch culture of *Escherichia coli* harboring a runaway-replication plasmid." *J. Chem. Eng. Japan*: **19**, 111–116 (1986).
12. Mizutani, S., Mori, H., Shimizu, S., Sakaguchi, K., Kobayashi, T., "Effect of amino acid supplement on cell yield and gene product in *Escherichia coli* harboring plasmid." *Biotechnol. Bioeng.*: **28**, 204–209 (1986).
13. Taniguchi, M., Kotani, N., Kobayashi, T., "High concentration cultivation of *Bifidobacterium longum* in fermenter with cross-flow filtration." *Appl. Microbiol. Biotechnol.*: **25**, 438–441 (1987).
14. Taniguchi, M., Kotani, N., Kobayashi, T., "High concentration cultivation of lactic acid bacteria in fermentor with cross-flow filtration." *J. Ferment. Technol.*: **65**, 179–184 (1987).
15. Kawai, S., Mizutani, S., Iijima, S., Kobayashi, T., "On-off regulation of tryptophan promoter in fed-batch culture." *J. Ferment. Technol.*: **64**, 503–510 (1986).
16. Iijima, S., Kai, K., Mizutani, S., Kobayashi, T., "Effects of nutrients on α-amylase production and plasmid stability in fed-batch culture of recombinant bacteria." *J. Chem. Tech. Biotechnol.*: **36**, 539–546 (1986).
17. Dairaku, K., Yamane, T., "Use of the porous teflon tubing method to measure gaseous or volatile substances dissolved in fermentation liquids." *Biotechnol. Bioeng.*: **21**, 1671–1676 (1978).
18. Mizutani, S., Iijima, S., Morikawa, M., Shimizu, K., Matsubara, M., Ogawa, Y., Izumi, R., Matsumoto, K., Kobayashi, T., "On-line control of glucose concentration using an automatic glucose analyzer." *J. Ferment. Technol.*: **65**, 325–331 (1987).
19. Nesmeyanov, A.N., In *Vapor pressure of the chemical elements*, p. 1–20. Elsevier, Amsterdam, 1963.
20. Cooper, C.M., Fernstrom, G.A., Miller, S.A., "Performance of agitated gas-liquid contactors." *Ind. Eng. Chem.*: **36**, 504–509 (1944).
21. Uzuka, Y., Naganuma, T., Tanaka, K., Odagiri, Y., "Effect of culture pH on the growth and biotin requirement in a strain of *Lipomyces starkeyi*." *J. Gen. Appl. Microbiol.*: **20**, 197–206 (1974).
22. Landwall, P., Holme, T., "Removal of inhibitors of bacterial growth by dialysis culture." *J. Gen. Microbiol.*: **103**, 345–352 (1977).
23. Taya, M., Ishii, S., Kobayashi, T., "Monitoring and control for extractive fermentation of *Clostridium acetobutylicum*." *J. Ferment. Technol.*: **63**, 181–187 (1985).
24. Lee, J.H., Woodard, J.C., Pagan, R.J., Rogers, P.L., "Vacuum fermentation for ethanol production using strains of *Zymomonas mobilis*." *Biotechnol. Lett.*: **3**, 177–182 (1981).
25. Mitsuoka, T., "Medical effect of lactic acid bacteria and a new field of utilization." *Nippon Shokuhin Kogyo Gakkaishi*: **31**, 285–296 (1984).
26. Aiba, S., Tsunekawa, H., Imanaka, T., "New approach to tryptophan production by *Escherichia coli*: Genetic manipulation of composite plasmids in vitro." *Appl. Environ. Microbiol.*: **43**, 289–297 (1982).

27. Uhlin, B.E. Nordstrom, K., "A runaway-replication mutant of plasmid Rldrd-19: Temperature-dependent loss of copy number control." *Mol. Gen. Genet.*: **165,** 167–179 (1978).
28. Remaut, E., Tsao, H., Fiers, W., "Improved plasmid vectors with a thermoinducible expression and temperature-regulation runaway replication." *Gene*: **22,** 103–113 (1983).
29. Nishimori, K., Shimizu, N., Kawaguchi, Y., Hidake, M., Uozumi, T., Beppu, T., "Expression of cloned calf prochymosin cDNA under control of the tryptophan promoter." *Gene*: **29,** 41–49 (1984).

IV-6
Bioreactor Design for Slurry Fermentation Systems

Murray Moo-Young and Yoshinori Kawase

Department of Chemical Engineering, University of Waterloo, Waterloo, Ontario, Canada, N2L 3G1

Slurry fermentation systems, which are characterized by fairly high concentrations of solid particles, are rapidly emerging in the bioprocess industries. They include so-called "dense cell cultures" and "solid-state fermentations." Compared with traditional dilute suspensions, these systems present challenging problems in the design, scale-up, and operation of bioreactors. Factors affecting the design of bioreactors for slurry fermentation systems are reviewed. In particular, mass transfer and mixing constraints are discussed for slurry fermentation systems with respect to several practical considerations, e.g., low-shear environment, foaming tendency, wall growth, and propensity for solids to settle out.

1. Introduction

Slurry fermentations involving solid-liquid dispersions are often encountered in the bioprocess industries. There is increasing interest in dense suspension systems, mainly because of the incentive for higher productivity per unit volume.

Traditional dilute suspensions of bacterial and yeast systems usually exhibit relatively low Newtonian viscosity. They rely on the use of carbon-source substrates in the form of dissolved liquid-phase sugar. In most cases, viscous solutions are not encountered so that mixing easily maintains a uniform concentration of dissolved oxygen in the bulk liquid phase and cell clumps are not formed.

For these systems, which can be treated as Newtonian two-phase (gas-liquid) systems, therefore, traditional design theories that have been fairly well established may be reasonably applicable (Moo-Young and Blanch, 1981). It should be noted that there are some cases in which unicellular broths exhibit highly viscous, non-Newtonian behavior, for example the production of an extracellular polysaccharide such as xanthan (Margaritis and Zajic, 1978).

On the other hand, slurry systems of fungal fermentations and of tissue cultures of plant or animal cells present challenging problems in the design and operation of bioreactors. In the cases of mycelial broths, the cells grow to sizes of several hundred microns and a mass of them causes their rheological complexity. They can no longer be assumed to be Newtonian two-phase systems. Such suspensions become very viscous and behave macroscopically as non-Newtonian fluids with an increase of the level of biomass. Therefore, the application of traditional chemical engineering principles may be limited for these

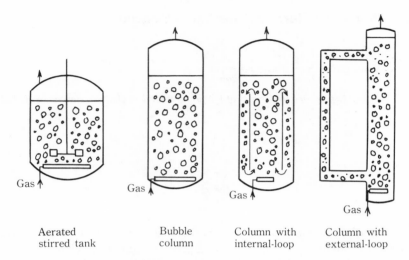

Aerated   Bubble   Column with   Column with
stirred tank   column   internal-loop   external-loop

FIG. 1. Basic geometric configurations for bioreactors.

systems and we have to develop new design strategies taking into account the constraints imposed by dense slurry systems. Despite the fact that the transport processes in fermentors are largely governed by the rheology of the fermentation broth (Aiba et al., 1973), little attention has been given to it. Furthermore, there appears to be considerable confusion and misinformation in the literature.

In addition, solid loading problems are often increased when support-immobilized bio-catalysts or water-insoluble solid substrates are involved.

It is clear that further understanding of bioreactor design with dense slurry broths is needed. An attempt is made to review the current state of the art of bioreactor design for slurry fermentation systems and to point out the gaps between our knowledge and real problems.

Correlations for the estimation of design parameters in slurry reactors, based on a stirred-tank, bubble column, and air-lift column (Fig. 1), are also presented.

2. Reactor Design

In many aerobic fermentation systems, oxygen supply is the limiting factor determining bioreactor productivity, especially at high cell densities when microbial growth is likely to be limited by the availability of oxygen in the liquid phase. Oxygen is transferred from air bubbles to suspended microbial cells via several mass transfer resistances, among which the liquid film resistance around bubbles usually controls the overall oxygen transfer rate (Fig. 2).

If good liquid mixing is not promoted and there are significant concentration gradients within the bulk liquid, oxygen transfer resistance in the bulk liquid may be an additional rate-controlling step. In some viscous fermentation broths, there are significant concentration gradients within the bulk liquid due to lack of perfect mixing. The liquid phase must be mixed by mechanical and/or pneumatic means to reduce the overall oxygen transfer resistance.

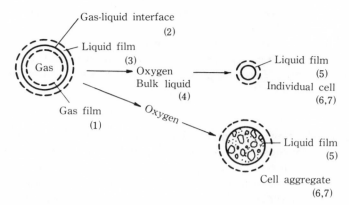

Fig. 2. Major mass transfer resistances for the transfer of oxygen from air bubble to microbial cells (Kargi and Moo-Young, 1985).

(1) gas film resistance between the bulk gas and gas-liquid interface.

(2) interfacial resistance at the gas-liquid interface.

(3) liquid film resistance between the interface and bulk liquid phase.

(4) liquid phase resistance for the transfer of oxygen to the liquid film surrounding single microbial cells.

(5) liquid film resistance around cells.

(6) intracellular or intrapellet resistance (in the case of microbial flocs or mycelial pellets).

(7) resistance due to consumption of oxygen inside a microbial cell.

If cell growth is in the form of pellets or clumps, intraparticle resistance may be important since oxygen must diffuse through to be available to the cells growing inside the clump. The size of the aggregate particles should be small enough to avoid anaerobic conditions. Cell aggregation is reduced by mixing.

However, bulk mixing sometimes results in damage to the organisms. Hydrodynamic forces influence the morphology of cell growth. Therefore, we have to consider avoiding shear damage to the cells while still providing enough shear to maintain good liquid-phase mixing or dispersion of air.

While bacterial adhesion to a surface is a potential advantage in fermentations using biofilm operation, it causes contamination or fouling due to wall growth of the microorganisms. The mechanism of the phenomenon, which may be related to liquid-phase mixing, is complicated and not well understood.

Foam is a major problem during many fermentations. Significant losses are incurred as a result of foam-outs. The severity of the problem depends on virtually all of the fermentation parameters ranging from medium composition to mechanical design and operation of the fermentors. As described above, mixing, mass transfer, and heat transfer in bioreactors are affected in multiple ways by the shear forces generated by mechanical and/or pneumatic mixing. Therefore, finding the optimal mixing is very important. Transport phenomena in bioreactors are largely governed by the rheology of the fermentation broth, the complexity of which always causes significant design problems.

At present, these are two basic approaches to discuss the transport phenomena in a dense slurry system.

a) The system is treated as a gas-liquid two-phase system. In other words, the cell suspension is assumed to be a pseudo-homogeneous liquid that exhibits non-Newtonian fluid behavior.

This approach may be subdivided into two cases.

a-1) In shear-dependent viscosity, the viscosity of the fermentation broth is represented as a function of shear rate. Viscosity, μ, is defined as the ratio of shear stress, τ, to shear rate, $\dot{\gamma}$,

$$\mu = \tau/\dot{\gamma}. \tag{1}$$

While the viscosity of a Newtonian fluid is independent of shear rate and is constant, that of a dense suspension depends on shear rate and is not constant. A power-law model, which is the simplest and most useful model of shear-dependent viscosity, has been widely used to represent the rheological property of the pseudo-homogeneous liquid. (Blakebrough et al., 1978; Reuss et al., 1982) (Fig. 3). According to this model, the viscosity of the suspension, μ_{sus}, is given as

$$\mu_{sus} = K\,\dot{\gamma}^{n-1}. \tag{2}$$

where n is the flow index and K is the consistency index. If $n < 1$, the liquid is shear thinning (pseudoplastic) and $n = 1$ is for a Newtonian fluid. Shear-thinning fluids are of primary importance in fermentation, as are fluids exhibiting a yield stress which will be discussed later.

a-2) In volume fraction-dependent viscosity, the density and viscosity of fermentation broths, ρ_{sus} and μ_{sus}, are expressed as functions of the volume fraction of solids, ϕ_s (Reuss et al., 1982; Sauer and Hampel, 1985):

$$\rho_{sus} = \phi_s \rho_s + (1 - \phi_s)\,\rho_l. \tag{3}$$

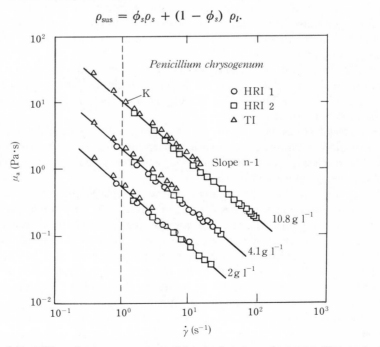

FIG. 3. Viscosity of *Penicillium chrysogenum* measured by a mixer type viscometer (Reuss et at., 1982).

$$\mu_{\text{sus}} = \mu_l \, (1 \, + \, 2.5 \, \phi_s \, + \, 10.05 \, \phi_s^2 \, + \, 0.00273 \, e^{-16.6\phi_s}). \tag{4}$$

Equation (4) is a semi-empirical correlation of Thomas' reduced data (Thomas, 1965).

b) The system is treated as a gas-liquid-solid three-phase system. In other words, the suspension is considered to be a heterogenous solid-liquid two-phase. This approach may be effective particularly for the cultures using water-insoluble solid substrates ("solid-state" fermentations). In this approach, classical chemical engineering principles for heterogeneous catalytic reactors are used. It should be noted, however, that most of them are empirical and based on the concepts for two-phase systems (Shah, 1979; Pandit and Joshi, 1984, 1986).

The shear-dependent viscosity approach (a-1) is the simplest to use and can be expected to provide a reasonable design theory except for systems with fairly high localized concentrations of solid particulates. In fact, the shear-dependent viscosity concept has been the most widely used to discuss the performance of bioreactors (Loucaides and McManamey, 1973; Godbole et al., 1984). The main problem in this approach is the current difficulty in finding reliable measurements of rheological properties of suspensions. There are several experimental arrangements to measure rheological properties. The most common viscometers are the concentric cylinder viscometer and the cone-and-plate viscometer. In these viscometers, the suspended solids tend to settle out. Furthermore, in the narrow gap between the concentric cylinders or the cone and plate the solids break and sometimes agglomerate. In other words, the shear in the viscometers often changes the morphology of cells in the test liquid or their rheological properties. An effective wall slip is also one of problems in measuring rheological properties of heterogeneous suspensions. Great care is needed to measure the rheological properties of suspensions. At present, a turbine mixer-type viscometer is generally recommended for these systems (Langer and Werner, 1981; Reuss et al., 1982). Using this type of viscometer, however, we can determine the rheological properties only at rather low shear rates which may be smaller than the characteristic shear rates in bioreactors (Moo-Young et al., 1986).

The indices in the power-law model are sometimes dependent on the shear rate range used to obtain flow curves. Therefore, they should be determined in the range of shear rate which characterizes the flow in bioreactors. In the case of stirred tanks, the average shear rate $\dot{\gamma}_a$ is estimated by (Metzner and Otto, 1957; Calderbank and Moo-Young, 1961b)

$$\dot{\gamma}_a = K_s N. \tag{5}$$

Typical K_s values that are dependent on geometries of tanks and impellers were summarized by Skelland (1967). This relationship for non-Newtonian fluids in unaerated stirred tanks has been directly applied to aerated stirred tanks.

The definition of average shear rate in bubble columns is a controversial issue. Many investigators (Nishikawa et al., 1977; Henzler, 1980; Capuder and Koloini, 1984; Schumpe et al., 1985) have used the following form

$$\dot{\gamma}_a = K_B U_{sg}. \tag{6}$$

where $K_B = 1500 - 5000$. However, this is only empirical and its applicability is questionable (Kawase and Moo-Young, 1986a).

The rheological properties of the fermentation broth (Charles, 1978) should be mentioned. Dispersed filamentous growth tends to cause shear-thinning; on the other hand, growth in pellet form tends to cause more Newtonian-like behavior. It has been observed that mycelial culture media have either shear-thinning fluids or fluids exhibiting a yield stress. The Bingham plastic model is a model for fluids having a yield stress:

$$\tau - \tau_0 = \mu_p \dot{\gamma} \tag{7}$$

where τ_0 is the yield stress and μ_p is the plastic viscosity. Note that the fluid will not flow until the applied shear stress exceeds the yield stress. Although the importance of the yield stress in reactor design has been pointed out, no correlation including the yield stress has been proposed for design parameters of bioreactors. Additional studies are required to shed further light on the effect of the yield stress on the performance of bioreactors.

2.1 Stirred-tank Reactor

The mechanically stirred, sparger-aerated slurry reactor is the most common design and a "standard" configuration following existing "conventional wisdom" is usually assumed. A standard stirred-tank fermentor may be satisfactory for traditional dilute suspension systems. However, it may not be reasonable for dense slurry systems. In a stirred tank the shear rate is highest near the impeller and decreases rather sharply with distance from it. If the fluid is shear-thinning, viscosity will be relatively low near the impeller but quite high at any position not close to it. Therefore, mixing and transport phenomena will be good only in the immediate region of the impeller. The viscosity gradients in non-Newtonian broths tend to cause most of the inlet air to flow toward the impeller which results in (i) lower gas residence time (channeling); (ii) increased bubble size; (iii) further deterioration of bulk mixing; and (iv) poor distribution of oxygen. This situation leads to losses in yield or productivity and difficulties in process control. Similar and even more pronounced problems can arise if the fluid has a high yield stress, and sometimes dead spaces are formed.

A vast amount of literature on the performance of aerated stirred tanks is available. However, in almost all previous studies the fluids under consideration have exhibited non-viscous Newtonian behavior, like water. Attempts have been made only in recent years to extend the study to non-Newtonian systems (Nienow and Ulbrecht, 1985).

There are two possible states: "complete suspension" in which all solids are in suspension, and "homogeneous suspension" in which the solid concentration is uniform throughout the reactor. The minimum agitation required to ensure complete suspension without aeration is estimated from the following equation (Zwietering, 1958)

$$N_m = \frac{a_1 \, d_p^{0.2} \, \mu_l^{0.1} \, g^{0.45} (\rho_s - \rho_l)^{0.45} \, w^{0.13}}{\rho_l^{0.55} \, D_I^{0.85}}. \tag{8}$$

The data indicate that the critical impeller speed increases in the presence of gas. This results from a reduction in the average liquid circulation velocity due to a reduction in power consumption in the presence of gas. Steiff and Weinspach (1982) and Wiedmann et al. (1985) also observed the same result and found that the power consumption per unit volume for complete suspension increases with increasing gas flow rate, particle diameter, solid mass fraction, and particle density. Wiedmann et al. (1985) presented a dimensionless correlation for complete suspension

$$\frac{N_m D_T^2}{\mu_l/\rho_l} = 5.58 \left\{ \left(\frac{U_{sg} D_I}{\mu_l/\rho_l} \right) + 1 \right\}^{0.083} \Psi_s^{0.12} \left(\frac{D_I^3 g}{\mu_l^2/\rho_l^2} \right)^{0.462} \left(\frac{d_p}{D_I} \right)^{0.26} \left(\frac{\rho_s - \rho_l}{\rho_l} \right)^{0.51}. \tag{9}$$

Power consumption is an important consideration in process economics. The bulk of the power input in aerobic fermentations is required to ensure adequate oxygen transfer. The need for very high power inputs imposes considerable constraints on fermentor design. This fact has led to a great number of reports on reducing power requirements and on more favorable consideration of non-standard stirred-tank fermentors.

The power required to agitate gassed liquid systems is less than that for ungassed liquids since the apparent density and viscosity of the liquid phase around the impeller decrease upon gassing. However, reduction in power on aeration is substantial only above a certain Reynolds number (Solomon et al., 1981; Nienow et al., 1983). Unfortunately, we still do not have reliable correlations for dense slurry systems, and further studies are needed. Multiple impellers are often required to obtain adequate gas redispersion in dense slurry systems (Solomon et al., 1981; Nienow et al., 1983; Hickman and Nienow, 1986). The choice of impeller systems is very important and difficult for dense slurry systems. Large, slow-moving impellers (e.g., the anchor or helical-ribbon types) have traditionally been used for highly viscous fluids but they are not suitable for gas dispersion. On the other hand, small, fast ones (e.g., turbine and paddle) provide satisfactory gas dispersion but produce high shear. One possible choice is a combination of a turbine into which the air is sparged and a propeller placed above the turbine to improve the overall mixing of the bulk liquid. Mixing time has been used to describe the mixing efficiency in stirred tanks. The correlation for Newtonian fluids may provide a first approximation. Joshi et al. (1982) proposed a correlation for the flat blade

$$Nt_M = 20.41 \left(\frac{H}{D_c} + 1 \right) \left(\frac{D_c}{D_I} \right)^{13/6} \left(\frac{W}{D_I} \right) \left(\frac{Q_g}{NV_l} \right)^{1/12} \left(\frac{N^2 D_I^4}{g W V_l^{2/3}} \right). \tag{10}$$

The following correlation for ε_g in non-Newtonian suspensions was proposed by Capuder and Koloini (1984).

$$\varepsilon_g = 0.059 \left(\frac{N^2 Q_g}{\sigma} \right)^{0.68} \left(\frac{\mu_{susM}}{\mu_w} \right)^{-0.26}. \tag{11}$$

Machon et al. (1980) found that increasing shear-thinning leads to decreasing gas hold-up due to an increasing proportion of large bubbles which have a short residence time.

No correlation for bubble size in slurry systems is available at present. Therefore, the correlation by Calderbank (1967) will be used as a rough estimation

$$d_{vs} = 4.15 \frac{\sigma^{0.6} \varepsilon_g}{\varepsilon^{0.4} \rho_l^{0.6}} + 0.09. \tag{12}$$

Calderbank (1967) proposed a number of equations for d_{vs} in stirred tanks. All equations based on Kolmogoroff's isotropic turbulence theory have the general structure

$$d_{vs} = C_1 \frac{\sigma^{0.6}}{\varepsilon^{0.4} \rho_l^{0.2}} \varepsilon_g{}^\alpha \left(\frac{\mu_g}{\mu_l} \right)^{0.25}. \tag{13}$$

It should be noted that this form is also applicable to bubble columns.

Lee et al. (1982) measured specific surface area in stirred tanks with non-Newtonian solid suspension and found that it decreases with increasing shear-thinning due to an increase in the solid fraction.

It was observed by Steiff and Weinspach (1982) that solid particles cause a decrease and an increase in specific surface area at low and high impeller speeds, respectively.

Yagi and Yoshida (1975) obtained the following correlation for CMC (carboxymethyl cellulose) solutions ($n = 0.95 - 0.66$)

$$\left(\frac{k_L a \, D_I^2}{D}\right) = 0.060 \left(\frac{\mu_{\text{susM}}}{\rho_l D}\right)^{0.5} \left(\frac{\rho_l \, D_I^2 \, N}{\mu_{\text{susM}}}\right)^{1.5} \left(\frac{D_I N^2}{g}\right)^{0.19} \left(\frac{\mu_{\text{susM}} \, U_{sg}}{\sigma}\right)^{0.6} \left(\frac{D_I \, N}{U_{sg}}\right)^{0.32}. \quad (14)$$

The correlation for CMC solutions proposed by Höcker et al. (1981) may be written as

$$\frac{k_L a}{Q_g/V_l} \cdot \left(\frac{\mu_{\text{susM}}}{\rho_l \, D}\right)^{0.3} = 0.105 \left\{ \frac{P/Q_g}{\mu_l^{1/3} (\mu_{\text{susM}} g)^{2/3}} \right\}^{0.59}. \quad (15)$$

Henzler (1980) obtained a very similar correlation (Fig. 4).

Nishikawa et al. (1981) measured gas absorption in aerated stirred tanks with CMC solutions and analyzed their data using a two-region model in which an agitation-control region and an aeration-control region are considered. Their correlation can be written as

$$\left(\frac{k_L a \, D_c^2}{D}\right)^2 = 0.115 \left(\frac{\mu_{\text{susM}}}{\rho_l \, D}\right)^{0.5} \left(\frac{D_I^2 \, N \, \rho_l}{\mu_{\text{susM}}}\right)^{1.5} \left(\frac{\mu_{\text{susM}} \, U_{sg}}{\sigma}\right)^{0.5} \left(\frac{D_I \, N^2}{g}\right)^{0.367} \left(\frac{N \, D_I}{U_{sg}}\right)^{0.167} \left(\frac{D_c}{D_I}\right)^2 N_p^{0.8}$$

$$+ \, 0.112 \frac{\varepsilon_{av}}{\varepsilon_a/N_p + \varepsilon_{av}} \left(\frac{\mu_{\text{susN}}}{\rho_l \, D}\right)^{0.5} \left(\frac{U_{sg}}{\sqrt{g \, D_c}}\right) \left(\frac{g \, D_c^2 \, \rho_l}{\sigma}\right)^{0.66} \left(\frac{g \, D_c^3 \, \rho_l^2}{\mu_{\text{susN}}}\right)^{0.42}. \quad (16)$$

This approach seems reasonable but the resulting correlation is too complicated.

Loucaides and McManamey (1973) measured mass transfer into suspensions of paper

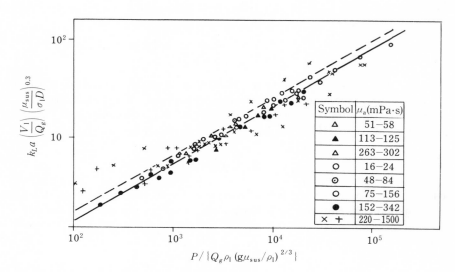

FIG. 4.   Mass transfer in aerated stirred tanks (Henzler, 1980).
-------- Höcker et al. (1981), ———— Henzler (1980).

pulp which has filamentous structure similar to many fermentation media and found that the increase in the $k_L a$ coefficient with power input was greater below a certain power input per unit volume.

Mass transfer to suspended particles in non-Newtonian fluids was analyzed by Kawase and Moo-Young (1986b). They extended the energy dissipation rate concept proposed by Calderbank and Moo-Young (1961a) to non-Newtonian systems. The resulting equation is

$$k_s = 0.075\, n^{1/3}\, \{\exp(1.37\, n + 1.71)\}^{4-n/6n\,(n+1)}\, D^{2/3}(K/\rho_l)^{-5n/6n\,(n+1)}\, \varepsilon^{4-n/6\,(n+1)}. \quad (17)$$

This equation can be applied to ungassed and gassed systems (Marrone and Kirwan, 1986).

In order to control the temperature of the fermentation process, reactors are usually equipped with either external jackets or internal coils for heating or cooling. In particular, heat transfer is one of major problems in designing large fermentors for aerobic operation. Although internal coils provide better heat transfer capabilities, the presence of internal coils causes microbial film growth on coil surfaces and alteration of mixing patterns.

Equation (17), where k_s and D are replaced by h and $(k/\rho_l C_p)$, respectively, is applicable to estimate the heat transfer rate in bioreactors. Steiff and Weinspach (1982) found that the heat transfer coefficients for aerated systems are generally lower than those for non-aerated systems. In many cases, the heat transfer rate decreases due to the addition of solids, which reduces the turbulence intensity (Kurpiers et al., 1985). The heat transfer data have been usually correlated using a dimensionless form based on Kast's (1962) approach (Fig. 5):

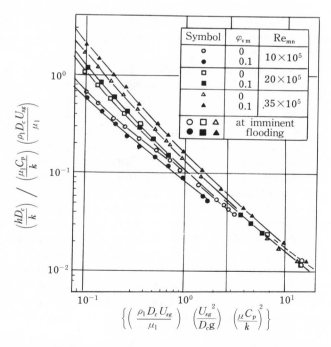

FIG. 5. Heat transfer in aerated stirred slurry reactors (Steiff and Weinspack, 1982).

$$\left(\frac{h\,D_c}{k}\right)\bigg/\left(\frac{\mu_l\,C_p}{k}\right)\left(\frac{\rho_l\,D_c\,U_{sg}}{\mu_l}\right)\propto\left\{\left(\frac{\rho_l\,D_c\,U_{sg}}{\mu_l}\right)\left(\frac{U_{sg}^2}{D_c g}\right)\left(\frac{\mu\,C_p}{k}\right)^2\right\}^{-1/3\ \text{or}\ -1/4}. \quad (18)$$

Keitel and Onken (1981) studied a stirred-loop reactor (stirred tank with a concentric draft tube) which shows a well-defined flow pattern and better gas dispersion. This reactor can be also regarded as a modification of the air-lift column (air-lift column with an impeller) and may be effective for dense slurry systems.

2.2   Bubble Column and Air-lift Column

Although aerated stirred tanks have been commonly used in the fermentation industry, recently bubble columns and air-lift columns have also begun to find applications in full-scale plants.

Due to the demand for high production capacity, some large-scale bubble columns or air-lift columns have been built. Examples are a single-cell protein unit in Billingham (total reactor volume of about 26,000 m³) and a wastewater treatment plant in Leverkusen Bayer (total reactor volume of about 20,000 m³).

Considerable construction difficulties arise with stirred-tank reactor volumes of 1,000 m³ and above. Another benefit of bubble and air-lift columns is avoiding shear damage due to the impeller. For example, Wase et al. (1985) cultured *Aspergillus fumigatis* in disc-turbine-agitated tanks and in an air-lift fermentor and found that the yield of cellulose in the stirred tank was reduced by shear damage to the mycelium and the use of an air-lift fermentor improved yields of the enzyme by about 20%. A possible additional benefit is a lower energy requirement compared with stirred tanks.

There have been numerous papers on the performance of bubble columns and they were extensively reviewed by Shah et al. (1982) and Heijnen and van't Riet (1984). However, attempts have been made only in recent years to extend previous studies on Newtonian fluids to non-Newtonian systems.

Although simple in construction, the design and scale-up of bubble columns are very difficult due to the complexity of their hydrodynamics. The rising gas bubbles induce liquid motion either by liquid entrainment or by density-driven circulation. The liquid rises with the bubbles in the center of the column and flows downward in the outer annular region. Liquid circulation strongly affects the performance of the bubble column.

We can usually observe three flow regimes: i) the bubbly flow regime, characterized by almost uniformly sized bubbles with equal radial distribution; ii) the churn-turbulent flow regime, characterized by large bubbles moving with high rising velocities in the presence of small bubbles in the environment; and iii) the slug flow regime, in which large bubbles are stabilized by the column wall leading to the formation of bubble slugs. These regimes occur in order of increasing gas flow rate. In the turbulent flow regime which is of practical interest in many commercial-scale bubble columns, a recirculating flow model (Ueyama and Miyauchi, 1979) and a circulation cell model (Joshi and Sharma, 1961) have been widely used to discuss transport phenomena in bubble columns. Recently, Kawase and Moo-Young (1986c) extended the recirculating flow model to non-Newtonian systems.

A minimum gas velocity for complete suspension (or the maximum amount of solid that can be kept in complete suspension for a given operating condition) can be estimated from (Roy et al., 1964)

$$\frac{w'_{\max}}{\rho_l} = 6.8 \times 10^{-4} \frac{b\,D_c\,U_{sg}\,\rho_g}{\mu_g}\left(\frac{\sigma\,\varepsilon_g}{U_{sg}\,\mu_l}\right)^{-0.23}\left(\frac{\varepsilon_g U_{tp}}{U_{sg}}\right)^{-0.18}\gamma^{-3}. \qquad (19)$$

According to Roy et al. (1964), the minimum gas flow rate increases with increasing solids concentration, while Imafuku et al. (1968) observed an opposite trend at lower concentrations. This contradiction may be due to the complex relationship between the solids fraction and hindered settling velocity.

The effects of rate processes on the bubble column performance depend on the nonideal mixing behavior of the phases. The axial dispersion coefficient is a single parameter in the model which has been used to characterize the nonideal mixing behavior in bubble columns.

Kawase and Moo-Young (1986c) analyzed liquid-phase mixing in bubble columns using an energy balance and the Prandtl mixing length theory. Their resulting equation is

$$\frac{E_z}{U_{sg}\,D_c} = 0.342\,n^{-8/3}\left(\frac{U_{sg}^2}{g\,D_c}\right)^{-1/3}. \qquad (20)$$

This equation indicates that the axial dispersion coefficient increases as the shear-thinning increases. The above correlation is in reasonable agreement with the data for axial dispersion coefficients for the liquid and the solid in the three-phase systems obtained by Kato et al. (1972).

Gas hold-up is one of the most important parameters characterizing the hydrodynamics of bubble columns; it depends mainly on gas velocity and often is very sensitive to the physical properties of the liquid.

An empirical correlation for gas hold-up in CMC solutions was obtained by Godbole et al. (1984).

$$\varepsilon_g = 0.207\,U_{sg}^{0.60}\,\mu_{\text{susN}}^{-0.19}. \qquad (21)$$

Schumpe and Patwari (1985) and Schumpe et al. (1985) proposed an empirical correlation based on the data in CMC, PAA (polyacrylamide), and xanthan gum

$$\varepsilon_g = 0.36\left(\frac{g\,D_c^2\,\rho_l}{\sigma}\right)^{-0.15}\left(\frac{g\,D_c^3\,\rho_l}{\mu_{\text{sus}}^2}\right)^{0.09}\left(\frac{U_{sg}}{\sqrt{g\,D_c}}\right)^{0.53}. \qquad (22)$$

Recently, Kawase and Moo-Young (1986d) proposed a theoretical correlation for ε_g

$$\varepsilon_g = 1.07\,n^{2/3}\left(\frac{U_{sg}^3}{g\,D_c}\right)^{1/3}. \qquad (23)$$

Agreement of this equation with the experimental data is comparable to that of their empirical correlation (Kawase and Moo-Young, 1986a). Although this is not a general form and the constant may change with the column diameter, Eq. (23) is in reasonable agreement with the data.

In general, the presence of solids does not affect the gas hold-up significantly and the correlations for two-phase systems are applicable to three-phase systems, particularly at high gas flow rates.

Sauer and Hempel (1985) obtained an empirical correlation for gas hold-up in suspensions based on the volume fraction-dependent viscosity approach (a-2)

$$\frac{\varepsilon_g}{1 - \varepsilon_g} = C_2\left(\frac{\sigma\,g}{\rho_g\,U_{sg}^4}\right)^{c_3}\left(\frac{\mu_{\text{sus}}^4\,g\,\rho_g}{\sigma^3\,\rho_{\text{sus}}^2}\right)^{c_4} \qquad (24)$$

In viscous non-Newtonian fluids, bubbles exist in two distinctive categories, a large number of very small ones with diameter less than 0.5 mm whose residence time is very long and large spherical cap bubbles with diameter larger than 10 mm.

Schumpe and Deckwer (1981) measured the specific interfacial area in CMC solutions and proposed an empirical correlation in the slug flow regime.

$$a = 0.487 \left(\frac{U_{sg}}{\mu_{susN}} \right)^{0.51}. \tag{25}$$

For churn-turbulent flow, Godbole et al. (1984) obtained a correlation

$$a = 19.2 \, U_{sg}{}^{0.47} \mu_{susN}{}^{-0.76}. \tag{26}$$

Capuder and Koloini (1984) measured specific interfacial area in non-Newtonian solid suspensions and proposed a correlation

$$aD_c = 1.67 \left(\frac{gD_c^2 \, \rho_l}{\sigma} \right)^{0.5} \left(\frac{gD_c^3 \, \rho_l^2}{\mu_{susck}^2} \right)^{0.021} \varepsilon_g^{1.13}. \tag{27}$$

Tuvekar and Sharma (1973) observed that the interfacial area in the presence of fine solid particles is a weaker function of gas velocity than the one in absence of solids.

Godbole et al. (1984) measured the volumetric mass transfer coefficients in CMC aqueous solutions and proposed the following empirical correlation.

$$k_L a = 8.35 \times 10^{-4} \, U_{sg}{}^{0.44} \, \mu_{susN}^{-1.01}. \tag{28}$$

Schumpe et al. (1985) also proposed an empirical correlation for $k_L a$

$$\left(\frac{k_L a \, D_c^2}{D} \right) = 0.021 \left(\frac{\mu_{susS}}{\rho_l \, D} \right)^{0.50} \left(\frac{g \, D_c^3 \, \rho_l}{\mu_{susS}} \right)^{0.60} \left(\frac{U_{sg}}{\sqrt{D_c \, g}} \right)^{0.49} \left(\frac{g \, D_c^2 \, \rho_l}{\sigma} \right)^{0.21}. \tag{29}$$

Recently, a theoretical correlation for $k_L a$ was derived using Higbie's penetration theory and Kolmogoroff's theory for isotropic turbulence by Kawase et al. (1986e). It may be written as

$$\left(\frac{k_L a \, D_c^2}{D} \right) = C_5 \left(\frac{K^{1/2} \, D_c^{(1-n)/2}}{\rho_l^{1/2} D^{1/2} \, U_{sg}^{(1-n)/2}} \right)^{1/2} \left(\frac{g \, D_c^2 \, \rho_l}{\sigma} \right)^{3/5}$$
$$\times \left(\frac{D_c^n \, U_{sg}^{2-n}}{K/\rho_l} \right)^{(2+n)/2(1+n)} \left(\frac{U_{sg}^2}{g \, D_c} \right)^{(2n-3)/10(1+n)} \varepsilon^{0.5}. \tag{30}$$

It can be conjectured that the particles affect mass transfer from the bubbles to the liquid by altering first the turbulence structure and then the thickness of the liquid film around the bubble. However, it is found in practice that the mass transfer coefficient k_L in three-phase systems is very similar to that in gas-liquid systems.

Kulkarni et al. (1983) found that for all solid concentrations the values of $k_L a$ were larger than those in the absence of solids but the $k_L a$ coefficient decreased with an increase in solid concentration in downflow bubble columns. In an upflow bubble column, Nguyen-Tien et al. (1985) observed higher and lower $k_L a$ values compared with those in the absence of solids at high liquid velocity and low gas velocity and at low liquid velocity and high gas velocity, respectively. They proposed the following equation:

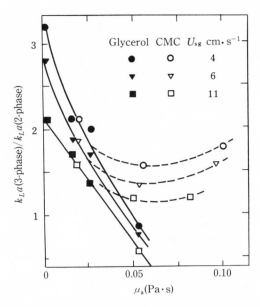

FIG. 6. Relative increase of volumetric mass transfer coefficients by particles as a function of liquid viscosity (Patwari et al. 1986).

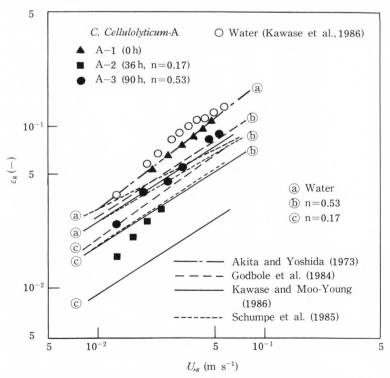

FIG. 7. Gas hold-up in *C. cellulolyticum* (Moo-Young et al., 1986).

$$k_L a = 0.394 \left(1 - \frac{\phi_s}{0.58}\right) U_{sg}^{0.67}. \tag{31}$$

Patwari et al. (1986) studied hydrodynamics and mass transfer in three-phase systems where the liquids were non-Newtonian. They reported that the gas hold-ups in three-phase systems were lower than those in gas-liquid two-phase systems. Figure 6 shows the relative increase of volumetric mass transfer coefficients for $d_p = 0.008$ m. It was found that the $k_L a$ coefficients for $d_p = 0.003$ m were lower than those in two-phase systems.

Loh et al. (1986) obtained the correlation for $k_L a$ in a three-phase circulating bed fermentor (based on approach a-2):

$$k_L a = 1.4 \times 10^{-4} \left(\frac{P}{V_l}\right)^{0.91} \frac{(1 - 2.5\phi_s)^{0.96}}{(1 - \phi_s)^{4.3}}. \tag{32}$$

Mass transfer to suspended particles in bubble columns is also discussed by Eq. (17). Needless to say, Eq. (17), in which k_s and D are replaced by h and $(k/\rho_l C_p)$, respectively, can be used to evaluate the heat transfer rate as well as the case of stirred tanks. Heat transfer in three-phase systems has been studied by Kato et al. (1981), Chiu and Ziegler (1983), and Kang et al. (1985).

Water-soluble polymer solutions (e.g., CMC, PAA, carboxypolymethylene) have been widely used to simulate fermentation broths. The correlations obtained using the data in those liquids are applicable to discussions of polysaccharide fermentations. However, their applicability to mycelial fermentation systems has not been discussed in detail. Re-

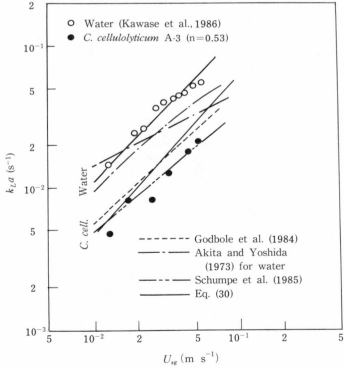

FIG. 8. Volumetric mass transfer in *C. cellulolyticum* (Moo-Young et al., 1986).

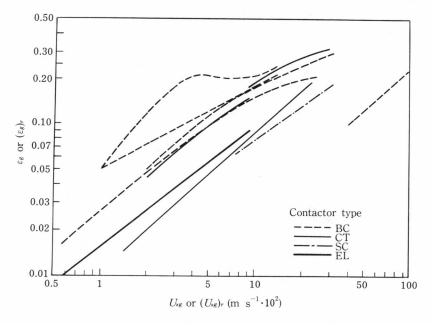

FIG. 9.   Gas hold-up in bubble columns (ε_g) and in the riser of air-lift columns (ε_g)$_r$. (Bello et al., 1985).
BC: bubble column, CT: concentric-tube air-lift column
SC: split cylindrical air-lift column
EL: external-loop air-lift column

cently, Moo-Young et al. (1986) discussed this problem using the data from mycelial fermentation systems for *Chaetomuim cellulolyticum* and *Neurospora sitophila*. It was found (Figs. 7 and 8) that the existing correlations obtained using the data in simulated media fit the experimental results for *C. cellulolyticum* reasonably. However, the existing correlations for ε_g and $k_L a$ lie considerably below the data for *N. sitophila*. It is conjectured that the rheological parameters determined at very low shear rates are responsible for this result. The flow index in the power-law model for *N. sitophila* determined using a turbine mixer-type viscometer instead of a concentric cylinder viscometer may be rather smaller than that characterizing the phenomena in the bioreactor. Furthermore, *N. sitophila* seems to have a yield stress.

André et al. (1985) studied the effect of paper pulp suspensions used to simulate a mycelial fermentation broth on the performance of a bubble column and found that $k_L a$ and ε_g values both decreased with increasing solid concentration. It should be emphasized that there is a great discrepancy in the results of various studies.

Considerable research has been devoted to aeration devices since they sometimes have a significant influence on the performance of bioreactors. In the case of dense suspension systems, clogging of spargers must be considered.

In order to improve the performance of bubble columns, a variant of the bubble column is rapidly gaining attention (Schügerl, 1982, 1985). Bubble columns with internal or external circulation loops, the so-called "air-lift towers," are two major examples. In an air-lift tower, there exists a significant liquid circulation that results in effective macroscale

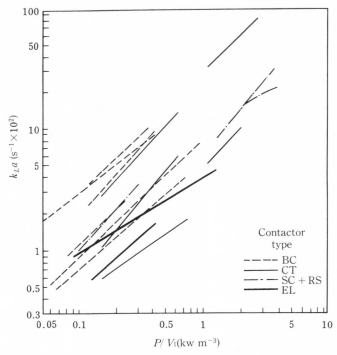

FIG. 10. Overall volumetric oxygen mass transfer coefficients in bubble columns and air-lift columns (Bello et al., 1985).
BC: bubble column, CT: concentric-tube air-lift column
SC: split cylindrical air-lift column
RS: rectangular split air-lift column
EL: external-loop air-lift column

mixing of the liquid phase as compared with the bubble column. Therefore, there has been an increasing number of studies of air-lift columns, some of which were summarized by Bello et al. (1985). However, most of them are restricted to non-viscous Newtonian fluids.

Self-generated liquid circulation improves heat transfer and mixing compared with bubble columns. However, the gas hold-up in the riser and the volumetric mass transfer coefficients have been reported to decrease as the circulating liquid velocity increases (Figs. 9 and 10).

Weiland (1984) recommended a diameter ratio between 0.8 and 0.9 for high oxygen transfer and efficient mixing and a ratio of 0.6 which has been widely used only to avoid sedimentation of large microbial aggregates (Table 1).

TABLE 1. Optimal diameter ratio in a column with internal-loop (Weiland, 1984).

Process parameter	Optimal diameter ratio	
	Coalescing systems	Non-coalescing systems
Gas hold-up	0.7 – 0.8	0.7 – 0.9
Mass transfer	0.7 – 0.8	0.9 – 1.0
Mixing time	0.9 – 1.0	0.9
Liquid velocity	0.6	0.6

Popovic and Robinson (1984) measured the gas hold-up and the volumetric mass transfer coefficient in a column with an external loop and proposed empirical correlations for them:

$$\varepsilon_g = 0.02 U_{sg}^{0.6504}(1 + A_d/A_r)^{-1.0516} \mu_{susN}^{-0.1039} \tag{33}$$

and

$$k_L a = 1.911 \times 10^{-4} U_{sg}^{0.525}(1 + A_d/A_r)^{-0.853} \mu_{susN}^{-0.89}. \tag{34}$$

Recently Schügerl et al. (1986) reported experimental data for gas hold-up, liquid circulation rate, gas residence time, and cell mass concentration in columns with an external loop. There is considerable uncertainty, and more attention must be devoted to design of air-lift columns.

2.3 Other Reactor Types

Packed beds as "fixed-film" bioreactors are becoming increasingly interesting for carrier-supported immobilized enzymes and cells. The immobilization techniques are numerous but the basic design approach is similar to that in conventional packed-bed chemical engineering. These reactors may also be operated in a countercurrent manner, e.g., for particle densities lower than that of the liquid.

In fluidized beds, the particles are fluidized mainly by the liquid rather than by the gas phase. Unlike a slurry bubble column in which small particles are fluidized by gas-induced liquid motion alone, momentum transfer in biofluidization where relatively large solid particle size may be used is carried out by the motion of liquid rather than the gas bubble motion. Results in traditional chemical reactor engineering of fluidized beds (Shah, 1979; Ramachandran and Chaudhari, 1983) are the basis of design for these bioreactor types.

Various types of reactor configurations for more dense slurry systems than considered above ($>15\%$ w/v solid loading) are used for composting, dump bioleaching, Koji, and other "solid-state" fermentations. Little quantitative information is available on these systems, however.

3. Concluding Remarks

For dilute homogeneous suspensions where solid particles are rather uniformly distributed throughout the reactor, as described above, there are relatively reliable design theories and design correlations. For systems with fairly high localized concentrations of solid particles, however, only first approximations can be obtained. Lack of homogeneity, segregational effects by foam formation, settling-out or wall growth, or sensitivity of cells to shear stress can be a limiting constraint in reactor design for slurry fermentation systems. The complexity of fermentation systems causes significant design problems and uncertainties. Improving our design theories and correlations is important. In particular, more innovative designs and operational procedures are required for high-density solid systems.

Nomenclature

A	cross-sectional area
a	specific surface area
a_1	constant in Eq. (8)
b	constant in Eq. (19)
C_1	constant in Eq. (13)
C_2, C_3, C_4	constants in Eq. (24)

C_5	constant in Eq. (30)
C_p	specific heat capacity
D_c	reactor diameter
D_I	impeller diameter
D	diffusion coefficient
E_z	axial dispersion coefficient
d_p	particle diameter
d_{vs}	bubble diameter
g	gravitational acceleration
H	clear liquid height
h	heat transfer coefficient
K	consistency index in power-law model
K_B	constant in Eq. (6)
K_s	constant in Eq. (5)
k	thermal conductivity
k_L	liquid film mass transfer coefficient
k_s	liquid-solid mass transfer coefficient
N	impeller speed
N_m	minimum impeller speed for complete suspension
N_p	power number
n	flow index in power-law model
P	power consumption
P_0	power consumption of gas-free liquid
Q	volumetric flow rate
t_M	mixing time
U_{sg}	superficial gas velocity
U_{tp}	terminal settling velocity of particle
V	volume
W	impeller blade width
w	percentage by weight of solids
w'_{max}	weight of solids/weight of suspension
α	exponent in Eq. (13)
γ	relative wettability in Eq. (19)
$\dot{\gamma}$	shear rate
ε	energy dissipation rate
ε_a	dissipation rate of energy by agitation
ε_{av}	dissipation rate of energy by aeration
ε_g	gas hold-up
μ	viscosity
μ_a	apparent viscosity
μ_p	plastic viscosity
μ_{susck}	$K (4000\ U_{sg})^{n-1}$
μ_{susM}	$K (11.83N)^{n-1}$
μ_{susN}	$K (5000 U_{sg})^{n-1}$
μ_{susS}	$K (2800 U_{sg})^{n-1}$
μ_w	viscosity of water

ρ	density

ρ density
σ interfacial tension
τ shear stress
τ_0 yield stress
ϕ_s volume fraction of solid
Ψ_s mass fraction of solid

Subscripts

a average
d downcomer
g gas
l liquid
r riser
s solid

sus   suspension

REFERENCES

1. Aiba, S., Humphrey, A.E., Millis, N.F., *Biochemical Engineering* 2nd. Ed. Academic Press, New York, 1973.
2. Akita, K., Yoshida, F., "Gas holdup and volumetric mass transfer coefficient in bubble columns." *Ind. Eng. Chem. Proc. Des. Dev.*: **12**, 76–80 (1973).
3. André, G., Moo-Young, M., Robinson, C.W., "Application of the Fourier transform to the measurement of K_La_L in non-mechanically agitated contactors." *Can. J. Chem. Eng.*: **63**, 202–211 (1985).
4. Bello, R.A., Robinson, C.W., Moo-Young, M., "Gas holdup and overall volumetric oxygen transfer coefficient in airlift contactors." *Biotechnol. Bioeng.*: **27**, 369–381 (1985).
5. Blakebrough, N., McManamey, W.J., Tart, K.R., "Rheological measurements on *Aspergillus niger* fermentation systems." *J. Chem. Tech. Biotechnol.*: **28**, 453–461 (1978).
6. Calderbank, P.H., "Gas absorption from bubbles." *Trans. Instn. Chem. Engrs.*: **45**, CE209–CE233 (1967).
7. Calderbank, P.H., Moo-Young, M.B., "The continuous phase heat and mass-transfer properties of dispersions." *Chem. Eng. Sci.*: **16**, 39–54 (1961a).
8. Calderbank, P.H., Moo-Young, M.B., "The power characteristics of agitators for the mixing of Newtonian and non-Newtonian fluids." *Trans. Instn. Chem. Engrs.*: **39**, 337–347 (1961b).
9. Capuder, E., Koloini, T., "Gas hold-up and interfacial area in aerated suspensions of small particles." *Chem. Eng. Res. Des.*: **62**, 255–260 (1984).
10. Charles, M., "Technical aspect of the rheological properties of microbial cultures." *Adv. Biochem. Eng.*: **8**, 1–62 (1978).
11. Chiu, T.-M., Ziegler, E.N., "Heat transfer in three-phase fluidized beds." *AIChE J.*: **29**, 677–685 (1983).
12. Godbole, S.P., Schumpe, A., Shah, Y.T., Carr, N.L., "Hydrodynamics and mass transfer in non-Newtonian solutions in a bubble columm." *AIChE J.*: **30**, 213–220 (1984).
13. Heijnen, J.J., Riet, K.V., "Mass transfer, mixing and heat transfer phenomena in low viscosity bubble column reactors." *Chem. Eng. J.*: **28**, 21–42 (1984).
14. Henzler, H.-J., "Begasen höherviskoser Flüssigkeiten." *Chem.-Ing.-Tech.*: **52**, 643–652 (1980).
15. Hickman, A.D., Nienow, A.W., "Mass transfer and hold-up in an agitated simulated fermentation broth as a function of viscosity." *Proc. Int. Conf. on Bioreactor Fluid Dynamics*, Cambridge (1986).
16. Höcker, H., Langer, G., Werner, U., "Mass transfer in aerated Newtonian and non-Newtonian liquids in stirred reactors." *Ger. Chem. Eng.*: **4**, 51–62 (1981).
17. Imafuku, K., Wang, T-Y., Koide, K., Kubota, H., "The behavior of suspended solid particles in the bubble column." *J. Chem. Eng. Japan*: **1**, 153–158 (1968).
18. Joshi, J.B., Pandit, A.B., Sharma, M.M., "Mechanically agitated gas-liquid reactors." *Chem. Eng. Sci.*:

37, 813–844 (1982).

19. Joshi, J.B., Sharma, M.M., "A circulation cell model for bubble columns." *Trans. Instn. Chem. Engrs.*: 57, 244–251 (1961).

20. Juvekar, V.A., Sharma, M.M., "Absorption of CO_2 in a suspension of lime." *Chem. Eng. Sci.*: 28, 825–837 (1973).

21. Kang, Y., Suh, I.S., Kim, S.D., "Heat transfer characteristics of three phase fluidized beds." *Chem. Eng. Commun.*: 34, 1–13 (1985).

22. Kargi, F., Moo-Young, M., "Transport phenomena in bioprocesses." In *Comprehensive Biotechnology* 2, M. Moo-Young (ed.), p. 5, Pergamon Press, Oxford, 1985.

23. Kast, W., "Analysis of heat transfer in bubble columns." *Int. J. Heat Mass Transfer*: 5, 329–336 (1962).

24. Kato, Y. Nishiwaki, A., Fukuda, T., Tanaka, S., "The behavior of suspended solid particles and liquid in bubble columns." *J. Chem. Eng. Japan*: 5, 112–118 (1972).

25. Kato, Y., Uchida, K., Kago, T., Morooka, S., "Liquid holdup and heat transfer coefficient between bed and wall in liquid-solid and gas-liquid-solid fluidized beds." *Powder Technol.*: 28, 173–179 (1981).

26. Kawase, Y., Moo-Young, M., "Influence of non-Newtonian flow behaviour on mass transfer in bubble columns with and without draft tubes." *Chem. Eng. Commun.*: 40, 67–83 (1986a).

27. Kawase, Y., Moo-Young, M., "Solid turbulent fluid heat and mass transfer—a unified model based on the energy dissipation rate concept." *Chem. Eng. J.*, in press (1986b).

28. Kawase, Y., Moo-Young, M., "Liquid phase mixing in bubble columns with Newtonian and non-Newtonian fluids." *Chem. Eng. Sci.*: 41, 1969–1977 (1986c).

29. Kawase, Y., Moo-Young, M., "Theoretical prediction of gas hold-ups in bubble columns with Newtonian and non-Newtonian fluid systems." *Ind. Eng. Chem. Fundam.*, in press (1986d).

30. Kawase, Y., Halard, B., Moo-Young, M., "Theoretical prediction of volumetric mass transfer coefficients in bubble columns: Newtonian and non-Newtonian fluids." *Chem. Eng. Sci.*, in press (1986e).

31. Keitel, G., Onken, U., "Gas-liquid mass transfer in a stirred loop Reactor." *Ger. Chem. Eng.*: 4, 250–258 (1981).

32. Kulkarni, A., Shah, Y., Schumpe, A., "Hydrodynamics and mass transfer in downflow bubble column." *Chem. Eng. Commun.*: 24, 307–337 (1983).

33. Kurpiers, P., Steiff, A., Weinspach, P.-M., "Heat transfer in stirred multiphase reactors." *Ger. Chem. Eng.*: 8, 48–57 (1985).

34. Langer, G., Werner, U., "Measurements of viscosity of suspensions in different viscometer flows and stirring systems." *Ger. Chem. Eng.*: 4, 226–241 (1981).

35. Lee, J.C., Ali, S.S. Tasakorn, P., "Influence of suspended solids on gas-liquid mass transfer in an agitated tank." *Proc. 4th European Conf. on Mixing*, H4, Noorduijkerhout (1982).

36. Loh, V.Y. Richards, S.R., Richmond, P., "Fluid dynamics and mass transfer in a three-phase circulating bed fermenter." *Proc. Int. Conf. on Bioreactor Fluid Dynamics*, Cambridge (1985).

37. Loucaides, R., McManamey, W.J., "Mass transfer into simulated fermentation media." *Chem. Eng. Sci.*: 28, 2165–2178 (1973).

38. Machon, V., Vlcek, J., "Some effects of pseudoplasticity on hold-up in aerated, agitated vessels." *Chem. Eng. J.*: 19, 67–74 (1980).

39. Margaritis, A., Zajic, J.E., "Mixing, mass transfer, and scale-up of polysaccharide fermentations." *Biotechnol. Bioeng.*: 20, 939–1001 (1978).

40. Marrone, G.M., Kirwan, D.J., "Mass transfer to suspended particles in gas-liquid agitated systems." *AIChE J.*: 32, 523–525 (1986).

41. Metzner, A.B., Otto, R.E., "Agitation of non-Newtonian fluids." *AIChE J.*: 3, 3–10 (1957).

42. Moo-Young, M., Blanch, H.W., "Design of biochemical reactors mass transfer criteria for simple and complex systems." *Adv. Biochem. Eng.*: 19, 1–69 (1981).

43. Moo-Young, M., Halard, B., Allen, D.G., Burrel, R., Kawase, Y., "Oxygen transfer to mycelial fermentation broths in an airlift fermentor." *Biotechnol. Bioeng.*, in press (1986).

44. Nguyen-Tien, K., Patwari, A.N., Schumpe, A., Deckwer, W.-D., "Gas-liquid mass transfer in fluidized particle beds." *AIChE J.*: 31, 194–201 (1985).

45. Nienow, A.W., Wisdom, D.J., Solomon, J., Machon, V., Vlcek, J., "The effect of rheological complexities on power consumption in an aerated, agitated vessel." *Chem. Eng. Commun.*: 19, 273–293 (1983).

46. Nienow, A.W., Ulbrecht, J.J., In *Mixing of liquids by mechanical agitation*, Gordon and Breach, New York, 1985.

47. Nishikawa, M., Kato, H., Hashimoto, K., "Heat transfer in aerated tower filled with non-Newtonian liquid." *Ind. Eng. Chem. Process Des. Dev.*: **16**, 133–137 (1977).
48. Nishikawa, M., Nakamura, M., Hashimoto, K., "Gas absorption in aerated mixing vessels with non-Newtonian liquid." *J. Chem. Eng. Japan*: **14**, 227–232 (1981).
49. Pandit, A.B., Joshi, J.B., "Three phase sparged reactors–some design aspects." *Reviews in Chem. Eng.*: **2**, 1–84 (1984).
50. Pandit, A.B., Joshi, J.B., "Mass and heat transfer characteristics of three phase sparged reactors." *Chem. Eng. Res. Des.*: **64**, 125–157 (1986).
51. Patwari, A.N., Nguyen-Tien, K., Schunmpe, A., Deckwer, W.-D., "Three-phase fluidized beds with viscous liquid: hydrodynamics and mass transfer." *Chem. Eng. Commun.*: **40**, 49–65 (1986).
52. Popovic, M., Robinson, C.W., "Estimation of some important design parameters for non-Newtonian liquid in pneumatically-agitated fermenters." CCES Meeting, Quebec (1984).
53. Ramachandra, P.A., Chaudhari, R.V., In *Three-phase catalytic reactors*. Gordon and Breach, London, 1983.
54. Reuss, M., Debus, D., Zoll, G., "Rheological properties of fermentation fluids." *Chem. Engr.*: **60**, 233–236 (1982).
55. Roy, N.K., Guha, D.K., Rao, M.N., "Suspension of solids in a bubbling liquid." *Chem. Eng. Sci.*: **19**, 215–225 (1964).
56. Sauer, T., Hempel, D.C., "Aufwirbelcharakteristik und relativer Gasgehalt in Suspensions-Blasensäulen." *Chem.-Ing.-Tech.*: **57**, 973–975 (1985).
57. Schügerl, K., "Characterization and performance of single- and multistage tower reactors with outer loop for cell mass production." *Adv. Biochem. Eng.*: **22**, 93–224 (1982).
58. Schügerl, K., "Nonmechanically agitated bioreactor systems." In *Comprehensive Biotechnology* **2**, M. Moo-Young (ed.), p. 99, Pergamon Press, Oxford, 1985.
59. Schügerl, K. Burschäpers, J., Czech, K., Frieling, M.V., Fröhlich, S., Gebauer, A., Lorenz, T., Lübbert A., Ross, A., Scheper, T., "Fluid-dynamic behavior of airlift tower loop reactors." *Proc. Int. Conf. on Bioreactor Fluid Dynamics*, Cambridge (1986).
60. Schumpe, A., Deckwer, W.D., "Determination of interfacial areas in non-Newtonian gas/liquid flow." ACS Meeting, Atlanta (1981).
61. Schumpe, A., Singh, C., Deckwer, W.-D., "Stoffübergangszahlen und effektives Schergefälle beim Belüften von Xanthan-Lösungen in Blasensäulen." *Chem.-Ing.-Tech.*: **57**, 988–989 (1985).
62. Schumpe, A., Patwari, A.N., "Hochviskose Medien in Blasensäulen: Hydrodynamik und Stoffaustausch." *Chem.-Ing. Tech.*: **57**, 874–875 (1985).
63. Shah, Y.T., In *Gas-Liquid-Solid Reactor Design*. McGraw-Hill, New York, 1979.
64. Shah, Y.T., Kelkar, B.G., Godbole, S.P., Deckwer, W.-D., "Design parameters estimations for bubble column reactors." *AIChE J.*: **28**, 353–379 (1982).
65. Skelland, A.H.P., In *Non-Newtonian flow and heat transfer*. Wiley, New York, 1967.
66. Solomon, J., Elson, T.P., Nienow, A.W., "Cavern sizes in agitated fluids with a yield stress." *Chem. Eng. Commun.*: **11**, 143–164 (1981).
67. Steiff, A., Weinspach, P.-M., "Fluid dynamics heat and mass transfer in agitated aerated slurry reactors." *Ger. Chem. Eng.*: **5**, 342–350 (1982).
68. Thomas, D.G., "Transport characteristics of suspension VIII." *J. Colloid Sci.*: **20**, 267–277 (1965).
69. Ueyama, K., Miyauchi, T., "Properties of recirculating turbulent two phase flow in gas bubble columns." *AIChE J.*: **25**, 258–266 (1979).
70. Wase, D.A.J., McManamey, W.J., Raymahasay, S., Vaid, A.K., "Comparisons between cellulase production by *Aspergillus fumigatus* in agitated vessels and in an air-lift fermentor." *Biotechnol. Bioeng.*: **27**, 1166–1172 (1985).
71. Weiland, P., "Influence of draft tube diameter on operation behaviour of airlift loop reactors." *Ger. Chem. Eng.*: **7**, 374–385 (1984).
72. Wiedmann, J.A., Steiff, A., Weinspach, P.-M., "Suspension behaviour of two- and three-phase stirred reactors." *Ger. Chem. Eng.*: **8**, 321–335 (1985).
73. Yagi, H., Yoshida, F., "Gas absorption by Newtonian and non-Newtonian fluids in sparged agitated vessels." *Ind. Eng. Chem. Process Des. Dev.*: **14**, 488–493 (1975).
74. Zwietering, T.N., "Suspending of solid particles in liquid by agitators." *Chem. Eng. Sci.*: **8**, 244–253 (1958).

Retrospect on Practical and Engineering Problems in Fermentation Industry

Juntang Yu*, Siliang Zhang*, and Baishan Fang[2]*

* *East China University of Chemical Technology, Shanghai, China*
[2]* *Huaqiao University, Quanzhou, China*

1. Dehumidification of Compressed Air for Producing Sterile Air

Compressed air is indispensable for aeration in the aerobic fermentation industry, and how to avoid wetting of the air filter medium, especially when a fibrous filter is used, is one of the most serious problems involved in realizing contamination-free operations.[1]

Water vapor contained in compressed air comes from moisture in the air. The absolute humidity, x (kg of water/kg of dry air), and relative humidity, ϕ (%) can be represented as follows:

$$x = 0.622 \frac{\phi p_s}{P - \phi p_s} \tag{1.1}$$

$$\phi = \frac{p_w}{p_s} \tag{1.2}$$

$$\phi = \frac{P}{p_s} \frac{x}{0.622 + x} \tag{1.3}$$

where

P is the total pressure of air; p_w the partial pressure of water vapor in a water-air mixture; and p_s the saturated water vapor pressure at the temperature of air.

When atmospheric air is compressed, the humidity will remain unchanged, while the relative humidity will change, depending on the temperature of the compressed air. From Eq. (1.3) we obtain:

$$\frac{\phi_2}{\phi_1} = \frac{P_2}{P_1} \frac{(p_s)_1}{(p_s)_2} \tag{1.4}$$

where subscripts 1 and 2 indicate the conditions of atmospheric and compressed air, respectively.

From Eq. (1.3) we can find that the dew point of compressed air is obviously higher than that of the atmospheric air, if x remains constant. For example, the dew point of an 80% atmospheric water-air mixture at 25°C is 21.3°C; when the mixture is compressed to 2 atm (gauge), the dew point will be increased to 33.1°C. Examining Eq. (1.4), we find that the relative humidity of compressed air will increase greatly as it is cooled to the same temperature as that of atmospheric air, so that it may reach or exceed the saturation point. As a result, water vapor will condense into mist or water drops. That is why compressed air condenses easily into liquid water at ordinary temperatures. According to our experience, it is preferable to control the compressed air to a relative humidity of about 60% and a temperature below 60°C before it enters the air filter.

The following example describes the pretreatment procedure suitable for air sterilization. If atmospheric air at 25°C and 80% humidity were compressed to 2 atm (gauge), the temperature of the compressed air leaving the compressor would be 110°C. This air is too hot to be admitted to the air filter. Cooling and some adjunct treatment are needed.

1. Cooling the compressed air down to 50°C (higher than the dew point) allows it to proceed to the air filter directly (Fig. 1). In this case, the relative humidity of the compressed air before entering the air filter would be

$$\phi_2 = \phi_1 \frac{(p_s)_1}{(p_s)_2} \frac{P_2}{P_1} = 0.80 \left(\frac{0.03126}{0.1217} \right) \left(\frac{2.8}{1} \right) = 57.5\%.$$

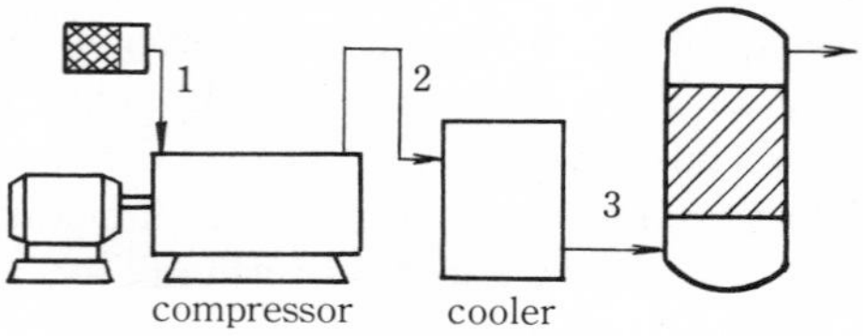

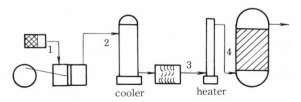

FIG. 1. Simple cooling procedure.

	Point 1	Point 2	Point 3
P (atm, absolute)	1	3	2.8[a]
t (°C)	25	110	50
ϕ (%)	80	5.31	57.5
x (kg water (kg dry air)$^{-1}$)	0.016	0.016	0.016

[a] Reduction in pressure due to friction loss in cooler.

2. Cooling the compressed air down to 30°C (lower than the dew point), and then heating it up to 40°C before it enters the air filter (Fig. 2) means that x after cooling would be

$$x = 0.622 \frac{0.0419}{2.8 - 0.0419} = 0.0095.$$

ϕ after cooling would be

$$\phi_2 = 1.00 \left(\frac{0.0419}{0.0734} \right) \left(\frac{2.7}{2.8} \right) = 55\%.$$

FIG. 2. Cooling followed by heating procedure.

	Point 1	Point 2	Point 3	Point 4
P (atm, absolute)	1	3	2.8	2.7[a]
t (°C)	25	110	30	40
ϕ (%)	80	5.31	100	55
x (kg water (kg dry air)$^{-1}$)	0.016	0.016	0.0095	0.0095

[a] Reduction in pressure due to friction loss in heater.

3. The hot compressed air can be divided into two parts; one part (80%) was cooled to 30°C, and the other (20%) was not cooled but was mixed with the cooled part. Then, the mixed air enters the air filter (Fig. 3). In this case, x after mixing would be

$$x = (0.016)\,(0.2) + (0.0095)\,(0.8) = 0.0108;$$

t after mixing would be

$$t = \frac{[(0.2)\,(0.2457)\,(110) + (0.8)\,(0.2425)(30)]\,(0.95)}{0.2432} = 43.9°C$$

where 0.2457, 0.2425, and 0.2432 are values of humid heat of air at x values of 0.016, 0.0095, and 0.0108, respectively. In addition a 5% heat loss is considered.
    ϕ after mixing would be

$$\phi = \frac{2.8}{0.0886}\,\frac{0.0108}{0.622 + 0.0108} = 53.9\%.$$

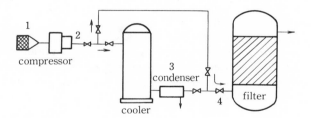

FIG. 3. Partial treatment procedure.

	Point 1	Point 2	Point 3	Point 4
P (atm, absolute)	1	3	2.8	2.8
t (°C)	25	110	30	43.9
ϕ (%)	80	5.31	100	53.9
x (kg water (kg dry air)$^{-1}$)	0.016	0.016	0.0095	0.0108

Except for hot and humid days, the first procedure is preferable, for it is simple as well as advantageous in terms of minimizing the equipment investment and maintenance cost. The third alternative is better than the second one, because it offers about 20% power saving.

2. Assessment of Agitator Speeds and Power Requirements in Production-Scale Fermentors

Up to now, the scale-up of fermentors has conventionally been carried out with semi-empirical methods. Generally, it is carried out on the basis of data obtained from a pilot fermentor, assuming geometrical similarity between the production and pilot fermentors. The criteria commonly used in scale-up are either the power consumption (under gassing) per unit volume of medium, P_g/V, or the volumetric oxygen transfer coefficient, k_La. The objective is to estimate the agitator speed required, superficial air velocity needed, and power requirement in the production fermentor.

Here, several convenient expressions to attain the above-mentioned aims will be re-written on the basis of a constant value of either P_g/V or k_La.[2]

The simplest method widely adopted in chemical reactor scale-up is based on $P/V =$ constant, i.e.,

$$P_2/V_2 = P_1/V_1, \qquad n_2^3 d_2^2 = n_1^3 d_1^2$$

(for turbulent regions).

Therefore,

$$n_2 = n_1 \left(\frac{d_1}{d_2}\right)^{2/3} \tag{2.1}$$

$$P_2 = P_1\left(\frac{V_2}{V_1}\right) = P_1\left(\frac{d_2}{d_1}\right)^3 \tag{2.2}$$

where

P is power consumption in the absence of aeration, kW;
V the volume of medium in fermentor, m^3;
n the agitator speed, s^{-1};
d the agitator diameter, m; and
subscript 1 is a pilot and 2 is a production-scale fermentor.

If a fermentation is carried out under aeration, the power requirement will decrease with an increase in the air flow rate. The relationship between power consumption and other parameters for gassed and ungassed systems has been presented by Michel and Miller[3]:

$$P_g \propto \left(\frac{P^2 n d^3}{Q^{0.56}}\right)^{0.45} \tag{2.3}$$

where

P_g is power consumption under aeration, kW; and
Q is air flow rate, $m^3 s^{-1}$.

Hence it is advisable to use $P_g/V =$ constant as a basis for scale-up, instead of $P/V =$ constant.

Since $P \propto n^3 d^5$ and $Q \propto d^2 v_s$, where v_s is the superficial air velocity, m s^{-1}, Eq. (2.3) can be modified as

$$P_g \propto [(n^3 d^5)^2 n d^3/(d^2 v_s)^{0.56}]^{0.45}$$

$$\propto n^{3.15} d^{5.35}/v_s^{0.252} \tag{2.4}$$

or

$$P_g/V \propto n^{3.15} d^{2.35}/v_s^{0.252}.$$

Remembering that $P_g/V = $ constant,

$$n_2 = n_1\left(\frac{d_1}{d_2}\right)^{0.746}\left(\frac{v_{s2}}{v_{s1}}\right)^{0.08} \tag{2.5}$$

$$P_2 = P_1\left(\frac{d_2}{d_1}\right)^{2.76}\left(\frac{v_{s2}}{v_{s1}}\right)^{0.24}. \tag{2.6}$$

Several dimensionless expressions have been published describing the relationship between volumetric oxygen transfer coefficient k_La and parameters, such as d, n, v_s as well as liquid characteristics μ (viscosity), D (diffusivity), ρ (medium density), σ (medium surface tension), e.g.,

$$k_La \propto \frac{D}{d^2}\left(\frac{d^2n}{\mu}\right)^a\left(\frac{dn^2}{g}\right)^b\left(\frac{\mu}{\rho D}\right)^c\left(\frac{\mu v_s}{\sigma}\right)^e\left(\frac{nd}{v_s}\right)^f$$

$$\propto n^x d^y v_s^z \rho^h D^i g^j \mu^k \sigma^l. \tag{2.7}$$

Taking for granted that liquid characteristics remain unchanged,

$$\frac{(k_La)_2}{(k_La)_1} = \left(\frac{n_2}{n_1}\right)^x\left(\frac{d_2}{d_1}\right)^y\left(\frac{v_{s2}}{v_{s1}}\right)^z = 1 \qquad n_2 = n_1\left(\frac{d_1}{d_2}\right)^{y/x}\left(\frac{v_{s1}}{v_{s2}}\right)^{z/x} \tag{2.8}$$

$$P_2 = P_1\left(\frac{d_2}{d_1}\right)^{5-3y/x}\left(\frac{v_{s1}}{v_{s2}}\right)^{3z/x}. \tag{2.9}$$

When Yagi's correlation ($H_L/D = 1$ and one turbine agitator) is used for Newtonian fluid,[4]

$$k_La \propto \frac{D}{d^2}\left(\frac{d^2n\rho}{\mu}\right)^{1.5}\left(\frac{dn^2}{g}\right)^{0.19}\left(\frac{\mu}{\rho D}\right)^{0.5}\left(\frac{\mu v_s}{\sigma}\right)^{0.6}\left(\frac{nd}{v_s}\right)^{0.32}$$

or

$$k_La \propto n^{2.2} d^{1.51} v_s^{0.28}\ldots\ldots$$

then

$$n_2 = n_1\left(\frac{d_1}{d_2}\right)^{0.686}\left(\frac{v_{s1}}{v_{s2}}\right)^{0.127} \tag{2.10}$$

$$P_2 = P_1\left(\frac{d_2}{d_1}\right)^{2.94}\left(\frac{v_{s1}}{v_{s2}}\right)^{0.381}. \tag{2.11}$$

When another correlation presented by Ye ($H_L/D = 1.5$ and two turbine agitators) is used for Newtonian fluid,[5]

$$k_La \propto \frac{D}{d^2}\left(\frac{d^2n\rho}{\mu}\right)^{1.65}\left(\frac{n^2d}{g}\right)^{0.127}\left(\frac{\mu}{\rho D}\right)^{0.5}\left(\frac{\mu v_s}{\sigma}\right)^{0.37}\left(\frac{nd}{v_s}\right)^{0.169}$$

or

$$k_La \propto n^{2.08} d^{1.6} v_s^{0.201}\ldots\ldots$$

then

$$n_2 = n_1\left(\frac{d_1}{d_2}\right)^{0.769}\left(\frac{v_{s1}}{v_{s2}}\right)^{0.097} \tag{2.12}$$

$$P_2 = P_1\left(\frac{d_2}{d_1}\right)^{2.69}\left(\frac{v_{s1}}{v_{s2}}\right)^{0.291}. \tag{2.13}$$

If the medium exhibits non-Newtonian behavior, the apparent viscosity μ_a will change with agitator speed n, i.e., μ_a is not constant in scale-up. No correlation suitable for this has been made available yet.

So far as the current situation is concerned, some comparisons will be illustrated next to appreciate which scale-up criterion of the two alternatives should be employed.

A pilot fermentor has a working volume $V_1 = 4.7$ m³, diameter of agitator $d_1 = 0.54$ m, agitator speed $n_1 = 240$ rpm, superficial air velocity $v_{s1} = 80$ m h⁻¹, and power consumption $P_1 = 7.5$ kW. If the production fermentor is 50 m³, i.e., 10.64-fold of scale-up, the assessed agitator speed n_2 and power requirement P_2 are listed below.

$$d_2 = 1.2 \text{ m}, \qquad d_2/d_1 = 2.2.$$

According to Aiba's suggestion,[6)]

$$\frac{Q_2/V_2}{Q_1/V_1} = \left(\frac{d_1}{d_2}\right)^{2/3}$$

then

$$v_{s2} = v_{s1}\left(\frac{d_2}{d_1}\right)^{1/3} = 104 \text{ mh}^{-1}, \quad \frac{v_{s2}}{v_{s1}} = 1.3.$$

Criterion for scale-up	Equations used	n_2, rpm	P_2, kW
P/V = const.	2.1, 2.2	142	80
P_g/V = const.	2.5, 2.6	136	70
$k_L a$ = const.	2.10, 2.11	135	69
	2.12, 2.13	128	58

From the results of assessment, it can be seen that there is no distinct difference between the two alternatives (especially when Yagi's correlation is used to evaluate $k_L a$). Both of them can be used as criteria for fermentor scale-up with a Newtonian fluid medium.

3. Estimation of Kinetic Parameters from Batch Cultivation Data

A method for estimating kinetic parameters (K_s, μ_{max}, $Y_{x/s}^{max}$, and m_s) from batch microbial cultivation data has been presented.[7)] The following kinetic equations are widely used.

$$\mu = \mu_{max}\frac{S}{K_s + S} \qquad \text{(Monod, 1942)} \tag{3.1}$$

$$Y_{x/s} = Y_{x/s}^{max}\left(\frac{D}{\mu_e + D}\right) \qquad \text{(Herbert, 1958)} \tag{3.2}$$

$$\frac{1}{Y_{x/s}} = \frac{1}{Y_{x/s}^{max}} + \frac{m_s}{\mu}$$

or

$$\frac{S}{X} = \frac{(S)_G}{X} + \frac{(S)_M}{X} \qquad \text{(Pirt, 1965)} \qquad (3.3)$$

$$\frac{-dS}{Xdt} = R_{max} \frac{S}{K_s + S} \qquad \text{(Knight, 1981)} \qquad (3.4)$$

where X is cell concentration; S limited substrate concentrations; $\mu = dX/(Xdt)$, specific cell growth rate; $Y_{x/s}$ yield coefficient of cell growth/substrate used; K_s saturation constant; $\mu_e = \mu$ based on endogenous metabolism; m_s maintenance coefficient; and $R_{max} = \mu_{max}/Y_{x/s}^{max}$ maximum specific substrate consumption rate.

From Eq. (3.3)

$$\frac{dX}{-dS} = Y_{x/s}^{max} \frac{\mu}{m_s Y_{x/s}^{max} + \mu}. \qquad (3.5)$$

Put Eq. (3.1) into Eq. (3.4) so that

$$\frac{dX}{-dS} = \frac{Y_{x/s}^{max} \mu_{max} S}{(Y_{x/s}^{max} m_s + \mu_{max}) S + K_s Y_{x/s}^{max} m_s} = \frac{AS}{S + B} \qquad (3.6)$$

where

$$A = Y_{x/s}^{max} \mu_{max}/(Y_{x/s}^{max} m_s + \mu_{max}) \qquad (3.7)$$

$$B = K_s Y_{x/s}^{max} m_s/(Y_{x/s}^{max} m_s + \mu_{max}). \qquad (3.8)$$

Integration of Eq. (3.6) with boundary condition $X_{s=s_0} = X_0$ (X_0, S_0: initial concentrations of cell and limited substrate, respectively) gives

$$X = A\left[(S_0 - S) - B\ln\left(\frac{S_0 + B}{S + B}\right)\right] + X_0. \qquad (3.9)$$

Whenever endogenous metabolism is taken into account, Eq. (3.4) becomes

$$\frac{-dS}{Xdt} = R_{max} \frac{S}{K_s + S} + m_s. \qquad (3.10)$$

Combining Eqs. (3.9) and (3.10)

$$-\int \frac{dS}{(R_{max} S/(K_s + S) + m_s) A\{(S_0 - S) + B\ln[(S + B)/(S_0 + B)] + X_0\}} = \int dt. \qquad (3.11)$$

Generally, K_s is rather small (on the order of 10^{-1}—10^{-4} kg m^{-3}) as compared with S_0 (on the order of 100 kg m^{-3}); so the value of B is very small. Consequently, $B \ll S_0$ and thus

$$\ln \frac{S + B}{S_0 + B} \approx \ln \frac{S + B}{S_0}.$$

If first term of expansion $\ln \dfrac{S + B}{S_0}$ is taken,

$$\ln\frac{S + B}{S_0} \approx 2\left(\frac{\dfrac{S + B}{S_0} - 1}{\dfrac{S + B}{S_0} + 1}\right) = 2\left(\frac{B + S - S_0}{B + S + S_0}\right) \approx 2\left(\frac{B + S - S_0}{S + S_0}\right)$$

then Eq. (3.11) becomes

$$-\int \frac{dS}{\left(R_{max}\dfrac{S}{K_s + S} + m_s\right)A\left[S_0 - S + 2B\left(\dfrac{B + S - S_0}{S + S_0}\right) + X_0\right]} = \int dt \quad (3.12)$$

or

$$\int\left(\frac{M}{ES + F} + \frac{NS + P}{S^2 + CS + D}\right)dS = A\int dt \qquad (3.13)$$

where

$$C = -(2B + X_0/A)$$

$$D = -(S_0 - 2B + X_0/A)S_0 - 2B^2 \approx -(S_0 - 2B + X_0/A)S_0$$

$$E = R_{max} + m_s$$

$$F = m_sK_s$$

$$M = \frac{(EK_s - F)(ES_0 - F)}{F^2 - CEF + DE^2}$$

$$N = (1 - M)/E$$

$$P = (S_0K_s - DM)/F.$$

If $S_{t=0} = S_0$, integration of Eq. (3.12) gives the relationship between S and t:

$$\frac{2M}{E}\ln(ES + F) + N\ln(S^2 + CS + D) + G\ln\frac{S + H}{S + I} - J = 2At \qquad (3.14)$$

where

$$G = \frac{2P - CN}{\sqrt{C^2 - 4D}}$$

$$H = \frac{C - \sqrt{C^2 - 4D}}{2}$$

$$I = \frac{C + \sqrt{C^2 - 4D}}{2}$$

$$J = \frac{2M}{E}\ln(ES_0 + F) + N\ln(S_0^2 + CS_0 + D) + G\ln\frac{S_0 + H}{S_0 + I}.$$

To estimate the value of kinetic parameters, we can combine Eqs. (3.7) and (3.8) with $R_{max} = \mu_{max}/Y_{x/s}^{max}$, then

$$\mu_{max} = AK_sR_{max}/(K_s - B) \qquad (3.15)$$

$$m_s = B\mu_{max}/AK_s. \qquad (3.16)$$

TABLE 1. Estimated and experimental x values.

t (h)		0	1	2	3	4	5	6	7
S (kg m^{-3})	exp.	10.0	9.92	9.76	9.44	8.79	7.49	4.89	0.062
X (kg m^{-3})	exp.	0.04	0.082	0.165	0.332	0.670	1.35	2.71	5.21
	est.	0.04	0.0818	0.164	0.332	0.672	1.35	2.71	5.21
Relative error (%)		0	−0.24	−0.61	0	0.30	0	0	0

$A = 0.5227$ kg kg^{-1}; $B = 9.20 \times 10^{-3}$ kg m^{-3}.

TABLE 2. Estimated and experimental t values.

S (kg m^{-3})	exp.	10	9.92	9.76	9.44	8.79	7.49	4.89	0.0062
t (h)	exp.	0	1	2	3	4	5	6	7
	est.	0	1.01	2.01	3.00	4.00	4.99	5.98	7.02
Relative error (%)		0	1	0.5	0	0	−0.2	−0.33	0.29

$K_s = 0.190$ kg m^{-3}; $R_{max} = 1.31$ h^{-1}.

It can be seen from Eqs. (3.15) and (3.16) that in order to obtain the values of μ_{max} and m_s, we must know the values of A, B, K_s, and R_{max} in advance. The values of A and B can be estimated from Eq. (3.9) by the trial-and-error method with a computer. To avoid blindness in the estimation, the simplex method can be used.

The values of K_s and R_{max} can be estimated from Eq. (3.14). There are eight parameters in Eq. (3.14), i.e., C, D, E, F, G, H, I, and J, but C, D, H, and I are functions of A and B; E, F, G, and J are functions of A, B, K_s, and R_{max} (with Eqs. (3.15) and (3.16), μ_{max} and m_s in E and F can be expressed in terms of K_s and R_{max}). A and B being known, there are only two parameters to be estimated, i.e., K_s and R_{max}, which can be estimated from Eq. (3.14) by a method similar to that for estimating A and B from Eq. (3.9). The four parameters can also be estimated simultaneously by the least squares method. Finally, the values of m_s, μ_{max} and $Y_{x/s}^{max}$ can be obtained after A, B, K_s, and R_{max} have been estimated.

To demonstrate the reliability of these estimation methods, we employed data from batch cultivation of *Klebsiella pneumoniae* presented by Esener et al.[8] to estimate the values of A, B, K_s, and R_{max}, with a microcomputer, then calculated the values of μ_{max}, m_s, and $Y_{x/s}^{max}$. The experimental values and estimated values of X obtained from Eq. (3.9) are given in Table 1.

Table 2 shows the experimental values and estimated values of t obtained from Eq. (3.14).

From the estimated values of A, B, K_s, and R_{max}, we obtain $\mu_{max} = 0.720$ h^{-1}; $m_s = 0.0666$ h^{-1}; and $Y_{x/s}^{max} = 0.549$ kg kg^{-1}.

4. Application of Kalman Filter to the Control of Microbial Processes

Since 1960, a variety of mathematical models for fermentation and its control has been suggested based on various viewpoints and data. Those models are typical under defined conditions and have served as the bases for a great deal of research.

In many cases, optimization of bioreaction processes is difficult, because too many parameters in the models and errors due to simplifications make the structured models impracticable. Instability of organisms used is grave enough to cause significant deviation from the models, particularly in the case of operation through multiple microbial generations. In addition, several kinds of stochastic noise in the measurement system will add to inaccuracy of state estimation.

Since 1970, a number of filter algorithms have been described. Svreck[9] reported that an extended Kalman filter could be used for estimating the state in microbial reaction systems and that it could overcome the noise in measurements.

We suggest that a "gray box" can be used to investigate microbial processes, and we studied yeast propagation with the Kalman filter.

4.1   Yeast Propagation Model

Data on cell mass, either observed or estimated, are essential for research and for process optimization. In our work, cell mass was estimated using the CO_2 evolution rate. The cell mass can be estimated with the following equation.[10]

$$\dot{X}_1 = \frac{dCO_2}{dt} = K_1 \frac{dX}{dt} + K_2 X. \tag{4.1}$$

Experimental data show that the growth of the organism follows a logistic model, i.e., the rate was independent of substrate concentration.

$$\dot{X}_2 = \frac{dX}{dt} = \mu_{max} X \left(1 - \frac{X}{X_{max}}\right). \tag{4.2}$$

It is difficult to predict the value of X_{max}, so we always assume X_{max} to be constant. Besides, the metabolic pathway of the organism can be changed with operation conditions, leading to deviation from the model.

4.2   State Estimation and System Identification of Yeast Propagation

4.2.1   State estimation based on the CO_2 evolution rate and dry weight of cells

From Eqs. (4.1) and (4.2), we obtain the process equation with state variables.

$$\left.\begin{array}{l} \dot{X}_1 = \dfrac{dCO_2}{dt} = K_1 \dfrac{dX}{dt} + K_2 X \\[2ex] \dot{X}_2 = \dfrac{dX}{dt} = \mu_{max} X \left(1 - \dfrac{X}{X_{max}}\right) \\[2ex] \dot{X}_3 = \dfrac{d\mu_{max}}{dt} = 0 \\[2ex] \dot{X}_4 = \dfrac{dK_1}{dt} = 0 \\[2ex] \dot{X}_5 = \dfrac{dK_2}{dt} = 0 \end{array}\right\}. \tag{4.3}$$

The observation equation is as follows:

$$H = \begin{pmatrix} 1 & 0 & 0 & 0 & 0 \\ 0 & 1 & 0 & 0 & 0 \end{pmatrix} \qquad X = (CO_2 \ X \ \mu_{max} \ K_1 \ K_2)^T$$

$$Y = HX. \tag{4.4}$$

Then state estimation and parameter identification can be carried out on a computer with Kalman filter.

Because the state estimation by the Kalman filter can be automatically adjusted in approaching the true system, determination of the initial values is arbitrary within a certain range. In our work, the observed values two hours after inoculation were taken as the initial values, i.e., CO_2, X_0, μ_{max}, K_1, and K_2 were 1.05 mmol h^{-1}, 3.0 g l^{-1}, 0.095 h^{-1}, 1.995 mmol g^{-1}, and 0.5 mmol g^{-1}h^{-1}, respectively.

Estimation of state can trace the real system with sensitivity depending on the value of Q/R. With higher Q/R, higher sensitivity of the adjustment can be achieved. But if too high a value of Q/R is taken, serious fluctuation in the estimation will occur. In this paper, a fairly high Q was taken to trace the true system quickly in the initial period, and then a reduced value of Q was used to estimate the state precisely.

The errors in CO_2 evolution rate and cell mass between the estimated and the measured values were 0.5 mmol h^{-1} and 0.2 g l^{-1}, respectively.

Let the diagonals of the R matrix be

$$R = \text{diag} [0.25 \quad 0.04]$$

where values are derived from CO_2 evolution and cell mass measurement data.

At the beginning, because of the low confidence of the model, let the ratio of information to noise Q/R equal 2, i.e.,

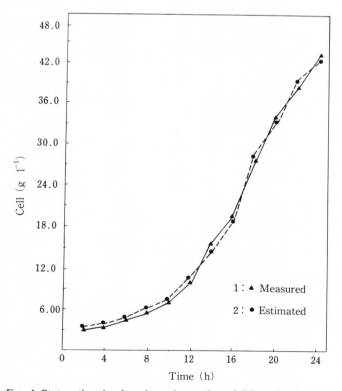

FIG. 4. State estimation based on observation of CO_2 and cell weight.

$$Q = \text{diag} [0.5 \quad 0.15 \quad 0.04 \quad 0.1 \quad 0.25].$$

After 10 h, as cells grow faster, let the diagonals of the Q matrix be

$$Q = \text{diag} [0.5 \quad 0.15 \quad 0.005 \quad 0.1 \quad 0.075].$$

In consideration of the algorithm of the computer and sampling interval, the step of filter interval is set at 0.5 h. Let the X_{max} be 50 g l^{-1}.

Figure 4 illustrates the results of the system identification, Curve 1, experimental cell mass, and Curve 2, the Kalman filter state estimation. In general, these variables are accurately tracked by the filter with an average error of less than 5%.

Figures 5, 6 and 7 illustrate parameter estimation for three batch data, in which the values of K_1 are maintained at about 2.0 mmol g cell^{-1} during the batch time. This indicates that the carbon dioxide evolution rate, which is related to cell mass, remains unchanged. The track of K_2 values is shown in Fig. 6. Because of the high activity of the organism and the sufficient supply of substrate in the early phase, the CO_2 evolution rate based on cell maintenance has a high value of K_2. When a certain cell concentration is reached, the activity is reduced, along with the CO_2 evolution rate and the K_2 value.

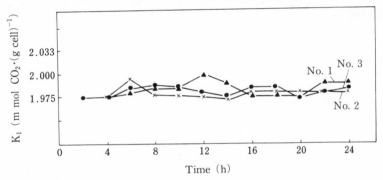

FIG. 5. Estimation of K_1.

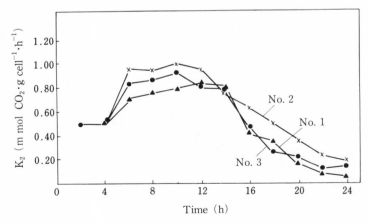

FIG. 6. Estimation of K_2.

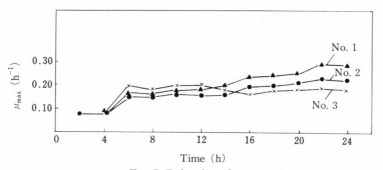

FIG. 7. Estimation of μ_{max}.

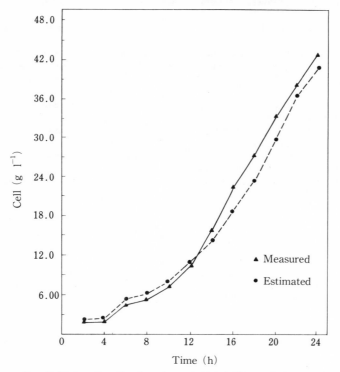

FIG. 8. State estimation based on observed CO_2 evolution rate.

4.2.2   Estimation of state based on CO_2 evolution

Similar to Eq. (4.4),

$$H = [1 \ 0 \ 0 \ 0 \ 0].   \tag{4.5}$$

Let $R = [0.25]$ and

$$Q = \text{diag } [0.5 \quad 0.15 \quad 0.04 \quad 0.1 \quad 0.25].$$

Ten hours later, Q becomes

$$Q = \text{diag} [0.5 \quad 0.15 \quad 0.0025 \quad 0.1 \quad 0.0757]$$
$$T = 0.5 \, h \qquad X_{max} = 50 \, g \, l^{-1}.$$

The results of the filter are shown in Fig. 8 and have an average error of estimated values of ca. 9.5%. This method can substitute for difficult cell mass measurements.

REFERENCES

1. Yu, J.T., Shen, Z.F., "An important step to avoid contamination of fermentation." *Chinese J. Antib.*: 2(3), 1–4 (1977).
2. Yu, J.T., "Scale-up of fermentors — A convenient method for estimating agitator speed under aeration." *Pharm. Eng. Des., China*: 4, 26–29 (1981).
3. Michel, B.J., Miller, S.A., "Power requirements of gas-liquid agitated systems." *AIChE J.*: 8, 263–266 (1962).
4. Yagi, H., Yoshida, F., "Gas absorption by Newtonian and non-Newtonian fluids in sparged agitated vessels." *Ind. Eng. Chem. Proc. Des. Dev.*: 14, 488–493 (1975).
5. Ye, Q., Li, Y.R., Yu, J.T., "Studies on the volumetric oxygen transfer coefficients for Newtonian and non-Newtonian fluids in aeration agitated tank." *J. East Chn. Inst. of Chem. Tech.*: 3, 313–326 (1982).
6. Aiba, S., Humphrey, A.E., Millis, N.F., In *Biochemical Engineering*, 2nd Ed., p. 196, Academic Press, New York, 1973.
7. Fang, B.S., "Estimation of kinetic and energetic parameters from batch cultivation data." *Chinese J. Biotechnol.*: 2(4), 54–60 (1986).
8. Esener, A.A., Roels, J.A., Kossen, N.W.F., "Theory and applications of unstructured growth models." *Biotechnol. Bioeng.*: 25, 2803–2841 (1983).
9. Svrcek, R.F., Elliott, R.F., Zajic, J.E., "The extended Kalman filter applied to a continuous culture model." *Biotechnol. Bioeng.*: 16, 827–846 (1974).
10. Alford, J.S., "Modeling cell concentration in complex media." *Biotechnol. Bioeng.*: 20, 1873–1881 (1978).

Enzyme Separation and Purification Using Improved Simulated-Moving-Bed Chromatography

Shih Yow Huang, Wen Shan Lee, and Chin Kuan Lin

Department of Chemical Engineering, National Taiwan University, Taipei, 10764, Republic of China

In the first enzyme separation and purification steps using simulated-moving-bed chromatography, high separation efficiency can be achieved. However, in successive operations of the simulated-moving-bed, an unavoidable pH lag occurs in adsorption and impurities in the crude enzyme solution lower the efficiency of desorption. In this paper, the chromatography process was further studied by adding more washing operations to the flow diagram. The additional washing was either to shift up the pH value of the flow before adsorption or to strip off impurities after adsorption but before desorption. Both the shift-up of pH value and purge of impurities by washing were effective in improving enzyme purification efficiency. Average values of specific activity and specific activity ratio in the product were 10,800 U (mg protein)$^{-1}$ and 13.9, respectively.

1. Introduction

Enzyme purification is one of the current areas of interest in biotechnological processes. Affinity chromatography has been considered one of the most effective techniques for purifying enzymes and other biochemical materials. Traditional affinity chromatography is carried out batchwise in a fixed-bed column. It usually consumes a large amount of eluant, which necessarily dilutes the product. We previously applied a pH-parametric pumping technique to the separation and purification of trypsin from hog pancreas as well as to bromelain from pineapple stem,[1] and achieved high separation efficiency. The specific activity ratio (SAR), i.e., the ratio of specific activity of product to that of feed, was 3.8; although the amount of eluant used in that method was much less than that used in traditional and batch chromatography, the product was considerably diluted.

In the late 1950s Universal Oil Products (UOP) developed a continuous adsorption process called simulated-moving-bed[2,3] in which the characteristics of the moving-bed were simulated by fixing the adsorption and by moving in succession the position of the feed line, elute line, and lines for withdrawal of raffinate and extract. The idea of the simulated-moving-bed originated in attempts to ameliorate disadvantages of the moving-bed, i.e., attrition of adsorbent particles, channeling, and non-uniformity of the bed, and was fueled by the successful recovery of n-alkane from its isomers. The process was successfully applied to the separation and purification of fructose from high-fructose syrup by Mitsubishi Chemical Industries Ltd. and Sanmatsu Industries Ltd.[4-6]

In our previous paper,[7] a continuous affinity chromatography a system with six columns was devised. The six columns in series constituted the six sections in a chromatographic column. In essence, adsorption and desorption proceeded simultaneously within the system, and both were carried out within two successive columns. One buffer column was installed after the adsorption and before the desorption section. The main purpose of this intermediate section was to buffer the neutralization of the two different liquids. Adsorption and desorption in each column were done alternately to enhance the utilization rate of the affinity adsorbent. The adsorbed enzyme in Columns 1 and 2, for example, was switched to desorption at a specific time in the operation, and at the same time columns 4 and 5 that had initially served as desorption were switched to adsorption. Positions of the inlet and outlet lines were changed intermittently by a time controller.[7] The feed position was automatically shifted alternately from columns 1 to 2 throughout. Positions of the desorbent inlet and the raffinate and extract outlets were changed in a similar fashion. Purification of trypsin from crude porcine pancreas by this process yielded a fairly high separation efficiency, and the average specific activity ratio of eluant was 3.9. Thus far, simulated-moving-bed chromatography has demonstrated the following definite advantages over traditional chromatography: smaller amounts of eluant are needed; simultaneous adsorption and desorption operations result in high productivity; products are less diluted because the extract is taken at its peak and raffinate is taken in its trough; and the recovery rate is high.

In applying simulated-moving-bed chromatography (SMBC) to the purification of a crude enzyme, a washing step must be inserted between the adsorption and desorption steps to strip impure protein from the adsorbent and to reduce the neutralization in adsorption.

This paper describes enzyme separation and purification using a modified SMBC. Our aim was to achieve high separation efficiency and column productivity. The modification was made by incorporating the washing steps for removing impure proteins and reducing neutralization. Experimental studies on separation and purification of crude trypsin from porcine pancreas were conducted to verify the effects of this modification.

2. Process Description and Experimental Apparatus

2.1 SMBC with Washing to Remove Impure Proteins

Figure 1 shows the configuration and operation mode of SMBC incorporating washing after desorption. The operation mode is illustrated based on a certain period shown in the same figure. Six columns were connected in a series, and each column of 16 mm ID × 40 cm Ht was loaded with 10 cm of high-affinity adsorbent (see below). Six steps, adsorption, recirculation, washing, recirculation, desorption, and recirculation constituted one cycle. After the operation had proceeded for a certain length of time, the adsorption column (Column 3 in Fig. 1) was switched to recirculation, column 2 was switched to washing, and so on. This kind of alternation creates a series of continuous chromatographic operations in which adsorption, stripping of impure proteins, and desorption are conducted simultaneously, with each column carrying out different operations in turn as programmed.

As the first step, six solenoid valves, designated by "1" (HA1, HB1, L1, HAo1, HBo1, and Lo1) in the figure were actuated and adsorption, washing, and desorption were done simultaneously in columns 3, 1, and 5. Similarly in the next step, another six solenoid valves, designated "2," were actuated and the operations shifted to (left-hand-side) colu-

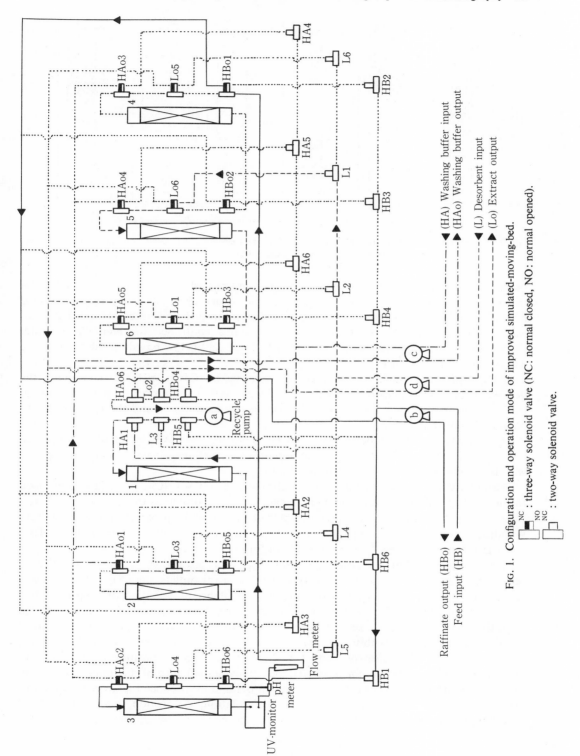

FIG. 1. Configuration and operation mode of improved simulated-moving-bed.

NC : three-way solenoid valve (NC: normal closed, NO: normal opened).
NO
NC : two-way solenoid valve.

Raffinate output (HBo) ▼
Feed input (HB) ▲

(HA) Washing buffer input ▼
(HAo) Washing buffer output ▲
(L) Desorbent input ▼
(Lo) Extract output ▲

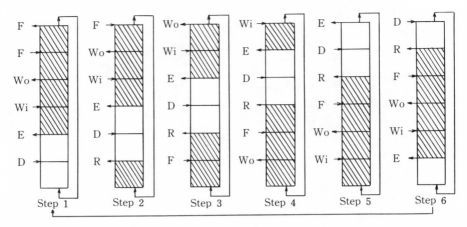

FIG. 2. Operating mode of improved simulated-moving bed system (F: feed; R: raffinate; D: desorbent; E: extract; Wi: washing buffer in; Wo: washing buffer out).

umns 4, 2, and 6. Thus, the six columns were continuously subjected to alternate chromatographic separations and the six steps completed one cycle (each step required 18–20 min of operation). Peristaltic pumps b, c, and d controlled flow rates of high pH feed, washing buffer, and desorbent, respectively. Pump a controlled the recirculation rate. Operations of the improved SMBC are described in detail below based on the time period corresponding to that in Fig. 1.

1) High pH feed buffer (0.1M Tris-HCl plus 0.05M CaCl$_2$, pH 8.0) containing the desired enzyme was pumped through solenoid valve HB1 to column 3 by peristaltic pump b. The effluent from column 3 was withdrawn partly as raffinate and the rest was fed to column 4 as recirculation.

2) High pH washing buffer (0.05M Tris-HCl plus 0.05M CaCl$_2$, pH 8.0) was charged into column 1 through HA1 and peristaltic pump c. The effluent from column 1 was partly withdrawn and the rest was recycled to column 2.

3) Similarly, the desorbent buffer (0.1M glycine plus 0.05M CaCl$_2$, pH 2.5) was pumped through L1 to column 5, and the effluent was partly withdrawn as extract, while the rest was recycled to column 6. Meanwhile, the raffinate, impure proteins, and extract were withdrawn continuously from the respective columns and discharged from lines HBo, HAo, and Lo, respectively (see bottom part of Fig. 1).

4) The periodic shifts to adsorption, washing, and desorption in each column were repeated. Changes in positions of inlets and withdrawal outlets were carried out using a set of time-controllers,[7] as illustrated in Fig. 2. The hatched area in the figure represents the high pH zone, while the blank area represents the low pH zone. The feed position is shifted from the first column to the second, then to the third column and so on, and positions of the desorbent inlet and withdrawal lines for raffinate and extract were similarly changed.

5) A combined glass electrode and UV monitor were set, in position at the outlet of column 3 for on-line measurement of pH and optical absorbance of the liquid throughout the cycle, under the assumption that changes in pH and absorbance monitored during one cycle (adsorption, recirculation, washing, recirculation, desorption, and recirculation) would represent those in all other columns in this chromatographic operation.

2.2 SMBC with Washing to Shift up pH before Adsorption

A high pH feed was pumped through solenoid valve HA1 to column 1 by peristaltic pump c, while high pH buffer (0.1M Tris-HCl plus 0.05M $CaCl_2$) was fed to column 3 through HB1 and b. Feed and high pH washing lines previously mentioned in *2.1* were interchanged each other.

2.3   Materials and Experimental Procedure

The enzyme purification systems demonstrated in this work were: (1) pure trypsin (porcine pancreas, Sigma Chemical Co., St. Louis, Mo) and (2) crude trypsin from porcine pancreas, for which the source and pretreatment are described elsewhere.[1]

The affinity adsorbent used was chicken-ovomucoid (CHOM) bound covalently on crab chitin by glutaraldehyde and reduced by sodium borohydride ($NaBH_4$). The details of preparation of this adsorbent have also been described previously.[1] Two particle sizes were used, 10–12 mesh (1.65–1.40 mm) and 40–60 mesh (0.35–0.246 mm). Trypsin activity was assessed by the method of Rick.[8] Protein concentration was determined by: C_f(mg ml^{-1}) $= 1.55A_{280} - 0.76A_{260}$.[9] The first term on the right-hand side represents absorbance of protein at 280 nm, while the second term is for nucleic acids at 260 nm.

The constituent of high pH feed buffer was determined in the previous work,[7] and that of the high pH washing buffer (0.005M Tris-HCl plus 0.05M $CaCl_2$, pH 8.0) as well as desorbent buffer (0.1M glycine plus 0.05M $CaCl_2$, pH 2.5) was determined by referring to the pKa values of Tris-HCl and glycine, respectively.

3. Results and Discussion

3.1   Washing to Elevate pH

The pH profile in one cycle (six steps) of the modified SMBC is shown in Fig. 3 (see the abscissa, Part 1). Washing was done before adsorption (feed). As seen in the figure, ad-

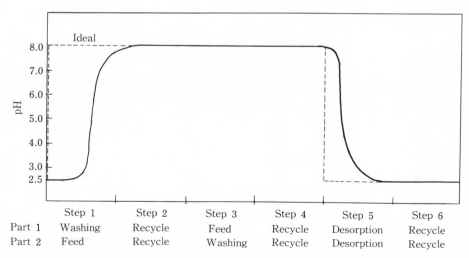

FIG. 3. pH profile for one cycle of modified simulated-moving-bed—(abscissa in Part 1: washing to elevate pH; in Part 2: washing to remove impure proteins).

sorption occurred at high pH, and decrease of pH during desorption was rapid. The pH profile improved considerably when compared with that obtained in our previous work.[7]

Figure 4 illustrates the performance of this modified SMB system in trypsin purification. Operation conditions, e.g., flow rates of feed, desorption, recycle, etc. were predetermined (see Figs. 4–5 and Figs. 8–9 later). Flow rates adopted here were within the range for stable operation of the simulated-moving-bed. Although the feed flow rate was high, the separation factor[10] obtained (2.35) was much higher than that (1.28) of the previous SMBC, which justifies pH shift-up before adsorption.

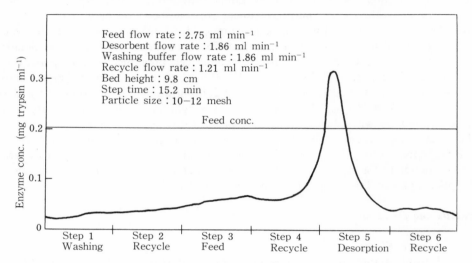

FIG. 4. Enzyme concentration profile (feed: pure trypsin solution, 0.210 mg ml⁻¹).

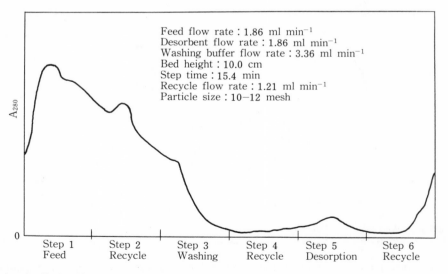

FIG. 5. A_{280} profile (purification of crude trypsin from porcine pancreas). Feed: 4,540 U ml⁻¹; SA: 782 U mg⁻¹.

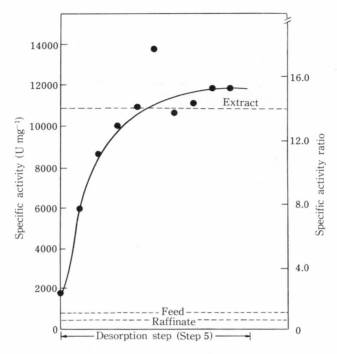

FIG. 6. SA and SAR profiles in desorption (see Fig. 5).

3.2 Washing to Remove Impure Proteins

Figure 5 shows the result for purification of crude enzyme solution (SA: 782 U (mg protein)$^{-1}$) with a feed rate of 1.86 ml min^{-1}. The rapid decrease in the A_{280} reading during washing in the figure indicates that impure proteins were removed efficiently, yielding high-purity trypsin. The average SAR value was 13.9 and the averaged SA value was 10,800 U (mg protein)$^{-1}$.

Subsequent SA and SAR values in desorption are shown in Fig. 6. Broken lines in the figure are averaged values during the desorption period. The extract and raffinate were collected from effluents of columns 5 and 1, respectively (see Fig. 6). Averaged SAR and SA values of extract were 14 and 13,900 U (mg protein)$^{-1}$, respectively, which are higher than those obtained in our earlier study.[1]

Trypsin purified in this modified SMBC was subjected to polyacrylamide gel electrophoresis (Fig. 7). Materials for molecular-weight standard were purchased from Pharmacia Fine Chemicals (Electrophoresis Calibration Kit, Uppsala, Sweden). The purified trypsin had a molecular weight of approximately 23,300, and was deemed acceptable.

3.3 Effect of Bed Height on Separation Efficiency

Figure 8 shows the changes in SA and SAR values during desorption when the bed height was increased to 16.2 cm. SA values at the early phase of desorption increased rather sluggishly compared to those shown in Fig. 6. This could be attributed to the increased amounts of contaminants on the adsorbent and to an increase in axial dispersion of

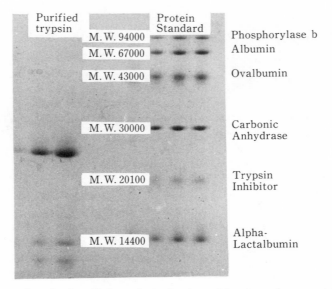

FIG. 7. Electrophoresis of purified trypsin (molecular weight of trypsin assessed as 23,300).

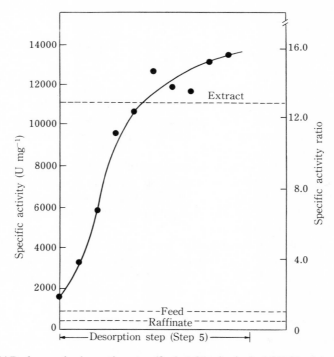

FIG. 8. SA and SAR changes in desorption step (feed: 1.86 ml min⁻¹; 4,860 U ml⁻¹; desorbent: 1.86 ml min⁻¹; washing buffer: 3.36 ml min⁻¹; recycle: 1.21 ml min⁻¹; step time: 24 min; bed height: 16.2 cm).

impure proteins as the height of the adsorbent bed increased. These adverse effects were minimized when the washing period was extended to 24 min from 15.4 min in Fig. 5. In the latter half of the desorption period, SA values (purity of trypsin) in Fig. 8 became higher than in Fig. 6. Thus an increase in adsorbent height results in a very limited improvement in SA values even if the washing period is considerably prolonged.

3.4 Effect of Particle Size on Separation Efficiency

Small particles with a relatively large surface area are expected to have higher adsorption capacity than large ones. Figure 9 shows the enzyme concentration profile when the adsorbent was from 40–60 mesh. Compared to that in Fig. 4 (particle size from 10–12 mesh), the use of smaller adsorbents had a rather flat concentration profile before reaching a peak steeper than that in Fig. 4. Clearly, smaller adsorbent particles exhibit better separation efficiency as well as higher product concentration.

The asymmetric peak in desorption in Fig. 9 indicates that even after adsorption had been completed, the column was not saturated with the desired enzyme. The second peak appearing in desorption was considered to be due to adsorption of impurities. It must be mentioned that particle size had a lower limit because of limitations in mechanical strength. A dense packaging of particles in a fixed-bed chromatographic column would result in lower enzyme separation and purification efficiency. From our experience, the particle size limit is around 40–60 mesh.

Generally, the incorporation of washing steps into our SMBC was effective in improving the efficiency of enzyme separation and purification. However, it must be remembered that the adsorption or desorption step was carried out exclusively in one column and thus the pH change in a column did not occur instantaneously during successive operations. Simplified operation would result in decreased separation and purification efficiency. Incidentally, if the number of columns for adsorption and desorption increase, the separation efficiency will be enhanced and, operational conditions could be with more freedom of choice.

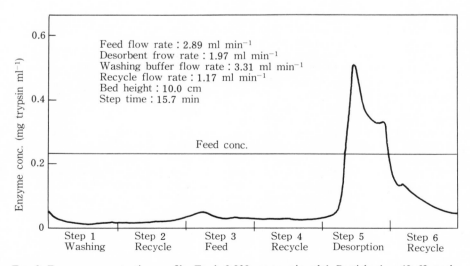

FIG. 9. Enzyme concentration profile. Feed: 0.230 mg trypsin ml^{-1}. Particle size: 40–60 mesh.

4. Conclusion

Two washing steps for different purposes were incorporated in simulated-moving-bed chromatography to enhance the efficiencies of enzyme separation and purification. One washing elevated the pH of purifying enzyme solution before proceeding to adsorption, and the other washed out the impurities contained in the raw enzyme solution after adsorption but before desorption. Experimental results showed that:

1. Both steps contributed to the improvement of enzyme purification; the maximum SA values reached 13,800 U (mg protein).$^{-1}$ This performance was better than that in our earlier report.

2. Increased adsorbent height was not as effective as had been expected in enhancing SA values. Nevertheless, washing *per se* was effective in achieving both objectives.

3. In order to reduce pH lag in successive SMBC operations, the number of columns for adsorption and desorption should be increased to obtain the ideal pH profile in the column.

Nomenclature

A_{260}, A_{280}	optical absorbance at 260 and 280 nm.
C_f	enzyme concentration in fluid mg ml^{-1}
SA	specific activity, i.e., enzyme activity (mg crude protein)$^{-1}$
SAR	specific activity ratio, i.e., ratio of SA of product to that of feed
SF	separation factor defined as ratio of enzyme concentration of extract to that of raffinate (for pure trypsin) or ratio of SA of extract to that of raffinate (for crude trypsin)

Acknowledgment

The authors would like to acknowledge financial support from the National Science Council, ROC (Grant Nos. NSC 73–0402–E002–11 and NSC75–0402–E002–11).

REFERENCES

1. Huang, S.Y., Lin, C.K., Juang, L.Y., "Separation and Purification of Enzymes by Continuous pH-Parametric Pumping." *Biotechnol. Bioeng.*: **27**, 1451–1457 (1985).
2. Broughton, D.B., "Molex: Case History of a Process." *Chem. Eng. Progress*: **64**, 60–65 (1968).
3. Broughton, D.B., Neuzil, R.W., Pharis, J.M., Breeley, C.S., "The Parex Process for Recovering Para-xylene." *Chem. Eng. Progress*: **66**, 70–75 (1970).
4. Mitsubishi Chemical Industries Ltd., Jap. Pat. 54–102288 (1979).
5. Mitsubishi Chemical Industries Ltd., Jap. Pat. 55–61903 (1980).
6. Sanmatsu Industries, Ltd., Jap. Pat. 53–149870 (1978).
7. Huang, S.Y., Lin, C.K., Chang, W.H., Lee, W.S., "Enzyme Purification and Concentration by Simulated Moving Bed Chromatography: An Experimental Study." *Chem. Eng. Commun.*: **45**, 291–309 (1986).
8. Rick, W., In *Methods of Enzymatic Analysis*, p. 800. H.U. Bergmeyer (ed.), Academic, New York, 1965.
9. Cooper, T.G., In *The Tools of Biochemistry*, p. 398–399. John Wiley & Sons, New York, 1983.
10. Chen, H.T., Wong, Y.W., Wu, S., "Continuous Fractionation of Protein Mixtures by pH Parametric Pumping: Experiment." *AIChE J.*: **25**, 320–327 (1979).

V. Environment

V-1

Prospects in Biotechnological Degradation of Xenobiotics During Waste Water Treatment

Hans-Juergen Rehm

Institut für Mikrobiologie, Universität Münster D-4400 Münster, Federal Republic of Germany

1. Introduction

Microorganisms play an important role in the natural cycles of compounds containing carbon, nitrogen, phosphorus, chlorine, sulfur, and others. Degradation of complex molecules to CO_2 and water, the so-called mineralization process, is termed biodegradation.

Until the beginning of this century all chemicals synthesized by plants, animals, men, and microorganisms could be degraded by microorganisms. Among these chemicals many chlorinated substances, e.g., chlortetracycline, chloramphenicol, and chlorine butyric acid, have been synthesized in nature by plants and microorganisms. Following this biosynthesis many dechloration mechanisms have been developed in evolution, because microorganisms had time enough in the course of evolution to develop mechanisms for degradation. The development during evolution of degradation mechanisms for each synthesized substance is of great importance, because all substances that cannot be degraded will accumulate in enormous amounts over time.

In the 20th century, we have synthesized many different chemicals including, among others, many xenobiotics. Microorganisms have not had enough time to develop degradation mechanisms for these substances. In the absence of such naturally evolved degradation mechanisms, men have had to resort to the construction of specific microorganisms that can degrade the new substances, and to developing processes to handle these microorganisms or their enzymes to break down xenobiotics at high speed.

This article will not review the degradation mechanisms of microorganisms for xenobiotics, but will examine some ideas for a microbial degradation of xenobiotics, especially in waste water treatment.

2. Some Important Xenobiotic Compounds

More than 65,000 chemical compounds are used every day (Maugh, 1978). Only some of them are important pollutants as, for instance, the organic priority pollutants according to the U.S. Environmental Protection Agency (Keith and Telliard, 1979) (Table 1).

These substances are recalcitrant or persistent against microbial degradation. Moreover, many compounds are recalcitrant because of the insolubility of the molecule, as in the case of synthetic polymers. The compounds listed in Table 1 are often of new chemical structures to which microorganisms have not been exposed during evolution.

TABLE 1. Organic priority pollutants according to the U.S. Environmental Protection Agency.

Chemical class	Number of compounds
Aliphatics	3
Halogenated aliphatics	31
Nitrosamines	3
Aromatics	14
Chloroaromatics (including 2, 3, 7, 8-tetrachloro-dibenzo-p-dioxin (TCDD))	16
Polychlorinated biphenyls (PCBs)	7
Nitroaromatics	7
Polynuclear aromatic hydrocarbons	16
Pesticides and metabolites (including 1, 1, 1-trichloro-2, 2-bis (p-chlorophenyl) ethane (DDT))	17

Aliphatics and aromatics are present in oil fractions and end up in the environment, for instance because of pipeline, refinery, or other industrial accidents or because of individual abuse. Halocarbons, for example, are used as industrial solvents for dry-cleaning textiles, as pesticides, and for many other applications. Developing the microbial metabolisms for the degradation of alkanes (Rehm and Reiff, 1981), aromatics (Lal and Saxena, 1982), and halocarbons (see Leisinger, 1983) and others has been the aim of many investigations (Chakrabarty, 1985).

Some degradation mechanisms are well known. From some degradation steps only the intermediates are known, but in many cases we need more knowledge about the degradation mechanisms, especially for chlorinated aromatics, or more information about the enzymes involved, especially for polynucleic aromatic hydrocarbons and halocarbons.

Degradation of microorganisms corresponds to their degradation mechanisms. On the one hand, there are the aerobic microorganisms, e.g., *Pseudomonas, Alcaligenes, Nocardia, Corynebacterium, Micrococcus* and some fungi, and on the other hand the anaerobic microorganisms, e.g., methanogenic bacteria and *Clostridium*.

Often mixed cultures of soil microorganisms that have not been well described are responsible for degradation of xenobiotics. Cooxidation is widespread among the degradation mechanisms. This short summary shows that much information is available on the degradation of xenobiotics but that more investigations are needed.

3. *Problems in Biotechnological Degradation of Xenobiotics*

Some of the problems in the application of microorganisms or enzymes to degrade xenobiotics are as follows:

Not all compounds can be degraded by microorganisms at present.

The rate of most degradation is not fast enough.

Specialized microorganisms are not stable enough for a long application.

Specialized microorganisms are washed out after inoculation in a continuous culture.

Specialized microorganisms are overwhelmed by other microorganisms in a mixed cultured wastewater plant.

Many substances are fully degraded only under very different conditions, e.g., DDT has been degraded in the first steps anaerobically, and in the last steps aerobically.

Many compounds are very toxic to microorganisms.

Many compounds are also very toxic to human beings so that only a few laboratories are able to investigate the conditions for degradation by microorganisms, e.g., some dioxins. In many cases close cooperation among microbiologists (geneticists included), bio-chemists, and technicians is necessary but not always achieved.

It would be very easy to add additional problems to this list, especially if the degradation of xenobiotics in polluted soil is considered.

In summary, the problems can be grouped as follows: on a scientific level, the study of the biology and biochemistry of microorganisms; on a genetic level, the construction of special degrading microorganisms; and on a technical level, the development of new methods for an application of new microorganisms.

4. Prospects for Isolation or Construction of New Microorganisms to Degrade Xenobiotics

4.1 Enrichment of New Active Strains

Methods for enriching new microorganisms have been developed recently especially by introducing continuous culture methods into enrichment techniques. Despite the fascinating new methods for constructing new strains of microorganisms by recombinant DNA methods, it will be necessary to isolate new strains with different degradation activities to obtain wild strains with new activities for new xenobiotics, to isolate sufficient strains to provide genetic material for constructing new strains, to obtain strains with high efficiency for anaerobic degradation of xenobiotics, and to create mixed cultures for degradation chains.

Harder (1981) has reviewed recent developments in the theory and practice of enrich-ment of microorganisms, and other workers have discussed critically the isolation of microorganisms with degradation potential. Sinton et al. (1986) have demonstrated that the metabolic activity for degrading 2,4-dichlorophenoxy acetic acid (2,4-D) is widely present in many microorganism species and often coupled with degradation of 2-methyl-4-chloro-phenoxy acetic acid (MCPA). But 2,4,5-trichlorophenoxy acetic acid (MCPA) is seldom degraded by those microorganisms.

Problems related to the handling of mixed cultures include the stability of the mixed culture, characterization of different strains, lack of knowledge of physiological and bio-chemical features, reproduction of the degradation results, and determination of kinetics with calculation of growth models.

4.2 Construction of New Strains by Recombinant DNA Methods

The recombinant DNA method has been used frequently to endow microorganisms with new degradation abilities (see Haas, 1983; Holloway, 1984; Chakrabarty, 1985). Catabolic pathways for degradation of xenobiotics are mostly encoded by plasmids (see Haas, 1983, and also Imanaka, 1986). Only a few metabolic pathways for xenobiotics are encoded by chromosomal genes as far as is known, e.g., in *Pseudomonas aeruginosa* the degradation of benzoate and p-hydroxylbenzoate.

Much important basic information about the encoding of catabolic pathways for degra-dation of xenobiotics is available. Some new strains have also been constructed, largely in Pseudomonadaceae. Plasmids play an important role in the evolution of biodegradation processes of new chemicals, especially chlorinated compounds. Also, highly chlorinated chemicals such as pentachlorophenol can be degraded by microorganisms (Watanabe, 1975; Suzuki, 1977; Brown et al., 1986).

Chakrabarty (1985) has shown some aspects of the development of suitable strains for degradation of 2, 3, 7, 8-tetrachlorodibenzo-*p*-dioxin (TCDD):

"This can be done by inoculating soil samples isolated from TCDD/pentachlorophenol contaminated areas with pAC25-containing microorganisms in a chemostat. Initially the chemostat may be operated with 3-Cba and pentachlorophenol as sole sources of carbon to allow rapid spread of pAC25 and putative pentachlorophenol degradative plasmids. With prolonged incubation, the concentration of PCP and 3-Cba will gradually be decreased while concentration of TCDD will be gradually increased. The rationale is that under increasing toxicity from TCDD and decreasing level of growth substrates, the mixed cultures in the chemostat harboring pAC25 and PCP-degradative plasmids, will be under strong selective pressure to undergo mutations in the plasmid-associated dechlorination genes so as to alter the substrate specificity of the enzymes to include TCDD and its metabolites. This may lead to total or partial degradation of TCDD."

It is doubtful however whether the new specialized microorganisms are resistant enough against environmental microflora in soil or in wastewater treatment plants. Therefore it should be the aim of genetic engineering to transfer the codes for catabolic degradation of xenobiotics to microorganisms normally present in great amounts in the environment that have a better growth rate there in comparison to other microorganisms.

4.3   Prospects for Techniques of Microbial Degrading of Xenobiotics

The detoxification and mineralization of recalcitrant compounds in industrial waste with the help of microorganisms is a central topic of our research. In the future, microbial strains that can efficiently degrade xenobiotics will be available. What are the prospects for applying such techniques using microorganisms, especially in wastewater treatment?

Many wastewater treatment specialists tend to regard any use of tailor-made microorganism cultures as useless in the effort to abate pollution. We cannot accept these opinions. New methods and ideas must be introduced in wastewater treatment or in the treatment of polluted soil by applying specialized microorganisms. Some industrial wastes that contain xenobiotics have a defined composition, pH, and other parameters. For this reason it may be possible to inoculate special microorganisms into the waste water before it is transported to a plant with other microorganisms. In this way a microbial process could be built up in which the conditions are so regulated that the specialized microorganisms can grow and degrade. Mixed cultures could also be established in such well-defined waste waters.

A good method for obtaining stable monocultures or mixed cultures of specialized microorganisms is the application of immobilized microorganisms. Immobilizing microorganisms has several advantages:

A better location in the wastewater plant so that the stability of the cultures is greater; higher availability in biomass concentration in the waste water; longer activity than free cells; reduced sensivity to toxic substances as with free cells; and under some conditions, the ability to spread by growth into the medium and to create a reservoir for continuous inoculation of waste water with new, special microorganisms.

Ca-Alginate- and polyacrylamide-entrapped cells of a special strain of *Pseudomonas putida* (P 8) are able to degrade much higher concentrations of phenol (3 g l^{-1} and more) in a shorter time than free cells (Bettmann and Rehm, 1984; Rehm, 1985). Degradation of 4-chlorophenol with Ca-immobilized cells with an entrapped *Alcaligenes* A 7-2 (see

TABLE 2. Comparison of degradation times of different phenolic concentrations by a mixed culture of *Pseudomonas putida* (P 8) and *Cryptococcus elinovii* (H 1) with degradation times of the mono-cultures, adsorbed on activated carbon (Moersen and Rehm, 1986).

Phenolic concentration (g l^{-1})	Degradation times of:		
	Monocultures of		Mixed culture of
	Cryptococcus elinovii (H1) (h)	*Pseudomonas putida* (P8) (h)	*Cryptococcus elinovii* (H1) + *Pseudomonas putida* (P8) (h)
5	44	44	10
10	95	112	80
15	220	not fully degraded	140
17	not fully degraded	not fully degraded	218
20	–	–	–

Knackmuss, 1981) is also possible in a continuous culture (Westmeier and Rehm, 1985). Not only in a sterile medium, but also in unsterile waste water degradation of phenols, *ortho-*, *meta-*, and *p*-cresols in combination with 4-chlorophenol is possible for more than three weeks in continuous culture (Bettmann and Rehm, 1985). Immobilized microorganisms in mixed cultures can have shorter degradation times for xenobiotics than immobilized monocultures (Table 2).

A mixed culture can moreover be adapted to a degradation of greater amounts of xenobiotics, e.g., phenols. This is due to Ca-alginate entrapped microorganisms (Zache and Rehm, 1986) and to microorganisms adsorbed on activated carbon (Moersen and Rehm, 1986).

The immobilized monocultures and mixed cultures remain active after being in use for some months or in storage for four months or longer. Activated carbon is better suited to applying adsorbed special microorganisms to waste water treatment. It has a great buffer capacity for xenobiotics which is an advantage at toxic or even at very low concentrations of xenobiotics. These low amounts can be enriched by activated carbon and then degraded by special microorganisms adsorbed on the surface and in the interior caves of the activated carbon. It has been demonstrated that a mixed culture of *Cryptococcus elinovii* and *Pseudomonas putida* was adsorbed not only on the surface but also in the interior caves of an activated carbon (Moersen and Rehm, 1986).

An additional prospective technique for detoxifying xenobiotics is the application of microbial enzymes. Johnson and Talbot (1983) have published a review article on this field in which future aspects for dehalogenation are presented. It would be desirable if new enzyme systems could be developed to dehalogenate aromatic compounds. Halidohydrolases have been described (Goldmann et al., 1968; Janke and Fritsche, 1979; Klages and Lingens, 1979), and many researchers are searching for new detoxification enzymes.

The application of detoxification enzymes may be restricted to crude enzymes that are immobilized, or to immobilized parts of microorganisms that contain these enzymes or enzyme systems. In the near future it can be expected that these immobilized detoxification enzymes, especially hydrolases, will be applied to special industrial effluents before these effluents are transported to microbial wastewater treatment plants.

The following illustration is a good synopsis of the application of specialized microorganisms and enzymes:

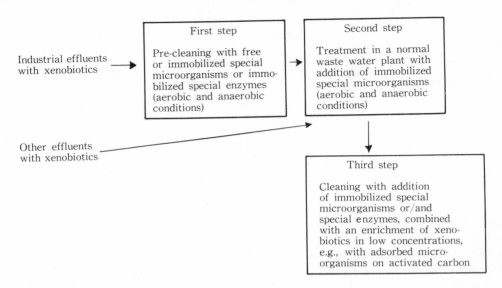

A great problem will be the microbial detoxification of chemicals, e.g., DDT, degraded partly in anaerobic and partly in aerobic steps. Here the experiences of nitrification and denitrification in wastewater plants should be useful.

It should be possible to devise a process that runs anaerobically at first and aerobically in a second step. The plant should contain both types of degradation microorganisms, perhaps in immobilized form: immobilized strains that can degrade DDT in the first steps anaerobically and additional immobilized strains that can degrade them in an aerobic way. A condition for this treatment is the survival of the special aerobic microorganisms during the anaerobic step and the survival of the anaerobic microorganisms during the aerobic step. This procedure can be perhaps guaranteed if the microorganisms involved are immobilized or if the same organisms can degrade anaerobically and aerobically, corresponding to the fermentation conditions (e.g., *Saccharomyces cerevisiae*).

Other strategies are also theoretically feasible. One of them is the addition of a large amount of microorganisms with plasmids for degrading xenobiotics. These plasmids can be captured by other microorganisms to promote xenobiotic degradation by these strains.

Another strategy may be to activate the microflora present by adding activated carbon, a method used recently with success. The reason is that adsorption of unknown degrading microorganisms and perhaps adsorption of toxic substances allows the microorganisms to develop better than without activated carbon. Many of the reasons for the successful application of activated carbons are unknown.

Techniques must be developed in the future suitable for wastewater treatment plants which consider primarily the behavior of the special microorganisms. With the close cooperation of microbiologists, chemists, and technicians, a new biotechnology for treatment of xenobiotic waste can be realized.

The problems will be much more difficult if a degradation of xenobiotics in soil is tried. Here many new methods must be developed that cannot be discussed here.

REFERENCES

1. Bettmann, H., Rehm, H.J., "Degradation of phenol by polymer entrapped microorganisms." *Appl. Microbiol. Biotechnol.*: **20**, 285–290 (1984).
2. Bettmann, H., Rehm, H.J., "Continuous degradation of phenol(s) by *Pseudomonas putida* P8 entrapped in polyacrylamide hydrazide." *Appl. Microbiol. Biotechnol.*: **22**, 389–393 (1985).
3. Brown, E.J., Pignatello, J.J., Martinson, M.M., Crawford, R.L., "Pentachlorophenol degradation: a pure bacterial culture and an epilithic microbial consortium." *Appl. Environ. Microbiol.*: **52**, 92–97 (1986).
4. Chakrabarty, A.M. (ed.), *Biodegradation and detoxification of environmental pollutants*. CRC Press, Boca Raton (1985).
5. Cook, A.M., Grossenbacher, H., Hütter, R., "Isolation and cultivation of microbes with biodegradative potential." *Experientia*: **39**, 1191–1198 (1983).
6. Goldmann, P., Milne, G.W.A., Keister, D.B., "Carbon-halogen bond cleavage. III. Studies on bacterial halidohydrolases." *J. Biol. Chem.*: **243**, 428–434 (1968).
7. Haas, D., "Genetic aspects of biodegradation by pseudomonads." *Experientia*: **39**, 1199–1213 (1983).
8. Harder, W., "Enrichment and characterization of degrading organisms." In *Microbial degradation of xenobiotics and recalcitrant compounds*, p. 77–96. T. Leisinger, A.M. Cook, R. Hütter, J. Nüesch, (eds.), Academic Press, London, 1981.
9. Holloway, B.W., "Pseudomonads." In *Genetics and breeding of industrial microorganisms*, p. 63–92. C. Ball (ed.), CRC Press, Boca Raton, 1984.
10. Imanaka, T., "Application of recombinant DNA technology to the production of useful biomaterials." *Adv. Biochem. Eng./Biotechnol.*: **33**, 1–27 (1986).
11. Janke, D., Fritsche, W., "Dechlorierung von 4-Chlorphenol nach extradioler Ringspaltung durch *Pseudomonas putida*." *Z. Allg. Mikrobiol.*: **19**, 139–141 (1979).
12. Johnson, L.M., Talbot, H.W., "Detoxification of pesticides by microbial enzymes." *Experientia*: **39**, 1236–1246 (1983).
13. Keith, L.H., Telliard, W.A., "Priority pollutants, a perspective view." *Environ. Sci. Technol.*: **13**, 416–423 (1979).
14. Klages, U., Lingens, F., "Degradation of 4-chlorobenzoic acid by a *Nocardia* species." *FEMS Microbiol. Lett.*: **6**, 201–203 (1979).
15. Knackmuss, H.J., "Degradation of halogenated and sulfonated hydrocarbons." In *Microbial degradation of xenobiotics and recalcitrant compounds*. p. 189–212. T. Leisinger, A.M. Cook, R. Hütter, J. Nüesch (eds.), Academic Press, London, 1981.
16. Lal, R., Saxena, D.M., "Accumulation metabolism and effects of organochlorine insecticides on microorganisms." *Microbiol. Rev.*: **46**, 95–127 (1982).
17. Leisinger, T., "Microbial degradation of environmental pollutants. I. General aspects." *Experientia*: **39**, 1183–1191 (1983).
18. Maugh, T.H., "Chemicals: how many are there?" *Science*: **199**, 162 (1978).
19. Moersen, A., Rehm, H.J., "Penolabbau durch eine an Aktivkohle adsorbierte Mischkultur." In *4. Jahrestagung der Biotechnologen*, Frankfurt, 3–4 June, 1986, Kurzfassung, p. 137. Detailed publication proposed in *Appl. Microbiol. Biotechnol.*, 1987.
20. Rehm, H.J., "Some results about multistep reactions with immobilized microorganisms." In *5th Congress of Yugoslav Microbiologists*, p. 30–41, Porec, 24–28 September, 1985.
21. Rehm, H.J., Reiff, I.: "Mechanisms and occurrence of microbial oxidation of long-chain alkanes." *Adv. Biochem. Eng.*: **19**, 175–215 (1981).
22. Sinton, G.L., Fan, L.T., Erickson, L.E., Lee, S.M., "Biodegradation of 2, 4-D and related xenobiotic compounds." *Enzyme Microb. Technol.*: **8**, 395–403 (1986).
23. Suzuki, T., "Metabolism of Pentachlorophenol by a Soil Microbe." *J. Environm. Sci. Health.*: **B12**, 113–127 (1977).
24. Watanabe, I., "Degradation of PCP in Soils. II. The Relationship between the Degradation of PCP

and the Properties of Soils, and the Identification of the Degradation Products of PCP." *Soil Sci. Plant Nutr.*: **21**, 405–414 (1975).

25. Westmeier, F., Rehm, H.J., "Biodegradation of 4-chlorophenol by entrapped *Alcaligenes* sp. A 7–2." *Appl. Microbiol. Biotechnol.*: **22**, 301–305 (1985).

26. Zache, G., Rehm, H.J., "Phenolabbau durch eine in Alginat eingeschlossene Mischkultur." In *4. Jahrestagung der Biotechnologen*, Frankfurt, 3–4 June, 1986. Kurzfassung, p. 150. Detailed publication proposed in *Appl. Microbiol. Biotechnol.*, 1987.

V-2
The Engineering of Biogeochemical Processes

Bernhard J. Ralph

Department of Mineral Processing & Extractive Metallurgy, School of Mines, University of New South Wales, Kensington, N.S.W. 2033, Australia

This essay describes the need for the greater involvement of biochemical engineers in the multidisciplinary teams required for the translation of biogeochemical phenomena into viable commercial processes. Some outstanding problems in the scaling-up of biogeochemical processes are discussed and some suggestions made as to possible solutions.

1. Introduction

Those areas of scientific and technological endeavour in which long-established disciplines overlap and interact have always been attractive, stimulating, and fruitful for research workers willing to participate in multidisciplinary activities. There is an excitement in looking at recalcitrant problems from new viewpoints and in applying well-established methodologies and conceptual frameworks in novel areas. In general terms, biotechnology is such an area of endeavour and embraces a wide range of specializations of which genetic manipulation and biochemical engineering are currently perhaps the most important.

Historically, biochemical engineering has been the bridge by which the fundamental discoveries of microbiology and biochemistry have been translated into the rationalization of long-established, empirically developed microbial processes as highly sophisticated technologies capable of large-scale, productive, and profitable operation. Biochemical engineering continues to be the vehicle by which the staggering range of new products resulting from the activities of both natural and genetically manipulated organisms can be exploited and brought to commercial fruition. Indeed, it is probable that the rate-limiting step in the development of the new biotechnology is the capability of the biochemical engineer to devise appropriate means for the up-scaling of processes shown to be feasible at the laboratory bench.

The main thrust of biochemical engineering over the past fifty years or so has been in the engineering of microbial transformations of organic raw materials and is manifest in the commercially successful operation of waste disposal processes, in the production of a wide range of food products and biocatalysts, and in making available very numerous compounds of medical, veterinary, and agricultural importance. While biological interventions in the transformations of inorganic compounds have been well known since the end of the 19th century, and the microbial role in geochemical phenomena is well understood, the

337

application of biochemical engineering to geobiological phenomena for the development of controlled and optimized processes has received much less attention.

2. *Historical Perspectives*

It is probable that the recovery of metallic copper from copper-sulphate-containing natural waters, arising from the chemical and biological degradation of copper sulphide minerals, was practiced in China as early as the second century B.C. (Yao Dun Pu, 1982), and certainly similar processes were well established in Spain by the 17th century A.D. (Taylor and Whelan, 1942). More recent developments of such early empirical technologies have been described by various authors (e.g., Ralph, 1979, 1985a, 1985b). However, no real understanding of the microbial role in such processes emerged until Colmer and Hinkle (1947) isolated and characterized the now well-known bacterium, *Thiobacillus ferro-oxidans*, with its capacity for greatly accelerating the oxidation of ferrous iron and sulphide ion, and unequivocally demonstrated the central importance of this organism in the generation of acidic mine drainage waters.

The past forty years have seen a great increase in fundamental understanding of geo-microbiology and biogeochemistry, stimulating some degree of rationalization of the traditional technology for the recovery of copper by bacterial leaching, and leading to a large number of demonstrations at the laboratory and pilot levels of novel biohydrometallurgical processes ranging from bacterial pretreatments for the enhancement of gold and antimony recovery to the treatment of lateritic nickeliferous ores. Currently, however, the major applications of biogeochemical phenomena embrace only the recovery of copper and uranium from low-grade ores by bacterial leaching, either from heaps or by *in situ* solution mining, the biocatalyzed regeneration of acidic ferric sulphate leaching reagent, and the related large-scale bacterial oxidation of ferrous sulphate in mine drainage waters as a part of waste water treatment programs (Ishikawa et al., 1983; Imaizumi, 1986).

3. *Some Impediments to the Development of Biogeochemical Processes*

It is clear from the literature that the potential for the development of useful biohydro-metallurgical processes and other applications of biogeochemical phemenena is large and extensive (Ralph, 1982a,b) and it is pertinent to enquire why commercially viable, large-scale processes have so far been slow in emerging (Lakshmanan, 1986). The advantages of biologically-mediated mining and metallurgical procedures have been claimed to be: i) the ability to deal with both high-grade and low-grade raw material; ii) high efficiency of recovery of metal values; iii) relatively low energy consumption; iv) lower capital investment in plant; v) greater flexibility with respect to complex ore types; and that they cause vi) less environmental damage.

The disadvantages of biohydrometallurgical processes, on the other hand, include: i) relatively slow reaction rates; ii) difficulties in optimization due to limited control over physicochemical parameters and microbial populations; iii) fragility of microbial agents; iv) particular difficulties in supplying adequate amounts of oxygen for aerobic processes; v) problems of recovery of solubilized metal from dilute leaching solutions; vi) undue influences of associated gangue materials; and vii) engineering difficulties in coping with the sheer scale of biohydrometallurgical processes as compared with those of conventional industrial microbial processes.

None of the listed disadvantages of biologically-mediated processes would appear unamenable to technical solution, even against the background of existing basic information and established engineering methodologies. The basic problem underlying the sluggish development of biotechnological processes in the mining and metallurgical industries most probably lies in the combination of the inhibiting influence of the current recessionary status of the resource industries in many countries and a lack of recognition by companies that conventional mining and metallurgical technologies may be approaching the limits of economical application. The general problem in developed countries has recently been perceptively analyzed and discussed by Spisak (1986).

It is the observation of the present author, in the Australasian context, that increasingly metallurgical processing problems arise that are unamenable to solution by conventional methodology based largely on physical separations and high-temperature chemistry. There is a growing need for new and imaginative approaches to such problems as the treatment of very finely mineralized ores, complex and recalcitrant mixed ores, more economical ways of exploiting some raw materials such as low-grade bauxites and lateritic minerals, the recovery of metals from dilute solution, and the long-term stabilization of wastes and residues. It is unlikely that all solutions to such problems will be via biologically-mediated processes; indeed, too vigorous an advocacy of exclusively biotechnological methods might be counter-productive. Rather, optimal solutions may lie in clever and imaginative integrations of chemical, biological, and pyrometallurgical processes.

In the Australasian context, the application of conventional mining and metallurgical procedures for the recovery of base and precious metals has been innovative and successful to date, but the mounting accumulations of tailings and other wastes, with estimated content of some billions of dollars worth of metal values and extensive deleterious environmental impact, suggests that an onslaught on the development of new technologies cannot be long delayed. In agreement with Spisak (1986) it is probable that "the greatest hurdle faced by micro-organisms in the mining industry may not lie in the field but rather in the board room." While conceding the importance of nontechnical factors in slowing the rate of development of mining biotechnology, the author is optimistic that the significant recent advances in Japan (Imaizumi, 1986), Canada (Bruynesteyn, 1986), Chile (Acevedo et al., 1983), and various European countries (Groudev and Groudeva, 1986; Rossi et al., 1983; Bosecker, 1986) to mention but a few, are likely to stimulate companies and agencies in other countries to renew efforts to find particular solutions to their indigenous problems.

4. Multidisciplinary Approaches

It is clear from the disadvantages of biologically-mediated processes that the majority require for their solution considerably more input than is likely to be derivable from any single discipline. A multidisciplinary approach is likely to be more successful, with varying contributions from the biological, geological, and engineering disciplines. The bringing together of a team of workers with such attributes is not easily achieved and the establishment of fruitful communication and collaboration is even more difficult. The parable of the stirred reactor is appropriate; when the impeller is still, the reactants segregate and the reaction rate declines to a low level, but when energy is fed into the system via the briskly moving impeller, interactions occur and a product results. In the multidisciplinary team context, the energy input is compounded of the organizational skills of the team leader, the enthusiasm and commitment of the team members, and the moral and financial support of the company or agency. The chemical engineering metaphor for the proper functioning of a multidisciplinary team has been deliberately chosen since, in the author's experience, the quantitative rigor of that discipline, its attention to the chemical feasibility of reactions and the physicochemical conditions under which they occur, and its insistence on the proper balancing of inputs and outputs, provides a framework within which the biological and geological disciplines function more effectively. The author's involvement with biogeochemical problems stems from the early 1960s and its subsequent course was greatly influenced by immediate biochemical engineering associates. Our early devising of continuous flow bacterial leaching of chalcopyrite concentrates (Moss and Andersen, 1968) owed a great deal to the employment of chemical engineering methodologies and concepts. An even more potent shaping factor was the author's four-month sojourn at the Institute of Applied Microbiology at the University of Tokyo in 1966, in the department of our esteemed colleague, Professor Shuichi Aiba. The dialog with Professor Aiba ranged from the design of some very exotic mineral bioreactor systems to the modelling of interrelationships in mixed microbial populations, but the most important outcome was the reinforcement of belief in the extreme potency of rigorous, quantitative approaches to the study of complex geobiological systems.

5. Problems of Large-scale Application of Biogeochemical Processes

Further development of biogeochemical processes is most likely to be applicable in the processing of three main types of raw material: ores whose metal grades lie below the currently economic cut-off point; ores of any grade whose mineralogical complexity defies economical treatment by existing technology; and tailings from metallurgical processes together with metal-rich residues from other industrial operations. It is anticipated that current technology will continue to be utilized for higher-grade materials but that, hopefully, overall planning for the exploitation of new ore deposits will take a longer view than is common at present and ensure that artificial impediments are not placed in the way of concurrent or subsequent processing by biohydrometallurgical techniques (or other novel technologies) of associated low-grade materials and metal-containing process residues. It is likely also that biogeochemical and other new technologies will need to be specifically applied, particularly in developed countries, to the very large quantities of waste materials and process residues which have accumulated from past mining and metallurgical operations, not only because of the very significant total content of metal values but also for environmental reasons.

While the biological agents involved in biogeochemical processes are central to such processes and while their particular characteristics define the operational parameters, solutions to the engineering problems concerned with the contriving of appropriate conditions for the maximizing of the biological roles probably remain the greatest impediments to the wider application of biotechnological procedures in the mining and metallurgical industries. Herein lies one of the great future opportunities for the biochemical engineer. The general engineering considerations involved in the large-scale application of biogeochemical processes have been discussed elsewhere (Ralph, 1985b). Some aspects merit further enlargement and in the following sections some attention is given to the problems associated with the particular nature of raw materials, the optimization of reaction configurations, the control and improvement of biological agents, and the recovery of products.

5.1 Raw Materials

In a typical mining and metallurgical operation, the materials handled fall into four categories: overburden containing no significant amount of metal values; low-grade ore material; ore material of a grade economically processable by existing technology; and a diversity of process residues. After primary extraction from the ore body, low-grade ore material may be separately stockpiled for future treatment but is not infrequently dumped with overburden. It is at this juncture that a number of problems may be generated from a conflict of quite legitimate imperatives. The mining engineer is driven by the need to dispose of very large quantities of material in as economical and expeditious a manner as possible and must consider such factors as the shortest possible transportation distances and disposition of the material in a way which ensures compactness and stability. Current procedure favors the use of valley features as receptacles, down-the-hillside dumping, or construction of truncated, pyramidal, free-standing heaps. In each of these modes, a maximum volume-to-surface area configuration is achieved. Consolidation of the dumped material by the passage of transport vehicles assists in mechanical stabilization.

If subsequent treatment of dumped material becomes desirable (by virtue of improved metal prices or for environmental reasons), the operator may be faced with a formidable array of self-inflicted problems. The dumped material, that can be in very large amounts (Piercy, 1982), could be in an unsuitable location for solution mining procedures (e.g., on a highly permeable underlying substratum), the mineralized material may be substantially diluted by overburden, the dumped material may have been consolidated to a point where permeability to aqueous solutions is very low, and the limited surface area may be a crucial inhibition to the penetration of air. It is probable that this sort of dilemma commonly arises from a lack of long-term planning, where the consideration of possible subsequent treatment of low-grade material by novel technologies is very much an after-thought, and the tilting of decision making toward short-term profitability rather than toward the achieving of a longer-term balance involving higher overall recovery and limitation of environmental impact. Alternate modes of approaching these problems are discussed in the next section.

The other major materials resulting from conventional metallurgical procedures are various process tailings. They may consist mainly of relatively inert gangue minerals. In the base metal industries, they commonly contain high proportions of low-value sulphide minerals such as pyrite and arsenopyrite. Some tailings, however, contain suites of minerals whose separation and recovery are beyond the capacity of orthodox physical separation procedures. All process tailings have in common a relatively small particle size, arising from the comminution procedures which generally precede physical separation processes such as flotation. They are commonly dispersed to tailings dams where they are dewatered and consolidated.

Process tailings of the first kind present no particular problems in disposal; they can be used as filling and are readily amenable to revegetation. Those rich in pyrite and arsenopyrite present recalcitrant problems in disposal since autogenous degradation and leaching, with release of sulphuric acid and solubilized metals, is inevitable if they are exposed to water and air. Sealing to prevent access of water and air has been a favored and generally successful mode of management.

It is the third type of process residue that is of greatest interest in the context of this essay, since such tailings may contain significant amounts of valuable metals such as gold and silver entrapped in a matrix of less valuable minerals. Complex mixtures of very fine-grained base metal sulphides may also occur. An example from the author's experience (Harris et al., 1983) will serve to illustrate how valuable metal may be recovered from such process residues by biohydrometallurgical procedures.

The principal tailings for the processing of the Renison Bell orebody near Zeehan on the west coast of Tasmania is a sulphide concentrate containing about 0.4% tin as the mineral cassiterite. The sulphide matrix is predominantly pyrrhotite with small amounts of pyrite, arsenopyrite, and other base metal sulphides. The residues in the tailings dams, which have been accumulating for about twenty years, are estimated to contain several hundreds of millions of dollars worth of tin at current prices. The particle size of the contained cassiterite does not exceed 150 μm in dimension and a substantial fraction is of submicron size. The consolidated tailings in the older dams are covered with a brown layer, 3–5 cm in thickness, that consists essentially of hydrated iron oxides, elemental sulphur (13–14%), and cassiterite. The tin mineral can be recovered from this oxidized layer by gravity separation and super-panning. The main bulk of the tailings material has remained unchanged over long periods. The tailings contain a mixed microbial population, the members of which include *Thiobacillus ferro-oxidans*, other higher-pH-tolerant *Thiobacilli*, and some unidentified bacterial species.

A number of experiments in 10 litre stirred reactors (Babij et al., 1980) were carried out in order to ascertain whether the rate of oxidation of the pyrrhotite matrix and consequent freeing of the contained cassiterite could be accelerated. Simple aeration showed a wide range of reactivity with different samples of the tailings material. Biologically-assisted oxidation using *T. ferro-oxidans, T. thio-oxidans*, and mixtures of these two species at pH 2.5 led to complete and rapid degradation of the sulphidic material with formation of ferric ions and sulphuric acid and freeing of the cassiterite. Other experiments using a consortium of the naturally occurring microorganisms at pH 4.5 yielded predominantly iron oxides and elemental sulphur and also freed cassiterite. The higher pH treatment yielded less environmentally damaging by products and the treated residue contained about 10% tin, at a recovery of 65% for the +6 μm fraction. Such a beneficiated residue could serve as a feedstock for a fuming process, and one can envisage a rationalized total recovery process, integrating orthodox metallurgical separation procedures, a biohydrometallurgical beneficiation stage, and a pyrometallurgical step leading to metallic tin as final product. Preliminary scaling-up investigations using a stirred tank reactor have been carried out, but the pressent uncertainty of the international tin market has inhibited further work on both the biological beneficiation step and the rationalization of the total process. Some of the problems associated with the scale-up of processes for the treatment of tailings are dealt with in the next section.

A similar area of current interest is the pretreatment of low-grade, auriferous pyritic ores or auriferous pyritic tailings to unlock the gold content and enhance recovery by cyanidation. Significant progress on this problem has been made by several workers (Livesey-Goldblatt et al., 1983; A. Bruynesteyn and B.C. Research–see Lakshmanan, 1986).

5.2   Optimization of Reaction Configurations

The area in which imaginative engineering innovation is most required is undoubtedly that of design of optimum systems for biologically catalyzed reactions, on a scale exceeding that of most commercial fermentations by several orders of magnitude and with extremely heterogeneous substrates, both in regard to physical characteristics and chemical composition. In this respect, the mineral substrates fall into two main categories, namely, those of large average particle size, and those that are finely divided.

Low-grade ore, after primary extraction, falls into the large-particle-size category, with individual aggregates ranging from boulder size to submicron dimensions. Ore bodies shattered explosively *in situ* have a similar particle size distribution. Such materials can be assembled in such a manner that they are readily permeable to percolating leaching solutions and the rates of degradation of the minerals they contain will be largely determined by the diffusion rates of reactants and products through the liquid films covering exposed mineral surfaces. The geometries of assesmblages and the modes of construction of masses of large-particle-size ores are crucial in obtaining maximum overall rates of metal solubilization.

Finely divided ore materials are typically process residues and tailings with average particle size not greater than 300–400 μm. *En masse*, they consolidate readily to states that are highly impermeable to aqueous solutions and to gases, and the potential advantage of enormously greater exposed mineral surface areas and correspondingly high metal solubilization rates is largely lost.

5.2.1 Configurations for the leaching of large-particle-size ore material

It was early established (Malouf and Prater, 1962) in the bacterial leaching of low-grade ores that the rate and extent of metal recovery in heap leaching operations are determined less by the inherent biochemical abilities of the complex microbial populations present than by the physicochemical constraints arising from the chemical nature of the raw materials and the manner in which its assemblage has been engineered. The experiences of many operators of successful bacterial leaching enterprises clearly show the advantages of aggregations for leaching purposes in which surface/volume ratios are maximized and in which consolidation during construction is avoided. The terraced leaching heaps of the Kennecott company in Utah, U.S.A. are a good example of this construction strategy.

Some interesting developments whereby the surface/volume ratio may be greatly increased for better aeration and distribution of leaching solutions by arranging graded ores in relatively thin layers or in "finger" heaps have been described by Robinson (1972) and indicate the scope for better and more imaginative engineering in the design of optimum configurations. Nevertheless, further development is impeded, as noted by Tuovinen (1986) by the lack to date of a comprehensive and thorough examination of all the physical, chemical, and microbiological parameters associated with dump, heap, and *in situ* leaching practices.

5.2.2 Leaching of finely divided ore materials

In order to overcome permeation problems it has been suggested that finely divided material such as process tailings might be first agglomerated or pelletized and then treated by heap leaching techniques but such procedures do not seem to have been practiced on a large scale. The more usual approach to the further treatment of such materials has been by the use of reactor systems and the equipment used has commonly been pachucas, other airlift devices, and stirred tanks. The raw material is characteristically dense and abrasive, and the power consumption for agitation and aeration is substantial. Recently, progress has been made with the design and operation of rotating drum devices for the hydrometallurgical treatment of base metal drosses by acid leaching (Robins, *personal communication*). Such devices yield high leach rates and excellent aeration, have relatively low power consumption and are amenable to considerable scale-up. The extrapolation of this technique for biohydrometallurgical purposes seems promising. The success of "barrel" leaching on a small scale for the bacterial pretreatment of auriferous arsenopyrite ores (Edvardsson, *personal communication*) seems to confirm the potential of rotating drum reactor systems for biohydrometallurgical purposes (see also, Vaseen, 1986).

Reactor systems have further inherent advantages. For example, they allow for tighter control of physicochemical factors, the use of more tightly defined biological agents, and operational flexibilities that include continuous flow operation and recycle of both organisms and substrates. More detailed treatments of reactor devices for biohydrometallurgical purposes have been published (e.g., Ralph, 1985b) and indicate the opportunities for development open to the enterprising biochemical engineer.

5.3   Control and Improvement of Biological Agents

In the dump, heap, and *in situ* leaching situations, the biological populations are extremely complex (Ralph, 1979) and variable in time (Goodwin et al., 1981). The possibility of improving such leaching situations by the introduction and establishment of superior strains of microorganisms has been the subject of some experimentation. The work of Groudev (1980) would seem to indicate that improved strains have great difficulty in establishing themselves and are rapidly displaced by wild strains. It is clear that if genetically modified leaching organisms are to play a significant role in such leaching operations, a range of characteristics other than those concerned specifically with the biodegradation mechanisms will need to be secured, before stable populations are established (Wichlacz, 1986). Spisak (1986) has suggested that improved organisms for biohydrometallurgical processes will most likely be developed through field selection and adaptive stress techniques rather than through genetic engineering (Holmes, 1986).

The microbial ecology of leaching heaps and associated environments is not particularly well developed and it is likely that the range of microbial types involved is considerably larger than has been suspected. The use of molecular probes (Yates et al., 1986; Yates and Holmes, 1986) for the identification and quantification of microorganisms found in mines and mine tailings has recently been described and should prove a powerful tool for improving knowledge of the microbial ecology of such situations.

It is in reactor systems for the treatment of finely divided metallurgical processing residues that new and improved organisms may find their greatest application. Reactor systems are more amenable to control of physicochemical and microbial variables than dumps or heaps and, given the successful development of appropriate mineral bioreactors, a very wide range of microbially catalyzed processes can be expected to be developed. The potential of highly metal-tolerant strains, strains with thermophilic and barophilic characteristics, combined with enhanced mineral biodegradation capabilities seems very wide. Exploitation of the full capabilities of special microbial strains will require a great deal of biochemical engineering expertise in the design and optimization of systems and processes.

5.4   Recovery of Products

A common economic limitation to the employment of biohydrometallurgical processes is the cost associated with the recovery of metals from dilute solution. Developments in this area are proceeding on a wide front with significant advances in membrane filtration technology, ion exchange methods, and improvements in solvent extraction procedures. Bioadsoportion offers an alternate technology, as does the possibility of recovery of some metals by the biogenesis of some of the extremely insoluble mixed metal sulphides (Robins, *personal communication*).

The possibilities of bioadsorption techniques have been reviewed recently by Shumate and Strandberg (1985) and it is clear that the potential for development of highly selective and quantitatively efficient biological reagents is high. As yet fundamental information on the nature of selective sites in the cell structures of organisms with notable metal accumulation abilities is rather scanty (Volesky, 1986); elucidation of the chemical nature of such sites could have implications for the synthesis of biomimetic materials useful in the design of novel ion-exchange and solvent-extraction reagents.

The biochemical engineering problems associated with the development of successful, large-scale bioadsorbent processes are likely to be challenging, and have been discussed by some of the authors previously cited. A prime problem would appear to be the production of bioadsorbent material with physical characteristics compatible with usage in column, counter-current, and fluidized bed devices, and properties related to sedimentation that allow ready separation from process streams. The author has suggested that sedimentation properties could be improved by linking the bioadsorbant with heavy metal or magnetite particles in the manner employed with other adsorbents in the CSIRO "Sirofloc" process (Ralph, 1983; Dixon, 1980; Weeks et al., 1981).

6. Conclusion

While there are a great many gaps in the microbiological, biochemical, and geochemical bases of the phenomena involved in biogeochemical processes, much of the fundamental information available has not been translated into viable, large-scale processes. The change in the nature of metalliferous resources and the numerous raw materials to which orthodox metallurgical technology is inapplicable suggest that the merits of biohydrometallurgical processes might be more intensively examined and pursued. It is probable that in a number of areas progress is limited by what are essentially engineering problems and it is imperative that the gaps be filled by much greater involvement of biochemical engineers in multidisciplinary teams.

REFERENCES

1. Acevardo, F., Gentina, J.C., Retamal, J., Godoy, A.M., Guerrero, L., "An experience on the bacterial leaching of Chilean copper ore." In *Recent Progress in Biohydrometallurgy*, (G. Rossi, A.E. Torma) (eds.), p. 201–212. Associazione Mineraria Sarda-09016-Italy, 1983.
2. Babij, T., Doble, R.B., Ralph, B.J., "A reactor system for mineral leaching investigations." In *Biogeochemistry of Ancient and Modern Environments*. (P.A. Trudinger, M.R. Walter, B.J. Ralph) (eds.), p. 563–572. Australian Academy of Science, Canberra, 1980.
3. Bosecker, K., "Bacterial metal recovery and detoxification of industrial waste." In *Biotechnology and Bioengineering Symposium Series No. 16*, (H.L. Ehrlich, D.S. Holmes) (eds.), p. 105–120. Interscience, John Wiley and Sons, New York, 1986.
4. Bruynesteyn, A., "Biotechnology: Its potential impact on the mining industry." In *Biotechnology and Bioengineering Symposium Series No. 16*, (H.L. Ehrlich, D.S. Holmes) (eds.), p. 343–350. Interscience, John Wiley and Sons, New York, 1986.
5. Colmer, A.R., Hinkle, M.E., "The role of micro-organisms in acid mine drainage." *Science*: **106**, 253–256 (1947).

6. Dixon, D.R., "Magnetic adsorbents: Properties and applications." *J. Chem. Tech. Biotechnol.*: **30**, 572–578 (1980).

7. Goodwin, A.E., Khalid, A.M., Ralph, B.J., "Microbial Ecology of Rum Jungle. Pt. 1. Environmental study of sulphidic overburden dumps, experimental heap-leach piles and tailings dam area." Australian Atomic Energy Commission. AAEC/E531 (1981).

8. Groudev, S., "Leaching of copper-bearing mineral substrates with wild microflora and with laboratory-bred strains of *Thiobacillus ferro-oxidans*." In *Biogeochemistry of Ancient and Modern Environments*, (P.A. Trudinger, M.R. Walter, B.J. Ralph) (eds.), p. 485–503. Australian Academy of Science, Canberra, 1980.

9. Groudev, S., Groudeva, V.I., "Biological leaching of aluminium from clays." In *Biotechnology and Bioengineering Symposium Series No. 16*, (H.L. Ehrlich, D.S. Holmes) (eds.), p. 90–99. Interscience, John Wiley and Sons, New York, 1986.

10. Harris, B., Khalid, A.M., Ralph, B.J., Winby, R., "Biohydrometallurgical beneficiation of tin process tailings." In *Recent Progress in Biohydrometallurgy*, (G. Rossi, A.E. Torma) (eds.), p. 596–616. Assosiazione Mineraria Sarda-09016-Italy, 1983.

11. Holmes, D.S., "The application of genetics and genetic engineering to biotechnology in the mining industry." In *Biotechnology and Bioengineering Symposium Series No. 16*, (H.L. Ehrlich, D.S. Holmes) (eds.), p. 299–300. Interscience, John Wiley and Sons, New York, 1986.

12. Imaizumi, T., "Some industrial applications of inorganic microbial oxidation in Japan." In *Biotechnology and Bioengineering Symposium Series No. 16*, (H.L. Ehrlich, D.S. Holmes) (eds.), p. 363–371. Interscience, John Wiley and Sons, New York, 1986.

13. Ishikawa, T., Murayama, T., Kawahara, I., Imaizumi, T., "A treatment of acid mine drainage utilizing bacterial oxidation." In *Recent Progress in Biohydrometallurgy*, (G. Rossi, A.E. Torma) (eds.), p. 393–407. Associazione Mineraria Sarda-09016-Italy, 1983.

14. Lakshmanan, V.I., "Industrial views and applications: Advantages and limitations of biotechnology." In *Biotechnology and Bioengineering Symposium Series No. 16*, (H.L. Ehrlich, D.S. Holmes) (eds.), p. 351–360. Interscience, John Wiley and Sons, New York, 1986.

15. Livesey-Goldblatt, E., Norman, P., Livesey-Goldblatt, D.R., "Gold recovery from arsenopyrite/pyrite ore by bacterial leaching and cyanidation." In *Recent Progress in Biohydrometallurgy*, (G. Rossi, A.E. Torma) (eds.), p. 627–641. Associazione Mineraria, Sarda-09016-Italy, 1983.

16. Malouf, E.E., Prater, J.D., "New technology for leaching waste dumps." *Min. Congr. J.*: **48**, 82–107 (1962).

17. Moss, F.J., Andersen, J.E., "The effects of environment on bacterial leaching rates." *Proc. Australas. Inst. Min. Metall.*: **225**, 15–25 (1968).

18. Piercy, P., "Waste dump leaching in a tropical environment." In *Interfacing Technologies in Solution Mining, Proceedings of 2nd SME-SPE International Solution Mining Symposium*, (W.J. Schlitt) (ed.), p. 241–250. Lucas Guinn Co., Hoboken, N.J., 1982.

19. Ralph, B.J., "Oxidative reactions in the sulfurcycle." In *Biogeochemical Cycling of Mineral-forming Elements*, (P.A. Trudinger, D.J. Swaine) (eds.), p. 369–400. Elsevier, Amsterdam, 1979.

20. Ralph, B.J., "Biological mining and biohydrometallurgy-prospects and opportunities." *The Australas. I.M.M. Conference*, p. 403–409, A.I.M.M., Melbourne (1982a).

21. Ralph. B.J., "The implications of genetic engineering for biologically-assisted hydrometallurgical processes." In *Proceedings of the Symposium on Genetic Engineering-Commercial Opportunities in Australia*, p. 82–102. Australian Government Publishing Service, Canberra (1982b).

22. Ralph, B.J., "Fermentation for Metals: Microbial Leaching of Minerals." In *Cells in Ferment*. (K.T.H. Farrer) (ed.), p. 30–42. Science and Industry Forum, Australian Academy of Science, Canberra, 1983.

23. Ralph, B.J., "Geomicrobiology and the new biotechnology, ONR Lecture, Society for Industrial Microbiology, Fort Collins, Colorado, August, 1984." *Devel. Ind. Microbiol.*: **26**, 411–447 (1985a).

24. Ralph, B.J., "Biotechnology applied to raw minerals processing." In *Comprehensive Biotechnology*, (M. Moo-Young, C.L. Cooney, A.E. Humphrey) (eds.), Vol. **4**, 201–234. Pergamon Press, Oxford, 1985b.

25. Robinson, W.I., "Finger dump preliminaries promise improved copper leaching at Butte." *Min. Eng.*: N9, 24 (1972).

26. Rossi, G., Torma, A.E., Trois, P., "Bacteria-mediated copper recovery from a cupriferous pyrrhotite ore: chalcopyrite/pyrrhotite interactions." In *Recent Progress in Biohydrometallurgy*, (G. Rossi, A.E. Torma) (eds.), p. 185–200. Associazione Mineraria Sarda-09016-Italy, 1983.

27. Shumate, S.E., Strandberg, G.W., "Accumulation of metals by microbial cells." In *Comprehensive Biotechnology*, (M. Moo-Young, C.L. Cooney, A.E. Humphrey) (eds.), Vol. **4**, 235–247., Pergamon Press, Oxford, 1985.
28. Spisak, J.F., "Biotechnology and the extractive metallurgical industries: Perspectives for success." In *Biotechnology and Bioengineering Symposium Series No. 16*, (H.L. Ehrlich, D.S. Holmes) (eds.), p. 331–341. Interscience, John, Wiley and Sons, New York, 1986.
29. Taylor, J.H., Whelan, P.F., "The leaching of cupreous pyrites and the precipitation of copper at Rio Tinto, Spain." *Trans. Inst. Min. Metall.*: **52**, 35–71 (1942).
30. Tuovinen, O.H., "Acid leaching of uranium ore materials with microbial catalysis." In *Biotechnology and Bioengineering Symposium No. 16*, (H.L. Ehrlich, D.S. Holmes) (eds.), p. 65–72. Interscience, John Wiley and Sons, New York, 1986.
31. Vaseen, V.A., "Biohydrometallurgy for gold recovery." In *Process Metallurgy 4. Fundamental and Applied Biohydrometallurgy* (R.W. Lawrance, R.M.R. Branion, H.D. Edner) (eds.), p. 481–482. Elsevier, Amsterdam, 1986.
32. Volesky, B., "Biosorbent materials." In *Biotechnology and Bioengineering Symposium Series No. 16*, (H.L. Ehrlich, D.S. Holmes) (eds.), p. 121–126. Interscience, John Wiley and Sons, New York, 1986.
33. Weeks, M.G., Munro, P.A., Spedding, P.L., "A new concept for rapid settling of yeast cells." *Chemica 81, Australasian Conference on Chemical Engineering*, Christchurch, N.Z. (1981).
34. Wichlacz, P.L., "Practical aspects of genetic engineering for the mining and mineral industries." In *Biotechnology and Bioengineering Symposium Series No. 16*, (H.L. Ehrlich, D.S. Holmes) (eds.), p. 319–325. Interscience, John Wiley and Sons, New York, 1986.
35. Yao Dun Pu, "The history and present status of practice and research work on solution mining in China." In *Interfacing Technologies in Solution Mining, Proceedings of 2nd SME-SPE International Solution Mining Symposium*, (W.J. Schlitt) (ed.), p. 13–20. Lucas Guinn Co., Hoboken, N.J., 1982.
36. Yates, J.R., Holmes, D.S., "Molecular probes for the identification and quantitation of micro-organisms found in mines and mine tailings." In *Biotechnology and Bioengineering Symposium Series No. 16*, (H.L. Ehrlich, D.S. Holmes) (eds.), p. 301–309. Interscience, John Wiley and Sons, New York, 1986.
37. Yates, J.R., Lobos, J.H., Holmes, D.S., "The use of genetic probes to detect micro-organisms in biomining operations." *J. Ind. Microbiol.*: **1**, 129–135 (1986).

Review on Effects of Some Physical and Chemical Pretreatments on Composition, Enzymatic Hydrolysis and Digestibility of Lemongrass, Citronella and Sugarcane Bagasse

Carlos Rolz

Applied Research Division, Central American Research Institute for Industry (ICAITI) P.O. Box 1552, Guatemala, C.A.

The following pretreatments were done on EX-FERMented sugarcane chips, citronella and lemongrass bagasse: sodium hydroxide, sodium carbonate plus calcium hydroxide, gaseous ammonia, SO_2, steam explosion, alkaline organosolv, aqueous-phenol and a full soda cook. Greater than 87% delignification was obtained in one of the organosolv sugarcane samples and the aqueous-phenol pretreatment. *In vitro* dry matter enzymatic digestibilities were increased 4.7- and 3.6-fold, respectively. The rate and extent of saccharification was higher for the steam-exploded and the nonwashed organosolv sugarcane samples. The general response to different pretreatments was similar with the lemongrass and citronella bagasse samples. However, the observed improvements were superior for lemongrass than for citronella, suggesting that these were, overall, species specific. Higher lignin losses were observed for the organosolv, the sulfur dioxide and the sodium hydroxide methods. Some of the results are in agreement with values reported for other raw materials like cereal straws. In the steam-exploded, the organosolv and the NaOH- and SO_2-pretreated materials, higher saccharification values were obtained at an earlier time. The *in vitro* enzymatic digestibilities were higher for the organosolv and the NaOH pretreatments, more than 5-fold for lemongrass and around 3.5-fold for citronella.

1. Introduction

Lignocellulosic biomass is relatively refractory to bioconversion. In natural ecosystems its biodegradation rate and extent of reaction are low; for example, degradation in the soil is an event that takes usually years to complete and in which a succession of fungi and aerobic and anaerobic bacteria in complex interactions bring about the required biochemical changes.[1-3] Polysaccharide degradation in the rumen takes place as the results of the action of a consortium of anaerobic bacteria, protozoa and phycomycete fungi and its bioconversion is far from complete.[4-8] The structural characteristics that have evolved in plants for protection against predation and biodegradation are also responsible for the limitations as forages for animal nutrition.[9]

Hence, in order to increase the rate and extent of holocellulose hydrolysis, pretreatment of the substrates is required in order to alter significantly the structural characteristics of the lignocellulosic matrix. Such pretreatment must enhance the close contact between microbe and fibers to have efficient enzyme action. There seems to be little doubt that the various pretreatments tested so far, alone or in combination (in a recent review[10] there was a list of 10 physical, 18 chemical and one biological treatments), enhance enzyme- or microbe-catalyzed holocellulose breakdown.[11-20] There seems to be, however, only a limited understanding of how these pretreatments enhance biodegradation.

The mode of holocellulolysis may be quite species specific, both in terms of substrate and microorganism. It is also an heterogeneous reaction and therefore influenced by the structural features of the substrate, the interaction between microbe and substrate surface and the complexity of the associated enzymatic system. Most of the research work done so far has employed enzymatic preparations of filamentous fungi grown in submerged culture and temperately grown cereal straws as substrates. It seems that the rate and extent of hydrolysis depends on the amount of substrate surface exposed, the proper adsorption of the biocatalyst, the degree of polymerization of the polysaccharide and minimum lignin, phenolic acid and acetyl contents.[21-39]

Most of the research in this field has been centered in wood and wood byproducts from temperate countries and in residues from annual crops like the straw (wheat, barley, etc.), beet pulp, corn stover and rice byproducts (straw and hulls). Sugarcane bagasse is an extremely important byproduct from crystallized sugar factories in tropical countries. It consists mainly of the lignocellulosic polymer matrix and small amounts of ash and water-soluble solids which are mostly sucrose. In the sugar industry mills most of the bagasse is used as fuel, and although the figure varies within each factory, energy efficient units can yield two tons of surplus bagasse on a dry basis for 100 tons of fresh cane processed. In independent fuel-ethanol distilleries part of this surplus is used for ethanol purification, making the overall system net energy positive, either for mechanized or for labor-intensive agriculture.[40-43] Nevertheless, even in this case, a surplus is possible if a new demand for bagasse is developed, either by applying energy saving options in ethanol separation or if new technologies are employed for processing cane chips directly into ethanol such as the EX-FERM process.[44-46] Especially important for bagasse use would be the hydrolysis of the holocellulose for monomer production or the separation of lignin or lignin derivatives. Henceforth research efforts to make bagasse more susceptible to biological processing are worth pursuing. In a separate article we are presenting work done to evaluate the effect of some physical and chemical pretreatments on the composition, susceptibility to enzymatic hydrolysis and *in vitro* enzymatic digestibilities of EX-FERMented sugarcane residue.[47] Some of these data will be commented on here.

Other agricultural residues found in tropical countries have not been studied at all. Such is the case of lemongrass and citronella bagasse. These are the lignocellulosic residues of steam distilling freshly cut lemongrass and citronella leaves for the recovery of the respective essential oils. The plants belong to the Gramineae family and have been classified as *Cymbopogon citratus* and *Cymbopogon winterianus*, respectively. The essential oil content is low, 0.3–0.7% by weight of fresh grass, and its recovery is not complete. After steam distillation, the bagasse is partially dried in the fields and a fraction is burned to generate steam for the stripping; the rest is left in the fields where natural biodegradation takes place.[48] Its use as a ruminant feed is limited due to animal rejection because of the residual aroma and flavor. Several attemps have been made to use it as a source of fiber for paper and board products but the process, at least in Guatemala, has only been carried out in pilot plant tests. Lemongrass oil production is around 1,000 tons per year, being produced by India, Guatemala, the Republic of China and Sri Lanka. Citronella oil production is around 3,000 tons per year produced by Indonesia, Sri Lanka, the Republic of China, Taiwan, Guatemala and Brazil.[49] This means that there is a possible worldwide availability of about 200,000 tons of dry bagasse per year that could be used as a source of lignocellulosic biomass. The material has already been subjected to low-pressure steam treatment which not only sterilizes it but induces physical and chemical changes in the lignocellulosic matrix. We have also in a separate article presented experimental results on the effects of some physical and chemical pretreatments on the composition, susceptibility to enzymatic hydrolysis and *in vitro* enzymatic digestibilities of these bagasses.[50]

2. Results

The analytical data for the raw bagasse samples are given in Table 1. The two *Cymbopogon* plants are similar. The holocellulose fraction, amenable to enzymatic hydrolysis to simple sugars, represented on dry basis 58.4% for lemongrass and 58.5% for citronella. The corresponding figures on an ash-free basis were 65.2% and 64.1%, respectively. Their crude protein content was lower than the values characteristic of hardwood, leaves, lucerne and ordinary grass hays, which are in the range of 8% to 23%[53,54]; however, it is higher than that of common straws[53,55] resembling instead the leaves (leaf sheath and blade) of such materials.[56] In terms of their NDF and ADF contents they show some similarity to the common straws and are of intermediate lignin content, again being close to the straws and sugarcane bagasse.[55,57,58] The data on Table 1 are, to our knowledge, the first of their kind reported for these materials; however, they must be taken with caution as individual components may vary within *Cymbopogon* species cultivated in different ecosystems and of course with plant maturity.

In Table 2 a summary of a comprehensive literature review on bagasse samples from many countries, cane maturity and variety is given. From these data the following pertinent observations can be made: a) With the exception of the hemicellulose content, all the other data for CAST in Table 1 fall within the values presented in Table 2. b) The data for permanganate lignin is similar in both tables; there exists, however, a large discrepancy between the permanganate and Klason values, the latter being about eight units higher. There is no easy explanation for this fact. For example, Hartley[99] mentioned that, contrary to what has been the case for sugarcane bagasse, the permanganate lignin values are some-

TABLE 1. Chemical analysis of raw materials.

	TEST[a]	CIST[a]	CAST[a]
Neutral detergent fiber (NDF)[b]	72.6	72.0	92.8
Acid detergent fiber (ADF)[b]	44.1	42.0	54.8
Permanganate lignin[b]	11.0	11.1	12.5
Cellulose[b]	29.9	28.5	43.3
Hemicellulose (NDF-ADF)	28.5	30.0	38.0
Ash[c]	11.0	9.3	2.4
Crude protein (Kjeldahl nitrogen × 6.25)[c]	5.1	4.5	1.2
Essential oil[c]	Traces	Traces	–
Water-soluble substances[d]	16.2	23.3	3.1

[a] TEST = lemongrass bagasse; CIST = citronella bagasse; CAST = EX-FERMented sugarcane residue.
[b] Determined according to the Van Soest techniques.[51]
[c] Determined according to the AOAC.[52]
[d] Determined in the controls of the *in vitro* dry and organic matter enzymatic digestibilities.

TABLE 2. Chemical analysis of sugarcane bagasse (Literature values[57-98]).

	NDF[a]	ADF[a]	Permanganate lignin	Klason lignin	Cellulose	Hemicellulose
1. Number of samples, n	15.0	14.0	18.0	35.0	41.0	48.0
2. Sample mean, x	87.0	59.2	12.5	20.5	41.7	26.2
3. Sample standard deviation, s	3.9	6.2	1.7	2.4	6.6	4.6
4. 95% confidence interval on population mean	±1.8	±3.0	±0.7	±0.7	±1.7	±1.1
5. Maximum value	94.5	68.3	15.5	27.5	53.1	34.2
6. Minimum value	81.6	49.2	10.0	14.3	26.6	10.0

[a] See Table 1.

times higher than those from the Klason method. However, as shown by Theander and Aman[53], this is not necessarily so for all plant species. c) As shown by Theander and Aman,[53] Windham et al.[100] and McAllan and Griffith,[101] the hemicellulose contents obtained by NDF-ADF usually are overestimated compared with the summation of hemicellulose constituent sugars. d) The presence of a lignin-carbohydrate complex (LCC) stabilized by chemical linkages has been proved in wood,[102] in gramineous plants this linkage might involve phenolic acids such as ferulic and p-coumaric acids;[103-107] the existence of LCCs in sugarcane bagasse has been reported[89,94,108,109] and Kato et al.[91] reported in detail the isolation of three fractions different from each other in contents of lignin, carbohydrate, phenolic acids and molecular weight. Usually the LCCs are not recovered in the acid detergent residue,[110] hence this could be an explanation of the low permanganate lignin values and the overestimated hemicellulose figures.

In vitro enzymatic digestibilities (IVDMED) are presented in Table 3 for the original three bagasses and for the pretreated samples. They are identified in a footnote and the interested reader can find the methodology employed for pretreatment in the corresponding references.[47,50] The digestibilities represent weight loss due to enzymatic action in terms of initial total dry matter but taking into account the weight loss of the control (no enzyme) samples. Hence, these figures represent true insoluble matter resistant to enzyme attack. All the pretreatments were effective; the observed changes for lemongrass were superior to the ones for citronella, probably due to the lower digestibility for the untreated sample of the former. However, in both cases the organosolv- and sodium-hydroxide-treated samples showed greater increases in digestibility: more than 5 times for lemongrass and around 3.5 for citronella. The same was true for cane bagasse, especially the organosolv variants, sodium hydroxide, the aqueous phenol and the steam-exploded samples. Lignin and hemicellulose contents decreased in all samples; the former was higher for the organosolv, aqueous phenol and steam-exploded samples for cane bagasse. Hemicellulose was fully hydrolyzed in the aqueous phenol and steam-exploded samples. In the case of lemongrass and citronella the sodium-hydroxide-treated sample also gave a high lignin loss. This was estimated employing the analytical data of the solid samples from each pretreatment, the corresponding solid yields and a mass balance. The solid yields were in all cases 100% with exception of TEEX, TEOA, CIEX, CIOA, CAEX, CAOK-1, CAOK-2, CAOA-1, CAOA-2, BCSODA and CABG which were: 75.4, 55.7, 75.8, 56.7, 77.9, 86.8, 80.5, 71.8, 67.4, 50.6 and 42.0%, respectively.

TABLE 3. *In vitro* dry matter enzymatic digestibilities (IVDMED) and lignin losses for pretreated samples.

Sample[a]	IVDMED %	Change[b]	Permanganate lignin content, %	Permanganate lignin loss, % of original
Lemongrass				
TENA	66.14	4.79	4.7	57.3
TECANA	40.80	2.96	6.9	37.3
TEAM	41.12	2.98	9.7	11.8
TESO	49.49	3.59	6.8	38.2
TEEX	47.76	3.46	9.5	34.9
TEOA	73.57	5.33	7.7	61.0
TEST	13.80 ± 1.60[c]	1.00	11.0	–
Citronella				
CINA	67.10	3.68	5.7	48.6
CICANA	37.87	2.08	7.8	29.7
CIAM	45.86	2.51	8.5	23.4
CISO	56.86	3.12	7.8	29.7
CIEX	47.12	2.58	11.8	19.5
CIOA	70.44	3.86	9.7	50.5
CIST	18.24 ± 2.02[c]	1.00	11.1	–
EX-FERMented sugarcane				
CANA	72.28	3.80	5.8	53.6
CACANA	33.43	1.76	8.6	31.2
CAAM	29.58	1.55	9.3	25.6
CASO	26.19	1.38	9.6	23.2
CAEX	56.68	2.98	5.7	64.5
CAOK-1	60.40	3.17	5.4	62.5
CAOK-2	64.88	3.41	6.4	58.8
CAOA-1	67.25	3.53	6.0	65.5
CAOA-2	89.80	4.72	2.4	87.1
BCSODA	98.64	5.18	0.7	97.2
CABG	69.02	3.63	3.5	88.2
CAST	19.04 ± 1.74[c]	1.00	12.5	–

[a] The identifications used were (first lemongrass): original (TEST, CIST), sodium hydroxide (TENA, CINA), sodium carbonate plus calcium hydroxide (TECANA, CICANA), ammonium (TEAM, CIAM), sulfur dioxide (TESO, CISO), steam explosion (TEEX, CIEX) and alkaline organosolv (TEOA, CIOA).[50] For the EX-FERMented sugarcane the identifications used were: sodium hydroxide (CANA); sodium carbonate plus calcium hydroxide (CACANA); ammonia (CAAM); sulfur dioxide (CASO); steam explosion (CAEX); alkaline organosolv (CAOK-1, CAOK-2, CAOA-1, CAOA-2); organosolv-phenol (CABG) and full soda cook (BCSODA).[47]

[b] Ratio between pretreated and unpretreated (TEST, CIST and CAST samples).

[c] 95% confidence interval.

The results of the susceptibility of the different materials to the hydrolytic action of a fungal cellulase enzyme are represented in detail as percentage of saccharification of the holocellulose fraction of each pretreated sample in the corresponding references.[47,50] The reducing sugars in the control (no enzyme) were employed to make the calculations, and hence these values represent true products of enzyme hydrolysis and do not take into account the soluble sugars produced during the pretreatment step. The percentage of saccharification and the enzymatic digestibility data are not equal as they are expressed on a different basis and have been obtained measuring unequal parameters and employing a different amount of enzyme. In Table 4 the saccharification data after 8 h and 72 h of re-

TABLE 4. Saccharification data.

Sample[a]	mg Reducing Sugars ml^{-1} 8 h	% Saccharification	mg Reducing Sugars ml^{-1} 72 h	% Saccharification
Lemongrass				
TENA	2.29	63.58	3.39	94.11
TECANA	1.56	32.05	2.46	50.55
TEAM	1.51	29.74	2.56	50.42
TESO	2.72	59.61	4.20	92.04
TEEX	2.14	61.61	5.39	100.00
TEOA	4.19	58.13	6.30	87.40
TEST	0.54	8.99	0.80	13.32
Citronella				
CINA	1.79	55.40	3.04	94.10
CICANA	0.33	44.27	1.29	65.53
CIAM	0.44	8.03	2.26	41.26
CISO	2.75	66.00	3.42	82.08
CIEX	0.94	27.56	3.89	100.00
CIOA	2.65	40.36	5.06	77.06
CIST	0.43	7.06	0.78	12.80
EX-FERMented sugarcane				
CANA	2.19	44.51	3.53	71.75
CACANA	1.54	24.84	2.05	33.06
CAAM	1.18	17.20	2.26	32.94
CASO	2.55	33.47	3.10	40.69
CAEX	4.00	79.78	4.33	86.36
CAOK-1	4.81	56.37	6.36	74.54
CAOK-2	1.85	19.66	3.69	39.22
CAOA-1	2.34	25.11	4.24	45.50
CAOA-2	3.64	37.80	6.82	70.83
BCSODA	2.46	24.70	6.05	60.75
CABG	3.31	34.77	3.39	35.61
CAST	0.77	8.93	0.93	10.79

[a] See footnote of Table 3.

action is presented for all samples. Again, there were more similarities than differences in the effects of the different pretreatments on the substrates. All the pretreatments were effective in enhancing the amount of holocellulose hydrolyzed by the enzyme. The steam-exploded, the organosolv- and the sodium hydroxide- and sulfur-dioxide pretreated materials were the ones more extensively attacked for lemongrass and citronella begasse and in which around 60 % saccharification values were obtained early in the digestion.

As expected, all the pretreated cane bagasse samples were hydrolyzed faster and more completely. Of the alkaline pretreated, the sample most susceptible to hydrolysis was the sample pretreated with NaOH under solid substrate conditions. This sample was even better than the one obtained with a full soda cook. The ammonia- and $Na_2CO_3 + Ca(OH)_2$-pretreated samples showed similar behavior but were inferior in the NaOH effect, as was found for the IVDMED data shown in Table 3. All organosolv treatments were effective in increasing the initial rate of hydrolysis and the extent of hydrolysis after 72 h, especially CAOK-1 and CAOA-2. As specified elsewhere,[47] the first one was not water washed after an alkaline (1 %) organosolv; in the second one a higher NaOH content (4 %) and anthraquinone as catalyst were employed and the sample was water washed. We do not know if washing the sample after pretreatment easily removed hydrolyzed complexes from it. This is a possible explanation for the fact that more carbohydrates were detected in the liquid phase in the CAOK-1 sample. The phenol-pretreated sample behaved similar to the CAOK-2 and CAOA-1 samples, and all of them were inferior to CAOK-1 and CAOA-2. Very similar increases were obtained for the IVDMED values, as mentioned above. The CASO sample showed a lower susceptibility to enzymatic hydrolysis than the NaOH and organosolv samples. However, the steam-exploded sample gave the highest values for the percentage of saccharification, of around 86%. This is 5.5 times the extent of hydrolysis for the untreated sample. Rao et al.[111] reported a 3.1-fold increase after steam explosion for 30 min at 0.68 MPa. Both of these values are much higher than those for the IVDMED parameter and summarized in Table 3. Again, these steam-exploded samples were not water washed after pretreatment and the possibility of the generation during steam explosion of easily hydrolyzable complexes might explain these results.

The percentage of saccharification and the enzymatic digestibility data are not equal as they are expressed on a different basis and have been obtained by measuring unequal parameters and employing different amounts of enzyme. Nevertheless, it is easy to transform one into the other. If the percentage of saccharification data at 72 h is converted into digestibility, it is relatively lower than the IVDMED values in Table 2. The fact that they are lower might reflect indirectly the hydrolyzing action of acid-pepsin in the two-enzyme digestibility assay. Dowman and Collins[112] commented that either the use of acid-pepsin or neutral detergent solution is recommended before the use of cellulases for determining forage digestibility, since in this way more insoluble material is hydrolyzed by the enzymatic action. The protein polymer might protect the holocellulosic fibers.

3. Discussion

The lignin losses for lemongrass and citronella shown in Table 3 are on the high side compared to literature values for other lignocellulosic residues where conflicting results have been reported: in wheat straw, a range of 12–30% decrease has been published.[113,114] Chesson[114] reported a 34% loss for barley straw, Taniguchi et al.[115] an 85% loss for rice straw and Ibrahim and Pearce[86] a 6% loss for pea straw. However, the same authors reported a 7% and 6% increase for barley straw and sunflower hulls, respectively; van Eenaeme et al.[116] showed a 9% increase for three different types of hays and Ben-Ghedalia and Shefet[117] an increase of 46% for cotton straw. The effects of sodium hydroxide on the *in vitro* dry-matter enzymatic digestibilities are extremely favorable. In the literature for other raw materials the *in vitro* ratios range from 1.2 to 2.5 times the original sample for wheat straw,[114,118,119] barley straw,[120–123] rice straw,[119] rye straw,[124] alfalfa straw,[120] sorghum stems,[119] cotton stems,[117] rice hulls,[125] sunflower hulls,[86,87] hays,[116] corn cobs[126] and screened cattle manure solids.[127] There is experimental evidence that both alkali-labile and alkali-resistant lignin-carbohydrate bonds exist and that their ratio in a particular plant material governs the decision to use alkali as a pretreatment to increase digestibility.[114] The alkali-labile bonds include the hemicellulose-phenolic acid and acetyl constituents of the cell walls,[99] both of which affect holocellulose hydrolysis.[105,128–134] In lemongrass and citronella bagasse these bonds seemed quite labile to alkaline conditions; since water-soluble compounds increased drastically and also the NDF values decreased from 72.6 and 72.0 to 48.2 and 44.7, respectively, as shown elsewhere.[50] Recently, Scalbert et al.[135] found for wheat straw that ferulic acid ethers cross-link between hemicelluloses and lignin. These are alkali labile and might explain the high solubility of straw lignins in soda. It could also be argued that the alkaline treatment modified the small amount of essential oils still present in the bagasse, hence removing any inhibitory effects upon the enzymes. In order to check this, bagasse samples were extracted with ether for 48 h, removing 2.13 g/100 g dry matter for citronella and 1.34 g/100 g dry matter for lemongrass. The corresponding IVDMED values were 19.45 for citronella and 16.06 for lemongrass, respectively. These were not statistically different so that the inhibition hypothesis does not seem valid. The ammonia pretreament was also typical and did not differ in the basic effects upon cell walls of gramineous plants as suggested by van Soest et al.[136]; however, it was different for the raw materials, illustrating again the variable response obtained with different residues.

The expected effects of the alkali treatments on the substrate were partially to dissolve hemicelluloses and lignin with a parallel increase in their *in vitro* or *in vivo* digestibilities.[57,137,138] The important chemical reaction in this pretreatment seems to be the saponification of esters of uronic acids associated with the xylan chains that induces an extensive swelling of the solid matrix,[138] promotes the penetration of microorganisms and makes possible a more favorable enzyme-fiber contact.

In Table 5 the data reported in the literature for sugarcane bagasse is summarized. There is a widespread variation in conditions used for the pretreatment which makes difficult any qualitative comparison; nevertheless, there are some general trends worth commenting on: a) our percentage IVDMED increases for NaOH are higher than those usually reported, with the exception of the data of Cabello et al.,[75] Awad and Abdel-Mottaleb,[79] Cabello and Conde[95] and Ismail et al.,[97] which are on the same order of magnitude; b) the one obtained for NH_3 compares well with those reported for NH_3 and NH_4OH pretreatments, with the exception of the data of Awad and Abdel-Mottaleb[79] which is rather high; c) our data for $Ca(OH)_2 + Na_2CO_3$ is lower than the one by Matei and Playne,[88] Playne[93] and Martin et al.[65]; d) we found higher digestibility values for NaOH than for NH_3 and $Ca(OH)_2 + Na_2CO_3$, which coincides with Ibrahim and Pearce[86]; e) as shown by Cabello et al.[75] IVDMED values are usually lower than IVDMD values, as a matter of fact those authors reported a linear relationship among the two; f) our data on lignin losses for NaOH were usually on the high side; they compare only with the data of Dekker and Richards[64] for cane bagacillo (fines) and of Ellenrieder and Castillo.[82] For ammonia our value is within values reported by Ibrahim and Pearce[86] and for calcium ours is higher.

Early work by Millet et al.[140] at the Forest Products Laboratory in Madison showed that after pretreatment with gaseous SO_2 under pressure for 2–3 h at a temperature of about 120°C samples of hardwoods and softwoods increased their enzymatic digestibility substantially. The Klason lignin values of the pretreated hardwoods decreased 67% from the initial content and 29.5% for the pretreated softwoods, indicating an extensive depolymerization of the original lignin. Lee et al.[141] also treated the lignocellulosic residue after hemicellulose acid-hydrolysis of a hardwood with gaseous SO_2 for 2–3 h at 150° and 170°C; they also showed an increase in the initial relative saccharification rate and in the extent of saccharification of the pretreated samples. Conner[142] treated pure cellulose and cotton linters and found that sulfur dioxide caused a dramatic decrease in the degree of polymerization but did not affect the enzymatic digestibility or the crystallinity of the substrate. Recent work with SO_2 pretreatment of wheat straw has been done at the Volcani Center in Israel.[143–145]

TABLE 5. Comparison of changes induced on bagasse by alkaline pretreatments (Figures expressed in % change over control; lignin is loss; digestibilities are increases).

Sample	Lignin	IVDMED
Randel[63]		
2% NaOH, 24 h, washing	–	60.2[a]
Dekker and Richards[64]		
7% NaOH, 1–1.5 h, washing	58.2	174.1[a]
Martin et al.[65]		
3–14% NaOH, 40 min, 90°C	–	181 –737[a]
4–14% NaOH, 30 min, 100°C	–	185 –577
8–16% Ca(OH)$_2$, 72 h	–	340 –378
Llamas et al.[69]		
2–6% NaOH, 45 days ensilage	14.9–22.1	–
Rexen[70]		
4.5% NaOH, Danish process	23.2	60.3
Oi et al.[72]		
14% NH$_4$OH, 24 h, washing	–	83.3
Cabello et al.[75]		
4–12% NaOH	–	30–280.7
Awad and Abdel-Mottaleb[79]		
2–10% NaOH	–	289.1–991.5
2–10% NH$_4$OH	–	230.5–406.0
Ellenrieder and Castillo[82]		
1% NaOH, 3 h, 80°C, washing	54.5	–
Molina et al.[94]		
2–6% NaOH	2.6–23.7	–
Ibrahim and Pearce[86]		
3–12% NaOH, 24 h	11.4–35.2	28.3– 66.9[a]
5–15% Ca(OH)$_2$, 24 h	1.1– 2.8	16.7– 28.9
3.6–7.7% NH$_4$OH, 30 days	1.1–14.8	0.4– 14.5
3.5–5.8% NH$_3$, 5–10 days	25.0–34.7	26.4– 31.7
Ibrahim and Pearce[87]		
6.0% NaOH, 24 h	–	103.7[a]
5.2% NH$_3$, 5 days	–	43.4
Ibraham and Pearce[139]		
6–2% NaOH, 24 h	–	48.5–116.1[a]
(spray, Beckman method and		
soak-and-press method)		
Matei and Playne[88]		
18% Ca(OH)$_2$ + Na$_2$CO$_3$, 1 h, 121°C	–	153.8
5% NH$_3$, 35 days	–	73.1
Playne[93]		
3–11% NaOH, 18 h	–	61.8–252.9
4–12% Ca(OH)$_2$ + Na$_2$CO$_3$, 18 h	–	96.1–243.1
3–11% NH$_3$, 18 h	–	47.0– 96.1
Cabello and Conde[95]		
2–6% NaOH	11.0–21.9	241.6–491.6
Ismail et al.[97]		
1–5% NaOH, 24 h, washing	–	400.5–566.7
(boiling, 4 h) washing	–	273.3–506.7

[a] Determined employing ruminant fluid as an enzyme source.

Our data are in agreement with them, although they used about seven times less chemical per unit dry weight of lignocellulosic residue. Ibrahim and Pearce[86,87] showed rather slight effects of SO_2 on barley and pea straws and cane bagasse. Although they tried several temperatures and holding times, it seems that they did not use the intermediate temperatures (about 70°C) and long holding times (3 days) necessary to induce the required benefits.

Steam pretreatment was originally developed in 1925 and has been extensively used in the manufacture of hardboard by the Masonite processes employing wood. In the late 1970s Iotech Corp. Ltd. and Stake Technology Ltd. in Canada started using this process for the production of feed for ruminants. Treatment of wood chips up to 185°C also is practiced commercially in the first stage of the manufacture of dissolving pulp by the prehydrolyzed Kraft process. Hence, wood steam-pretreament studies have shown that the lignocellulosic matrix is modified drastically and that the remaining solids are more susceptible to enzymatic hydrolysis. The structural changes and chemical reactions taking place seem to be a function of temperature and time of the pretreament, and can be summarized as follows: a) during short prehydrolysis (about 30–60 sec) taking place at temperatures of 220°–240°C lignin was extensively depolymerized to low-molecular-weight polymers that were soluble in aqueous alkali or ethanol-water solutions and a small amount to low-molecular-weight phenol-related compounds, although recondensation reactions could eventually take place; b) hemicelluloses were easily hydrolyzed into mono- and oligosaccharides of low molecular weights, and in addition, some degradation products were formed that apparently condensed with lignin, thereby increasing the lignin content; c) cellulose was the least hydrolyzed component, maintaining its degree of crystallinity but reducing its degree of polymerization; and d) the rapid steam-decompression or -explosion de-fabricates the cellular structure and caused a large increase in accessibility of the cellulose to enzymatic hydrolysis.[146-155] The resulting solids contained inhibitors which interfered with the growth of microorganisms and anaerobic fermentations.[156-160] Washing the solids with water or dilute alkali removes the inhibitors. However, it seems that anaerobic mixed cultures are less prone to inhibition, both in *in vitro* systems,[160] as also in *in vivo* ruminant feeding studies.[74,161-165] However, the chemical nature of the inhibitors is unknown,[166] as also is any antidotal mechanism existing in mixed cultures of anaerobic bacteria.

The solid yields obtained for steam-exploded lemongrass and citronella at 240°C (3.35 MPa) and 60 sec which were about 75% were within expected values; somewhat lower yields have been reported for other materials processed for longer time periods, from 65–68% for wheat, barley and oat straws (8 min at 1.1 MPa),[167] from 60–82% for rice straw and 78–99% for sorghum straw (30–60 min at 0.7 MPa),[81] 61% for sunflower hulls (5 min at 200°C with 6.9 MPa nitrogen pressure),[168] 62–70% for wheat straw (5 min at 200°C with 3.5 MPa carbon dioxide pressure).[38,169] The yield should drop with exposure time at constant pressure as has been shown for aspen wood by Saddler and Brownell[170] due to the formation of pyrolysis gases, acetic acid from acetyl groups, formic acid, furfural and other compounds which are stripped with water during decompression. There is a lack of information in this respect for annual crop residues. Taniguchi et al.[171] have shown that this is the case for rice straw and husk and peanut husk when pressure was increased for a fixed time.

The solid yield for cane bagasse after steam explosion at 240°C (3.35 MPa) and 60 sec was about 78%. This figure was within expected values: Cheong et al.[172] found that yield decreased with exposure time at fixed pressure and with increasing pressure at a constant exposure time, with the minimum yield being 70.7% at 0.16 MPa (199°C). Hart et al.[71] found higher values, 92–95% at 2.07 MPa for 5 min and 84–94% at 2.76 MPa for 0.5–1.5 min. A minimum yield of 79.2% has been reported by Rangnekar et al.[81] for 60 min at 0.88 MPa, and he found the same yield dependence with time and pressure. Morjanoff et al.[173,174] explored various steam-pretreatment techniques; the yield was between 69% to 74% after 1 h at 160–180°C. The solid fraction remaining was treated for 5 min at 208°C and the corresponding yield was 87.6%. Hence a combined minimum yield of 60.4% was obtained. Dekker and Wallis[168] reported a yield of 58% after a pretreatment for 4–20 min at 200°C and 6.9 MPa by nitrogen pressurization. The yield varied from 63.5% to 71.9% by 5 min at 200°C and a pressure from 3.45 to 13.8 MPa employing pure CO_2 or in mixtures with N_2.[60,169] Lower yield values can be obtained if an explosion soda pulping is employed; for example, Mamers et al.[175] reported a figure of 52.5% for 5 min at 3.4–13.8 MPa with 7% Na_2O.

In Table 4 it is shown that for both lemongrass and citronella steam-explosion solids were 100% enzymatically saccharified, reflecting an improvement of 10 times on the cellulose susceptibility to enzymatic action. Values of 6.2 times and 6.3 to 7.7 times have been reported for sunflower hulls by Dekker and Wallis[168] and for wheat straw by Linden et al.[176] and Vallander and Eriksson,[177] respectively. Taniguchi et al.[171] reported a 6-fold increase for rice straw and 5.6 and 1.1 for rice husk and peanut husk, respectively. Mac-Donald and Mathews[178] reported an improvement of 5.5 times for aspen wood. In Table 4 the improvements in the enzymatic digestibilities are not as good as those obtained with sodium hydroxide, sulfur dioxide and organosolv; nevertheless, they are within what has been observed for other steam-treated materials.[60,80,81,87,167,169,173,176,179–182] This improvement is not only related to a partial delignification[183–187] but more importantly to a decrease in the degree of polymerization of the cellulose[60] and an increase in surface area.[152] The cellulose crystallinity remains unchanged[171] and seems in this pretreatment not to be the main determinant influencing saccharification rates, although for other chemical pretreatments this is an important parameter.[25] Due to the partial hemicellulose and lignin (and probably LCCs) hydrolysis and loss of soluble monomers caused by the acid medium from the cleavage of acetyl groups, there is an actual crystallinity increase of the remaining solids. Hence it seems that, as Gharpuray et al.[33] have suggested, the enzymatic hydrolysis is mainly affected by the substrate surface area, lignin content and cellulose crystallinity in that order.

The modified organosolv pretreatments either with ethanol or phenol caused the greatest delignification of all methods tested, accompanied by a substantial loss of hemicelluloses. This behavior is typical in this process as clearly shown, among others, by Phillips and Humphrey[188,189] for poplar wood, in which regardless of the treatment conditions used, the lignin removal was always accompanied by hemicellulose removal in a linear correlation among the two. The IVDMED increases were very high, from 3.2 times for CAOK-1 to 4.7-fold for CAOA-2. The sample treated with phenol (CABG) responded similarly to CAOK-1. There were increases from 4- to 5-fold for lemongrass and citronella. These figures are the highest of all pretreatments. To our knowledge, there are no data for this parameter for other organosolv-pretreated lignocellulosic materials. Susceptibility to enzymatic attack on the residual solids has been reported by other authors: Avgerinos and Wang[190] found pretreated corn stover more easily degraded by a mixed culture of *Clostridium thermocellum* ATCC 31924 and *Cl. thermosaccharolyticum* ATCC 31925; Neilson et al.[191] reported an increase of total sugars produced per g of cotton-wood dry matter of 50 times after 6 h of enzymatic hydrolysis; the glucose yield was increased by approximately 3.7 times for ethanol- or butanol-pretreated rice straw[192]; about 90% hydrolysis has been reported in 24 h for poplar wood by Holtzapple and Humphrey[193] and for birch and spruce by Sakakibara et al.[194]

4. Conclusion
4.1   For Lemongrass and Citronella
The three alkaline pretreatments, the SO_2, steam explosion and organosolv fractionation, were quite effective for inducing the necessary physical and chemical changes in lemongrass and citronella bagasse that substantially increased their susceptibility to enzymatic hydrolysis. The solids yields were around 75% and 55% for steam explosion and organosolv, respectively. Higher delignifications were observed for organosolv, SO_2 and the sodium-hydroxide methods. The steam-exploded, organosolv-, NaOH- and SO_2-pretreated materials were more extensively attacked by enzymatic action. The observed changes were, in general, better for lemongrass than the ones for citronella; however, both were superior to the ones so far reported for common cereal straws and more lignified materials like hulls or even wood. Results show that pretreatment effects are, indeed, species specific, and clearly point out that the enhancement of enzymatic hydrolysis is also not only due to the extent of delignification of the raw material but also to the effect on physical properties of the holocellulose.

4.2   For Sugarcane Bagasse
All the pretreatments were effective in increasing the IVDMED, the initial rate of holocellulose hydrolysis and the saccharification extent. The best pretreatments were the organosolv variants, steam explosion, sodium hydroxide and aqueousphenol. Lignin contents decreased in all samples. Delignification was higher for the organosolv, aqueous phenol and steam-exploded samples. Hemicellulose was fully hydrolyzed in the aqueous-phenol and steam-exploded samples. Extensive comparison with literature values showed quite a large variation in pretreatment effects. It is quite possible that an explanation for this fact might be the structural differences found in the lignocellulosic matrix of the various cane varieties processed in the tropical sugarcane growing regions of the world.

Acknowledgments
The author wishes to thank Dr. J.N. Saddler of the Eastern Laboratory of Forintek Canada Corporation, Ottawa, for the steam-exploded sample and Dr. J.P. Sachetto from BATTELLE-Geneve for the aqueous phenolic pretreatment. Support for the project came from contract No. 936–5542–G–00–2041–00 from US-AID's Program in Science and Technology Cooperation (PSTC). We sincerely appreciate their encouragement and support.
It is a great honor to have been invited to contribute to a technical publication commemorating the retirement of Prof. Shuichi Aiba. As a human being, he has been a true friend; as a teacher, an inspiration.

REFERENCES
1. Bazin, M.J., Saunders, P.T., Prosser, J.I., "Models of microbial interactions in the soil." *CRC Crit.*

Reviews Microbiol.: **4**, 463–498 (1976).

2. Lynch, J.S., "Straw residues as substrates for growth and product formation by soil micro-organism." In *Straw decay and its effects on disposal and utilization.* p. 47–56. E. Grossbard (ed.), John Wiley & Sons, Chichester, N.J., 1979.

3. MacCubbin, A.E., Hodson, R.E., "Mineralization of Detrital Lignocellulose by Salt Marsh Sediment Microflora." *Appl. Environ. Microbiol.*: **40**, 735–740 (1980).

4. Stewart, C.S., Dinsdale, D., Chang, E.J., Paniagua, C., "The digestion of straw in the rumen." In *Straw decay and its effects on disposal and utilization.* p. 123–130. E. Grossbard (ed.), John Wiley & Sons, Chichester, N.J., 1979.

5. Morrison, I., "The degradation and utilization of straw in the rumen." In *Straw decay and its effects on disposal and utilization.* p. 237–245. E. Grossbard (ed.), John Wiley & Sons, Chichester, N.J., 1979.

6. Hobson, P.N., "Polysaccharide degradation in the rumen." In *Microbial polysaccharides and polysaccharases.* R.C.W. Berkeley, G.W. Gooday, D.C. Ellwood (eds.), p. 377–397. Academic Press, London, 1979.

7. Demeyer, D.I., "Rumen microbes and digestion of plant cell walls." *Agr. Environ.*: **6**, 295–337 (1981).

8. Russell, J.B., Hespell, R.B., "Microbial rumen fermentation." *J. Dairy Sci.*: **64**, 1153–1169 (1981).

9. Van Soest, P.J., "Limiting factors in plant residues of low biodegradability." *Agr. Environ.*: **6**, 135–143 (1981).

10. Fan, L.T., Lee, Y.-H., Gharpuray, M.M., "The Nature of Lignocellulosics and Their Pretreatments for Enzymatic Hydrolysis." *Adv. Biochem. Eng.*: **23**, 157–187 (1982).

11. Mandels, M., Hontz, L., Nystrom, J., "Enzymatic Hydrolysis of Waste Cellulose." *Biotechnol. Bioeng.*: **16**, 1471–1493 (1974).

12. Millett, M.A., Baker, A.J., Feist, W.C., Mellenberger, R.W., Satter, L.D., "Modifying wood to increase its *in vitro* digestibility." *J. Anim. Sci.*: **31**, 781–788 (1970).

13. Millett, M.A., Baker, A.J., Satter, L.D., "Pretreatments to enhance chemical, enzymatic, and microbiological attack of cellulosic materials." *Biotechnol. Bioeng. Symp. Series No. 5*, 193–219 (1975).

14. Dunlap, C., Thomson, J., Thomson, J., Chiang, L.C., "Treatment processes to increase cellulose microbial digestibility." *Am. Inst. Chem. Eng. Symp. Series*: **72**, 58–63 (No. 158) (1976).

15. Lipinsky, E.S., "Perspectives on preparation of cellulose for hydrolysis." *Adv. Chem. Series*: **181**, 1–23 (1979).

16. Horton, G.L., Rivers, D.B., Emert, G.H., "Preparation of cellulosics for enzymatic conversion." *Ind. Eng. Chem. Pro. Res. Dev.*: **19**, 422–429 (1980).

17. Datta, R., "Energy Requirements for Lignocellulose Pretreatment Processes." *Process Biochem.*: **16**, 16–19 (June/July) (1981).

18. Rexen, F., "Principles for pre-treatment of cellulose substances." In *Production and Feeding of SCP.* p. 2–14. M.P. Ferranti, A. Fiechter (eds.), Applied Science Publishers, London, 1983.

19. Hartley, R.D., Keene, A.S., "Pre-treatment of cereal straws and poor quality hays." In *Production and Feeding of SCP.* p. 90–92. M.P. Ferranti, A. Fiechter (eds.), Applied Science Publishers, London, 1983.

20. Ladisch, M.R., Lin, K.W., Voloch, M., Tsao, G.T., "Process considerations in the enzymatic hydrolysis of biomass." *Enzyme Microb. Technol.*: **5**, 82–102 (1983).

21. Cowling, E.B., Brown, W., "Structural features of cellulosic materials in relation to enzymatic hydrolysis." *Adv. Chem. Series*: **95**, 152–187 (1969).

22. Cowling, E.B., "Physical and chemical constraints in the Hydrolysis of Cellulose and Lignocellulosic materials." *Biotechnol. Bioeng. Symp. Series No. 5*, 163–181 (1975).

23. Cowling, E.B., Kirk, T.K., "Properties of Cellulose and Lignocellulosic Materials as substrates for Enzymatic Conversion Processes." *Biotechnol. Bioeng. Symp. Series No. 6*, 95–123 (1976).

24. Tsao, G.T., Ladisch, M., Ladisch, C., Hsu, T.A., Dale, B., Chou, T., "Fermentation substrates from cellulosic materials: Fermentable sugars from cellulosic materials." *Ann. Reports Ferment. Processes*: **2**, 1–21 (1978).

25. Sasaki, T., Tanaka, T., Nanbu, N., Sato, Y., Kainuma, K., "Correlation between X-ray diffraction measurements of cellulosic crystalline structure and the susceptibility to microbial cellulase." *Biotechnol. Bioeng.*: **21**, 1031–1042 (1979).

26. Fan, L.T., Lee Y.-H., Beardmore, D.H., "Major chemical and physical features of cellulosic materials as substrates for enzymatic hydrolysis." *Adv. Biochem. Eng.*: **14**, 101–117 (1980).

27. Fan, L.T., Lee, Y.-H., Beardmore, D.H., "Mechanism of the enzymatic hydrolysis of cellulose: Effects

of major structural features of cellulose on enzymatic hydrolysis." *Biotechnol. Bioeng.*: **22**, 177–199 (1980).

28. Fan, L.T., Lee, Y.-H., Beardmore, D.H., "The influence of major structural features of cellulose on rate of enzymatic hydrolysis." *Biotechnol. Bioeng.*: **23**, 419–424 (1981).

29. Fan, L.T., Gharpuray, M.M., Lee, Y.-H., "Evaluation of pretreatments for enzymatic conversion of agricultural residues." *Biotechnol. Bioeng. Symp. Series No. 11*, 29–45 (1981).

30. Losyakova, L.S., Serebrennikov, V.M., Kozhemyakina, O.P., Boihkavera, N.G., "Enzymatic hydrolysis of sawdust of various types of wood in relation to the method of pretreatment used." *Appl. Biochem. Microbiol.*: **16**, 305–312 (1980).

31. Lee, S.B., Shin, H.S., Ryu, D.D.Y., Mandels, M., "Adsorption of cellulase on cellulose: Effect of physicochemical properties of cellulose on adsorption and rate of hydrolysis." *Biotechnol. Bioeng.*: **24**, 2137–2153 (1982).

32. Lee, S.B., Kim, I.H., Ryu, D.D.Y., Taguchi, H., "Structural properties of cellulose and cellulase reaction mechanism." *Biotechnol. Bioeng.*: **25**, 33–51 (1983).

33. Gharpuray, M.M., Lee, Y.-H., Fan, L.T., "Structural modification of lignocellulosics by pretreatments to enhance enzymatic hydrolysis." *Biotechnol. Bioeng.*: **25**, 157–172 (1983).

34. Sinitsyn, A.P., Klesov, A.A., "Influence of pretreatment on the effectiveness of the enzymatic conversion of cotton linter." *Appl. Biochem. Microbiol.*: **17**, 506–517 (1982).

35. Lin, K.W., Ladisch, M.R., Schaefer, D.M., Noller, C.H., Lecthenberg, V., Tsao, G.T., "Review on effect of pretreatment on digestibility of cellulosic materials." *Amer. Inst. Chem. Eng. Symp. Series*: **77**, 102–106 (1981).

36. Ryu, D.D.Y., Lee, S.B., "Enzymatic hydrolysis of cellulose: Effects of structural properties of cellulose on hydrolysis kinetics." In *Enzyme Engineering* **6**, p. 325–333. I. Chibata, S. Fukui, L.B. Wingard (eds.), Plenum Press, New York, 1982.

37. Gilbert, I.G., Tsao, G.T., "Interaction between solid substrate and cellulase enzymes in cellulose hydrolysis." *Ann. Reports Ferment. Processes*: **6**, 323–358 (1983).

38. Marsden, W.L., Gray, P.P., "Enzymatic hydrolysis of cellulose in lignocellulosic materials." *CRC Crit. Reviews Biotechnol.*: **3**, 235–276 (1986).

39. Grethlein, H.E., "Pretreatment for enhanced hydrolysis of cellulosic biomass." *Biotech. Advs.*: **2**, 43–62 (1984).

40. da Silva, J.G., Serra, G.E., Moreira, J.R., Goncalves, J.C., Goldemberg, J., "Energy balance for ethyl alcohol production from crops." *Science*: **201**, 903–906 (1978).

41. Chambers, R.S., Herendeen, R.A., Joyce, J.J., Penner, P.S., "Gasohol: Does it or doesn't it produce positive net energy?" *Science*: **206**, 789–795 (1979).

42. Hopkinson, C.S., Day, J.W., "Net energy analysis of alcohol production from sugarcane." *Science*: **207**, 302–304 (1980).

43. Polack, J.A., Birkett, H.S., West, M.D., "Sugar cane: Positive energy source for alcohol." *Chem. Eng. Progr.*: **77**, 62–65 (1981).

44. Rolz, C., de Cabrera, S., Garcia, R., "Ethanol from sugar cane: EX-FERM concept." *Biotechnol. Bioeng.*: **21**, 2347–2349 (1979).

45. Rolz, C., "A new technology to ferment sugar cane directly: The EX-FERM Process." *Process Biochem.*: **15**, 2–6 (August/September) (1980).

46. Rolz, C., "Ethanol from sugar crops." *Enzyme Microb. Technol.*: **3**, 19–23 (1981).

47. Rolz, C., de Arriola, M.C., Valladares, J., de Cabrera, S., "Effects of some physical and chemical pretreatments on the composition, enzymatic hydrolysis and digestibility of lignocellulosic sugar cane residue." *Process Biochem.*: **22**, 17–23 (February) (1987).

48. Guenther, E., *The Essential Oils*. D Van Nostrand, Toronto, Vol. 4, p. 20–131 (1950).

49. Robbins, S.R.J., "Selected markets for the essential oils of lemongrass, citronella and eucalyptus." Tropical Products Institute Report G171, London (1983).

50. Rolz, C., de Arriola, M.C., Valladares, J., de Cabrera, S., "Effects of some physical and chemical pretreatments on the composition and enzymatic hydrolysis and digestibility of lemongrass and citronella bagasse." *Agr. Wastes*: **18**, 145–161 (1986).

51. van Soest, P.J., Robertson, J.B., "Systems of analysis for evaluating fibrous feeds." In *Standarization of Analytical Methodology for Feeds*. International Development Research Centre Publication No. IDRC-134e, p. 49–60, Ottawa, 1980.

52. AOAC. *Official methods of analysis.* 12th ed., Association of Official Analytical Chemists, Washington (1975).

53. Theander, O., Åman, P., "Chemical composition of some forages and various residues from feeding value determinations." *J. Sci. Food Agric.*: **31**, 31–37 (1980).

54. Ciszuk, P., Murphy, M., "Digestion of crude protein and organic matter of leaves by rumen microbes *in vitro.*" *Swedish J. Agric. Res.*: **12**, 35–40 (1982).

55. Klopfenstein, T.J., "Increasing the nutritive value of crop residues by chemical treatment." In *Upgrading and byproducts for animals.* J.T. Huber (ed.), p. 39–60. CRC Press Inc., Boca Raton, 1981.

56. Aman, A., Nordkvist, E., "Chemical composition and *in vitro* digestibility of botanical fractions of cereal straw." *Swedish J. Agrc. Res.*: **13**, 61–67 (1983).

57. Jackson, M.G., "Review article: The alkali treatment of straws." *Anim. Feed Sci. Technol.*: **2**, 105–130 (1977).

58. Ohlde, G., Becker, K., "Suitability of cell-wall constituents as predictors of organic matter digestibility in some tropical and subtropical by-products." *Anim. Feed. Sci. Technol.*: **7**, 191–199 (1982).

59. Paturau, J.M., *Byproducts of the cane sugar industry.* 2nd ed., p. 9, Elsevier, Amsterdam (1982).

60. Puri, V.P., "Effect of crystallinity and degree of polymerization of cellulose on enzymatic saccharification." *Biotechnol. Bioeng.*: **26**, 1219–1222 (1984).

61. Knapp, S.B., Watt, R.A., Wethern, J.D., "Sugarcane bagasse as a fibrous papermaking material. I. Chemical composition of Hawaian bagasse." *TAPPI*: **40**, 595 (1957).

62. Chang, C.D., Kononenko, O.K., Herstein, K.M., "The ammoniation of sugar cane bagasse." *J. Sci. Food Agric.*, **12**, 687–693 (1961).

63. Randel, P.F., "A comparison of the digestibility of two complete rations containing either raw or alkali-treated sugarcane bagasse." *J. Agr. Univ. Puerto Rico*: **56**, 18–25 (1972).

64. Dekker, R.F.H., Richards, G.N., "Effect of delignification on the *in vitro* rumen digestion of polysaccharides of bagasse." *J. Sci. Food Agric.*: **24**, 375–379 (1973).

65. Martin, P.C., Cribero, T.C., Cabello, A., Elias, A., "The effect of sodium hydroxide and pressure on the dry matter digestibility of bagasse and bagasse pith." *Cuban J. Agric. Sci.*: **8**, 21–28 (1974).

66. Preston, T.R., "Sugar cane as the basis for the intensive animal production in the tropics." In *Proc. Conf. Animal Feeds of Tropical and Subtropical Origin*, p. 69–83. (Ministry Overseas Development, Tropical Products Institute, London), 1975.

67. Meade, G.P., Chen, J.C.P., "Bagasse and its uses." In *Cane Sugar Handbook, 10th ed.*, G.P. Meade, J.C.P. Chen (eds.), p. 102–105, John Wiley & Sons, New York, 1977.

68. Swakon, D.H.D. "Quantitative and qualitative digestion of different tissues in sugarcane stem by rumen microorganisms *in vitro.*" MS Thesis, Univ. Florida, College of Agriculture (1977).

69. Llamas, G., Shimada, A.S., Castellanos, S., Merino, H., "Estudio del valor alimenticio de subproductos de la caña de azúcar con bovinos en corral." *Tec. Pec. Mex.*: **36**, 59–64 (1979).

70. Rexen, F., "Low-quality forages improve with alkali treatment." *Feedstuffs*: **51**, 33–34 (1979).

71. Hart, M.R., Walker, Jr., H.G., Graham, R.P., Hanni, P.J., Brown, A.H., Kohler, G.O., "Steam treatment of crop residues for increased ruminant digestibility. I. Effects of process parameters." *J. Anim. Sci.*: **51**, 402–408 (1980).

72. Oi, S., Yamanaka, H., Yamamoto, T., "Methane fermentation of bagasse and some factors to improve the fermentation." *J. Ferment. Technol.*: **58**, 367–372 (1980).

73. Ibrahim, M.N.M., Pearce, G.R., "Effects of gamma irradiation on the composition and *in vitro* digestibility of crop by-products." *Agr. Wastes*: **2**, 253–259 (1980).

74. Taylor, J.D., Esdale, W.J., "Increased utilization of crop residues as animal feed through autohydrolysis." *Proc. Bioenergy*: **80**, 285–286 (1980).

75. Cabello, A., Conde, J., Otero, M.A., "Prediction of the degradability of sugarcane cellulosic residues by indirect methods." *Biotechnol. Bioeng.*: **23**, 2737–2745 (1981).

76. Dahiya, D.S., Veeramani, H. "Effect of bagasse treatment on microbial growth." *Inst. Sugar J.*: **83**, 263–264 (1981).

77. Dekker, R.F.H., Wallis, A.F.A., "Autohydrolysis-explosion of lignocellulose as pretreatment for enzymic saccharification." *Proc. 5th Australian Biotechnol. Conf.*, p. 145–148 (1982).

78. Han, Y.W., Ciegler, A., "Use of nuclear wastes in utilization of lignocellulosic biomass." *Process Biochem.*: **17**, 32–38 (Jan./Feb.) (1982).

79. Awad, W.I., Abdel-Mottaleb, F.T., "Enzymatic saccharification of some Egyptian cellulosic wastes." *Proc. 5th Int. Alcohol Fuel Technol. Symp. I*, p. 89–93 (1982).

80. Ibrahim, M.N.M., Pearce, G.R., "The effects of boiling and of steaming under pressure on the chemical composition and *in vitro* digestibility of crop by-products." *Agr. Wastes*: **4**, 443–452 (1982).

81. Rangnekar, D.V., Badve, V.C., Kharat, S.T., Sobale, B.N., Joshi, A.L., "Effect of high-pressure steam treatment on chemical composition and digestibility in vitro of roughages." *Anim. Feed Sci. Technol.*: **7**, 61–70 (1982).

82. Ellenrieder, G., Castillo, J.J., "Evaluation of pretreatments favouring enzymatic hydrolysis of sugar cane bagasse." *Lat. am. j. chem. eng. appl. chem.*: **13**, 199–214 (1983).

83. Molina, E., Boza, J., Aguilera, J.F., "Nutritive value for ruminants of sugar cane bagasse ensiled after spray treatment with different levels of NaOH." *Anim. Feed Sci. Technol.*: **9**, 1–17 (1983).

84. Puri, V.P., "Ozone pretreatment to increase digestibility of lignocellulose." *Biotechnol. Lett.*: **5**, 773–776 (1983).

85. Han, Y.W., Catalano, E.A., Ciegler, A., "Chemical and physical properties of sugarcane bagasse irradiated with gamma rays." *J. Agric. Food Chem.*: **31**, 34–38 (1983).

86. Ibrahim, M.N.M., Pearce, G.R., "Effects of chemical pretreatments on the composition and *in vitro* digestibility of crop by-products." *Agr. Wastes*: **5**, 135–156 (1983).

87. Ibrahim, M.N.M., Pearce, G.R., "Effects of chemical treatments combined with high-pressure steaming on the chemical composition and *in vitro* digestibility of crop by-products." *Agr. Wastes*: **7**, 235–250 (1983).

88. Matei, C.H., Playne, M.J., "Production of volatile fatty acids from bagasse by rumen bacteria." *Appl. Microbiol. Biotechnol.*: **20**, 170–175 (1984).

89. Crosthwaite, C., Ishihara, M., Richards, G.N., "Acid-ageing of lignocellulosics to improve ruminant digestibility—application to bagasse, wheat and rice straw and oat hulls." *J. Sci. Food Agric.*: **35**, 1041–1050 (1984).

90. Kumakura, M., Kaetsu, I., "Effect of radiation pretreatment of bagasse on enzymatic and acid hydrolysis." *Biomass*: **3**, 199–208 (1983).

91. Kato, A., Azuma, J.-I., Koshijima, T., "Lignin-carbohydrate complexes and phenolic acids in bagasse." *Holzforschung*: **38**, 141–149 (1984).

92. Neely, W.C., "Factors affecting the pretreatment of biomass with gaseous ozone." *Biotechnol. Bioeng.*: **26**, 59–65 (1984).

93. Playne, M.J., "Increased digestibility of bagasse by pretreatment with alkalis and steam explosion." *Biotechnol. Bioeng.*: **26**, 426–433 (1984).

94. Molina, O.E., Perotti de Gálvez, N.I., Frigerio, C.I., Córdoba, P.R., "Single cell protein production from bagasse pith pretreated with sodium hydroxide at room temperature." *Appl. Microbiol. Biotechnol.*: **20**, 335–339 (1984).

95. Cabello, A., Conde, J., "Evaluation of newer methods of pretreatment for biological utilization of cellulosic residues." *Acta Biotechnol.*: **5**, 191–196 (1985).

96. Trivedi, S.M., Ray, R.M., "Saccharification of cellulosic wastes by high-strength cellulase system of mix-shake cultivated *Scytalidium lignicola* and *Trichoderma longibrachiatum*." *J. Ferment. Technol.*: **63**, 299–304 (1985).

97. Ismail, A.M.S., Hamdy, A.H.A., Naim, N., El-Refai, A.M.H., "Enzymatic saccharification of Egyptian sugar-cane bagasse." *Agr. Wastes*: **12**, 99–109 (1985).

98. Krishna, G., "Nylon bag dry-matter digestibility in agroindustrial by-products and wastes of the tropics." *Agr. Wastes*: **13**, 155–158 (1985).

99. Hartley, R.D., "Chemical constitution, properties and processing of lignocellulosic wastes in relation to nutritional quality for animals." *Agr. Environ.*: **6**, 91–113 (1981).

100. Windham, W.R., Barton, II, F.E., Himmelsbach, D.S., "High-pressure liquid chromatographic analysis of component sugars in neutral detergent fiber for representative warm- and cool-season grasses." *J. Agric. Food Chem.*: **31**, 471–475 (1983).

101. McAllan, A.B., Griffith, E.S., "Evaluation of detergent extraction procedures for characterising carbohydrate components in ruminant feeds and digests." *J. Sci. Food Agric.*: **35**, 869–877 (1984).

102. Obst, J.R., "Frequency and alkali resistance of lignin-carbohydrate bonds in wood." *TAPPI*: **65**, 109–112 (April) (1982).

103. Hartley, R.D., "*p*-Coumaric and ferulic acid components of cell walls of ryegrass and their relationship with lignin and digestibility." *J. Sci. Food Agric.*: **23**, 1347–1354 (1972).

104. Hartley, R.D., "Carbohydrate esters of ferulic acid as components of cell-walls of *Lolium multiflorum.*" *Phytochem.*: **12**, 661–665 (1973).

105. Chesson, A., Gordon, A.H., Lomax, J.A., "Substituent groups linked by alkali-labile bonds to arabinose and xylose residues of legume, grass and cereal straw cell walls and their fate during digestion by rumen microorganisms." *J. Sci. Food Agric.*: **34**, 1330–1340 (1983).

106. Gordon, A.H., Lomax, J.A., Chesson, A., "Glycosidic linkages of legume, grass and cereal straw cell walls before and after extensive degradation by rumen microorganisms." *J. Sci. Food Agric.*: **34**, 1341–1350 (1983).

107. Tanner, G.R., Morrison, I.M., "Phenolic-carbohydrate complexes in the cell walls of *Lolium perenne.*" *Phytochem.*: **22**, 1433–1439 (1983).

108. Nagaty, A., El-Sayed, O.H., Ibrahim, S.T., Mansour, O.Y., "Chemical and spectral studies on hemicelluloses isolated from bagasse." *Holzforschung*: **36**, 29–35 (1982).

109. du Toit, P.J., Olivier, S.P., van Biljon, P.L., "Sugar cane bagasse as a possible source of fermentable carbohydrates. I. Characterization of bagasse with regard to monosaccharide, hemicellulose, and amino acid composition." *Biotechnol. Bioeng.*: **26**, 1071–1078 (1984).

110. Gaillard, B.D.E., Richards, G.N., "Presence of soluble lignin-carbohydrate complexes in the bovine rumen." *Carbohydr. Res.*: **42**, 135–145 (1975).

111. Rao, M., Seeta, R., Deshpande, V., "Effect of pretreatment on the hydrolysis of cellulose by *Penicillium funiculosum* cellulase and recovery of enzyme." *Biotechnol. Bioeng.*: **25**, 1863–1871 (1983).

112. Dowman, M.G., Collins, F.C., "The use of enzymes to predict the digestibility of animal feeds." *J. Sci. Food Agric.*: **33**, 689–696 (1982).

113. Braman, W.L., Abe, R.K., "Laboratory and *in vivo* evaluation of the nutritive value of NaOH-treated wheat straw." *J. Anim. Sci.*: **46**, 496–505 (1977).

114. Chesson, A., "Effects of sodium hydroxide on cereal straws in relation to the enhanced degradation of structural polysaccharides by rumen microorganisms." *J. Sci. Food Agric.*: **32**, 745–758 (1981).

115. Taniguchi, M., Tanaka, M., Matsuno, R., Kamikubo, T., "Evaluation of chemical pretreatment for enzymatic solubilization of rice straw." *Eur. J. Appl. Microbiol. Biotechnol.*: **14**, 35–39 (1982).

116. Van Eenaeme, C., Istasse, L., Lambot, O., Bienfait, J.M., Gielen, M., "Effect of sodium hydroxide treatment on chemical composition and *in vitro* and *in vivo* digestibility of hay." *Agr. Environ.*: **6**, 161–170 (1981).

117. Ben-Ghedalia, D., Shefet, G., "The effect of sodium hydroxide and ozone treatments on the composition and *in vitro* organic matter digestibility of sodium straw." *Nut. Rep. Inst.*: **20**, 179–182 (1979).

118. Wilson, R.K., Pigden, W.J., "Effect of a sodium hydroxide treatment on the utilization of wheat straw and poplar wood by rumen microorganisms." *Can. J. Anim. Sci.*: **44**, 122–123 (1964).

119. Chandra, S., Jackson, M.G., "A study of various chemical treatments to remove lignin from coarse roughages and increase their digestibility." *J. Agric. Sci. Camb.*: **77**, 11–17 (1971).

120. Ololade, B.G., Mowat, D.N., Winch, J.E., "Effect of processing methods on the *in vitro* digestibility of sodium hydroxide treated roughages." *Can. J. Anim. Sci.*: **50**, 657–662 (1970).

121. Carmona, J.F., Greenhalgh, J.F.D., "The digestibility and acceptability to sheep of chopped or milled barley straw soaked or sprayed with alkali." *J. Agric. Sci. Camb.*: **78**, 477–485 (1972).

122. Rexen, F., Thomsen, K.V., "The effect on digestibility of a new technique for alkali treatment of straw." *Anim. Feed Sci. Technol.*: **1**, 73–83 (1976).

123. Raininko, K., Heikkilä, T., Lampila, M., Kossila, V., "Effect of chemical and physical treatment on the composition and digestibility of barley straw." *Agr. Environ.*: **6**, 261–266 (1981).

124. Anderson, D.C., Ralston, A.T., "Chemical treatment of ryegrass straw: *in vitro* dry matter digestibility and compositional changes." *J. Anim. Sci.*: **37**, 148–152 (1973).

125. Hutanuwatr, N., Hinds, F.C., Davis, C.L., "An evaluation of methods for improving the *in vitro* digestibility of rice hulls." *J. Anim. Sci.*: **38**, 140–148 (1974).

126. Berger, L., Klopfenstein, T., Britton, R., "Effect of sodium hydroxide on efficiency of rumen digestion." *J. Anim. Sci.*: **49**, 1317–1323 (1979).

127. Armentano, L.E., Rakes, A.H., "Effects of sodium hydroxide treatment on nutritive value of screened manure solids fed to dairy cattle." *J. Dairy Sci.*: **65**, 390–395 (1982).

128. Jung, H.-J. G., Fahey, Jr., G.C., "Effect of phenolic compound removal on *in vitro* forage digestibility." *J. Agric. Food Chem.*: **29**, 817–820 (1981).

129. Jung, H.G., Fahey, Jr., G.C., "Nutritional implications of phenolic monomers and lignin: a review." *J. Anim. Sci.*: **57**, 206–219 (1983).

130. Jung, H.-J. G., Fahey, Jr., G.C., "Interactions among phenolic monomers and *in vitro* fermentation." *J. Dairy Sci.*: **66**, 1255–1263 (1982).

131. Jung, H.-J. G., Fahey, Jr., G.C., Merchen, N.R., "Effects of ruminant digestion and metabolism on phenolic monomers of forages." *British J. Nutr.*: **50**, 637–651 (1983).

132. Jung, H.G., Fahey, Jr., G.C., Garst, J.E., "Simple phenolic monomers of forages and effects of *in vitro* fermentation on cell wall phenolics." *J. Anim. Sci.*: **57**, 1294–1305 (1983).

133. Bacon, J.S.D., Chesson, A., Gordon, A.H., "Deacetylation and enhancement of digestibility." *Agr. Environ.*: **6**, 115–126 (1981).

134. Theander, O., Uden, P., Aman, P., "Acetyl and phenolic acid substituents in timothy of different maturity and after digestion with rumen microorganisms or a commercial cellulase." *Agr. Environ.*: **6**, 127–133 (1981).

135. Scalbert, A., Monties, B., Lallemand, J.-Y., Guittet, E., Rolando, C., "Ether linkage between phenolic acids and lignin fractions from wheat straw." *Phytochem.*: **24**, 1359–1362 (1985).

136. Van Soest, P.J., Ferreira, A.M., Hartley, R.D., "Chemical properties of fibre in relation to nutritive quality of ammonia-treated forages." *Anim. Feed Sci. Technol.*: **10**, 155–164 (1983/84).

137. Capper, B.S., Morgan, D.J., Parr, W.H., "Alkali-treated roughages for feeding ruminants: a review." *Trop. Sci.*: **19**, 73–88 (1977).

138. Tarkow, H., Feist, W.C., "A mechanism for improving the digestibility of lignocellulosic materials with dilute alkali and liquid ammonia." *Adv. Chem. Series*: **95**, 197–218 (1969).

139. Ibrahim, M.N.M., Pearce, G.R., "A soak- and -press method for the alkali treatment of fibrous crop residues. Laboratory studies on aspects of the procedure." *Agr. Wastes*: **8**, 195–213 (1983).

140. Millett, M.A., Baker, A.J., Satter, L.D., "Physical and chemical pretreatments for enhancing cellulose saccharification." *Biotechnol. Bioeng. Symp. Series No. 6*, 125–153 (1976).

141. Lee, Y.Y., Lin, C.M., Johnson, T., Chambers, R.P., "Selective hydrolysis of hardwood hemicellulose by acids." *Biotechnol. Bioeng. Symp. Series No. 8*, 75–88 (1978).

142. Conner, A.H., "Effects of aqueous sulfur dioxide on cellulose." *Biotechnol. Lett.*: **2**, 439–444 (1980).

143. Ben-Ghedalia, D., Miron, J., "Effect of sodium hydroxide, ozone and sulphur dioxide on the composition and *in vitro* digestibility of wheat straw." *J. Sci. Food Agric.*: **32**, 224–228 (1981).

144. Ben-Ghedalia, D., Miron, J., "The effect of combined chemical and enzyme treatments on the saccharification and *in vitro* digestion rate of wheat straw." *Biotechnol. Bioeng.*: **23**, 823–831 (1981).

145. Ben-Ghedalia, D., Miron, J., "The response of wheat straw varieties to mild sulphur dioxide treatment." *Anim. Feed Sci. Technol.*: **10**, 269–276 (1983/84).

146. Stanek, D.A., "A study of the low-molecular weight phenols formed upon the hydrolysis of Aspenwood." *TAPPI*: **41**, 601–609 (October) (1958).

147. Lora, J.H., Wayman, M., "Delignification of hardwoods by autohydrolysis and extraction." *TAPPI*: **61**, 47–50 (June) (1978).

148. Wayman, M., Lora, J.H., "Delignification of wood by autohydrolysis and extraction." *TAPPI*: **62**, 113–114 (September) (1979).

149. Jurasek, L., "Enzymic hydrolysis of pretreated Aspenwood." *Dev. Ind. Microbiol.*: **20**, 177–183 (1979).

150. Marchessault, R.H., St. Pierre, J., "A new understanding of the carbohydrate system." In *Future sources of organic raw materials*. p. 613–625. L.E. St. Pierre, G.R. Brown (eds.), Pergamon, Oxford, 1980.

151. Barnet, D., Dupeyre, D., Excoffier, G., Gagnaire, D., Nava Saucedo, J.E., Vignan, M.R., "Flash hydrolysis of aspen wood: characterization of water soluble compounds." In *Energy from Biomass*. p. p. 889–893. A. Strub, P. Chartier, G. Schleser (eds.), Applied Science, London (1983).

152. Schultz, T.P., Biermann, C.J., McGinnis, G.D., "Steam explosion of mixed hardwood chips as a Biomass pretreatment." *Ind. Eng. Chem. Prod. Res. Div.*: **22**, 344–348 (1983).

153. Schultz, T.P., Templeton, M.C., Biermann, C.J., McGinnis, G.D., "Steam explosion of mixed hardwood chips, rice hulls, corn stalks, and sugar cane bagasse." *J. Agric. Food Chem.*: **32**, 1166–1172 (1984).

154. Biermann, C.J., Schultz, T.P., McGinnis, G.D., "Rapid steam hydrolysis/extraction of mixed hardwoods as a biomass pretreatment." *J. Wood Technol.*: **4**, 111–128 (1984).

155. Saddler, J.N., Brownell, H.H., Clermont, L.P., Levitin, N., "Enzymatic hydrolysis of cellulose and various pretreated wood fractions." *Biotechnol. Bioeng.*: **24**, 1389–1402 (1982).

156. Nesse, N., Wallick, J., Harper, J.M., "Pretreatment of cellulosic wastes to increase enzyme reactivity." *Biotechnol. Bioeng.*: **19**, 323–336 (1977).

157. Sinitsyn, A.P., Clesceri, L.S., Bungay, H.R., "Inhibition of cellulases by impurities in steam-exploded wood." *Appl. Biochem. Biotechnol.*: **7**, 455–458 (1982).

158. Sinitsyn, A.P., Bungay, H.R., Clesceri, L.S., "Enzyme management in the Iotech process." *Biotechnol. Bioeng.*: **25**, 1393–1399 (1983).

159. Saddler, J.N., Mes-Hartree, M., Yu, E.K.C., Brownell, H.H., "Enzymatic hydrolysis of various pretreated lignocellulosic substrates and the fermentation of the liberated sugars to ethanol and butanediol." *Biotechnol. Bioeng. Symp. Series No. 13*, 225–238 (1983).

160. Khan, A.W., Asther, M., Giuliano, C., "Utilization of steam- and explosion-decompressed aspen wood by some anaerobes." *J. Ferment. Technol.*: **62**, 335–339 (1984).

161. Garrett, W.N., Walker, Jr., H.G., Kohler, G.D., Hart, M.R., Graham, R.P., "Steam treatment of crop residues for increased ruminant digestibility. II. Lamb feeding studies." *J. Anim. Sci.*: **51**, 409–413 (1980).

162. Sharma, H.R., Forsberg, N.E., Guenter, W., "The nutritive value of pressure-steamed aspen (*Populus tremuloides*) for mature sheep." *Can. J. Anim. Sci.*: **59**, 303–312 (1979).

163. Sharma, H.R., Guenter, W., Devlin, T.J., Ingalls, J.R., McDaniel, M., "Comparative nutritional value of corn silage and steamed aspen in the diet of wintering calves and finishing beef steers." *Can. J. Anim. Sci.*: **60**, 99–106 (1980).

164. Sharma, H.R., Ingalls, J.R., Guenter, W., "Evaluation of steamtreated wheat straw as a roughage source in dairy cow rations." *Can. J. Anim. Sci.*: **62**, 181–190 (1982).

165. Calder, F.W., "Effect of steam-treated grass silage on animal gain." *Can. J. Plant Sci.*: **62**, 89–94 (1982).

166. Mes-Hartree, M., Saddler, J.N., "The nature of inhibitory materials present in pretreated lignocellulosic substrates which inhibit the enzymatic hydrolysis of cellulose." *Biotechnol. Lett.*: **5**, 531–536 (1983).

167. Dietrichs, H.H., Sinner, M., Puls, J., "Potential of steaming hardwood and straw for feed and food production." *Holzforschung*: **32**, 193–199 (1978).

168. Dekker, R.F.H., Wallis, A.F.A., "Autohydrolysis-Explosion as pretreatment for the enzymic saccharification of sunflower seed hulls." *Biotechnol. Lett.*: **5**, 311–316 (1983).

169. Puri, V.P., Mamers, H., "Explosive pretreatment of lignocellulosic residues with high-pressure carbon dioxide for the production of fermentation substrates." *Biotechnol. Bioeng.*: **25**, 3149–3161 (1983).

170. Saddler, J.N., Brownell, H.H., "Steam-explosion pretreatment for enzymatic hydrolysis." *Biotechnol. Bioeng. Symp. Series No. 14*, 55–68 (1984).

171. Taniguchi, M., Tanaka, M., Goto, T., Matsuno, R., Kamikubo, T., "Effect of puff cooking on structure and chemical properties of several lignocellulosic agrowastes." *Agric. Biol. Chem.*: **49**, 1243–1249 (1985).

172. Cheong, Y.W.Y., d'Espaignet, J.T., Deville, P.J., Sansoucy, R., Preston, T.R., "The effect of steam treatment on cane bagasse in relation to its digestibility and furfural production." *Proc. 15th Cong. Int. Sugarcane Tech.* p. 1887–1894 (1974).

173. Morjanoff, P., Dunn, N.W., Gray, P.P., "Improved enzymic digestibility of sugar cane bagasse following high temperature autohydrolysis." *Biotechnol. Lett.*: **4**, 187–192 (1982).

174. Morjanoff, P., Gray, P.P., "Optimization of steam explosion as a method for increasing susceptibility of sugar cane bagasse to enzymatic saccharification." *Biotechnol. Bioeng.*: **29**, 733–741 (1987).

175. Mamers, H., Yuritta, J.P., Menz, D.J., "Explosion pulping of bagasse and wheat straw." *TAPPI*: **64**, 93–96 (1981).

176. Linden, J.C., Murphy, V.G., Moreira, A.R., "Wheat straw autolysis." In *Advances in Biotechnology*, Vol. II. p. 41–46. M. Moo-Young, C.W. Robinson (eds.), Pergamon Press, Toronto, 1981.

177. Vallander, L., Eriksson, K.-E., "Enzymic saccharification of pretreated wheat straw." *Biotechnol. Bioeng.*: **27**, 650–659 (1985).

178. MacDonald, D.G., Mathews, J.F., "Effect of steam treatment on the hydrolysis of aspen by commercial enzymes." *Biotechnol. Bioeng.*: **21**, 1091–1096 (1979).

179. Kaufmann, V.W., Sinner, M., Dietrichs, H.H., "Zur verdaulichkeit von stroh und holz nach aufschluss mit gesättigtem wasserdampf bei höheren temperaturen sowie extraktion mit wasser und verdünnter natronlauge." *Z. Tierphysiol. Tierenährg. u. Futtermittelkde*: **40**, 91–96 (1978).

180. Buchholz, K., Puls, J., Gödelmann, B., Dietrichs, H.H., "Hydrolysis of cellulosic wastes." *Process Biochem.*: **16**, 37–43 (Dec./Jan.) (1980/81).

181. Puri, V.P., Pearce, G.R., "Alkali-explosion pretreatment of straw and bagasse for enzymic hydrolysis." *Biotechnol. Bioeng.*: **28**, 480–485 (1986).

182. Clemente, A., Menaia, J., Roda-Santos, M.L., Seabra, J., Ramalho Ribeiro, M.J.C., Fernandez, T.H., *3rd Med. Congress Chem. Eng.* p. 175 (1984).

183. Cross, H.H., Smith, L.W., DeBarth, J.V., "Rates of *in vitro* forage fiber digestion as influenced by chemical treatment." *J. Anim. Sci.*: **39**, 808–812 (1974).

184. Barton, II., F.E., Akin, D.E., "Digestibility of delignified forage cell walls." *J. Agric. Food Chem.*: **25**, 1299–1303 (1977).

185. Darcy, B.K., Belyea, R.L., "Effect of delignification upon *in vitro* digestion of forage cellulose." *J. Anim. Sci.*: **51**, 798–803 (1980).

186. Akin, D.E., Burdick, D., "Relationships of different histochemical types of lignified cell walls to forage digestibility." *Crop Sci.*: **21**, 577–581 (1981).

187. Belyea, R.L., Foster, M.B., Zinn, G.H., "Effect of delignification on *in vitro* digestion of alfalfa cellulose." *J. Dairy Sci.*: **66**, 1277–1281 (1983).

188. Phillips, J.A., Humphrey, A.E., "Process technology for the biological conversion of lignocellulosic materials to fermentable sugars and alcohols." In *Wood and Agricultural Residues: Research on use for feed, fuels and chemicals.* p. 503–527. J. Sotles (ed.), Academic Press, New York, 1983.

189. Phillips, J.A., Humphrey, A.E., "An overview of process technology for the production of liquid fuels and chemical feedstocks via fermentation." In *Organic chemicals from biomass.* p. 294–304. D.L. Wise (ed.), The Benjamin/Cummings Pub. Co., Menlo Park, 1983.

190. Avgerinos, G.C., Wang, D.I.C., "Selective solvent delignification for fermentation enhancement." *Biotechnol. Bioeng.*: **25**, 67–83 (1983).

191. Neilson, M.J., Shafizadeh, F., Azig, S., Sarkanen, K.V., "Evaluation of organosolv pulp as a suitable substrate for rapid enzymatic hydrolysis." *Biotechnol. Bioeng.*: **25**, 609–612 (1983).

192. Ghose, T.K., Pannir Selvam, P.V., Ghosh, P., "Catalytic solvent delignification of agricultural residues: Organic catalysts." *Biotechnol. Bioeng.*: **25**, 2577–2590 (1983).

193. Holtzapple, M.T., Humphrey, A.E., "The effect of organosolv pretreatment on the enzymatic hydrolysis of poplar." *Biotechnol. Bioeng.*: **26**, 670–676 (1984).

194. Sakakibara, A., Edashige, Y., Sano, Y., Takeyama, H., "Solvolysis pulping with cresol-water system." *Holzforschung*: **38**, 159–165 (1984).

Subject Index